Palgrave Series in Indian Ocean World Studies

Series Editor

Gwyn Campbell
McGill University
Montreal, Canada

This is the first scholarly series devoted to the study of the Indian Ocean world from early times to the present day. Encouraging interdisciplinarity, it incorporates and contributes to key debates in a number of areas including history, environmental studies, anthropology, sociology, political science, geography, economics, law, and labor and gender studies. Because it breaks from the restrictions imposed by country/regional studies and Eurocentric periodization, the series provides new frameworks through which to interpret past events, and new insights for present-day policymakers in key areas from labor relations and migration to diplomacy and trade.

More information about this series at
http://www.springer.com/series/14661

Greg Bankoff • Joseph Christensen
Editors

Natural Hazards and Peoples in the Indian Ocean World

Bordering on Danger

palgrave
macmillan

Editors
Greg Bankoff
University of Hull
Hull, UK

Joseph Christensen
Murdoch University
Perth, Australia

Palgrave Series in Indian Ocean World Studies
ISBN 978-1-349-94856-7 ISBN 978-1-349-94857-4 (eBook)
DOI 10.1057/978-1-349-94857-4

Library of Congress Control Number: 2016942672

Printed on acid-free paper

This Palgrave Macmillan imprint is published by Springer Nature
The registered company is Nature America Inc. New York

Acknowledgements

The editors would like to acknowledge a number of people whose assistance and support has made this volume possible. In particular, we thank Professor James Warren, Convenor, and the participants at the *Dimensions of the Indian Ocean World Past* conference, Western Australian Maritime Museum, 12–14 November 2012; Professor David Starkey and the Maritime Historical Studies Centre group at the University of Hull; Professor Kevin Hewison and the staff and students of the Asia Research Centre, Murdoch University; and Professor Gwyn Campbell, Dr Erik van Dongen and Lori Callaghan at the Indian Ocean World Centre, McGill University. Unless otherwise indicated, all maps that appear in this volume were produced by Dr Alexander Brown, Murdoch University, and Karina Pelling, ANU College of Asia and the Pacific. Finally, we would like to extend our thanks to Linda Rossi, Anna Chapman, Ashleigh Forrest and Bas van Maren for their support of the volume and the projects upon which it is based.

Contents

1 **Bordering on Danger: An Introduction** 1
Greg Bankoff and Joseph Christensen

2 **Revisiting Southeast Asian History with Geology: Some Demographic Consequences of a Dangerous Environment** 31
Anthony Reid

3 **'The Sea Becomes Mulberry Fields and Mulberry Fields Become the Sea': Dikes in the Eastern Red River Delta, c.200 BCE to the Twenty-First Century CE** 55
Li Tana

4 **'The Most Horrible of Evils': Social Responses to Drought and Famine in the Bombay Presidency, 1782–1857** 79
George Adamson

5 **Philippine Typhoons Since the Seventeenth Century** 105
James Francis Warren

6 **Bushfire in Madagascar: Natural Hazard, Useful Tool, and Change Agent** 143
Christian A. Kull

7 **Emperor Tự Đức's 'Bad Weather': Interpreting Natural Disasters in Vietnam, 1847–1883** 169
Kathryn Dyt

8 **Storm over San Isidro: Repeated 'disasters' and Civic Community Culture in the Nineteenth-Century Philippines** 199
Greg Bankoff

9 **Disaster Management and Colonialism in the Indonesian Archipelago, 1840–1920** 225
Alicia Schrikker

10 **Cyclones, Drought, and Slavery: Environment and Enslavement in the Western Indian Ocean, 1870s to 1920s** 255
Matthew S. Hopper

11 **Their Inescapable Portion? Cyclones, Disaster Relief, and the Political Economy of Pearlshelling in Northwest Australia, 1865–1935** 283
Joseph Christensen

Index 313

Notes on Contributors

George Adamson works at the boundary of environmental history, historical geography and climatology, focusing on climate variability and human responses in colonial India. Adamson is Lecturer in Geography at King's College London, having undertaken doctoral and postdoctoral research at the University of Brighton and University of Sussex. He is an affiliate of the Centre for World Environmental History.

Greg Bankoff has worked and published extensively on the historical dimension of how societies adapt to risk as well as engaged with contemporary civil defence and emergency management practices in Asia, Australasia and more recently Europe. His most recent publications include co-authoring The Red Cross's *World Disaster Report 2014: Culture and Risk* and a companion co-edited volume entitled *Cultures and Disasters: Understanding Cultural Framings in Disaster Risk Reduction* (2015). He is Professor of Modern History at the University of Hull, UK.

Joseph Christensen holds a BA and a PhD from the University of Western Australia. He is a postdoctoral fellow at the Asia Research Centre, Murdoch University, where he works in the fields of maritime and environmental history. He is co-editor of *Historical Perspectives of Fisheries Exploitation in the Indo-Pacific* (2014).

Kathryn Dyt is a historian with an interest in the environment and religious and cultural practices in Vietnam. Her current PhD research at the Australian National University, funded by a Vice-Chancellor's Scholarship for Doctoral Study, explores the relationship between the environment, emotion and kingship in nineteenth-century Vietnam.

Matthew S. Hopper is an associate professor in the History Department at California Polytechnic State University, San Luis Obispo. He holds a PhD in History from UCLA (2006) and has held fellowships at Yale University (2009), the Institute for Advanced Study (2015), and the University of Cambridge (2016). His writing has appeared in *Annales*, *Itinerario*, and the *Journal of African Development*. His book *Slaves of One Master: Globalization and Slavery in Arabia in the Age of Empire* was published in 2015.

Christian A. Kull researches the political ecology of resource management issues like fire, introduced plants, peasant agriculture and forest management, particularly in Madagascar and around the Indian Ocean rim. Trained in the USA, he has taught in Canada and Australia, and, since 2015, in Switzerland.

Tana Li holds a BA and an MA from Peking University and a PhD from Australia National University, where she is currently a senior fellow in the School of Culture, History and Language. Li is also Director of the Centre for the Study of the Chinese Southern Diaspora.

Anthony Reid is Emeritus Professor of Southeast Asian History at the Australian National University. His recent books include *An Indonesian Frontier: Acehnese and Other Histories of Sumatra* (2004), *Imperial Alchemy: Nationalism and Political Identity in Southeast Asia* (2010), *To Nation by Revolution: Indonesia in the 20th Century* (2011) and *A History of Southeast Asia: Critical Crossroads* (2015).

Alicia Schrikker is Lecturer in Colonial and Global History at the Institute for History at Leiden University. She wrote her PhD on colonial transitions in Sri Lanka and has been involved in two research projects on natural disasters in history: one examining the interaction between government and society in the wake of natural disasters in colonial Indonesia in the nineteenth and early twentieth centuries and the other focusing on the representation of disaster in colonial and postcolonial contexts in Southeast Asia and the Caribbean.

James Francis Warren is Professor of Southeast Asian Modern History at Murdoch University. He is an award-winning historian who has published numerous monographs, journal articles and book chapters. His books include *Iranun and Balangingi: Globalisation, Maritime Raiding and the Birth of Ethnicity* (2002), *The Sulu Zone 1768–1898: The Dynamics of External Trade, Slavery and Ethnicity in the Transformation of a Southeast Asian Maritime State* (2007) and *Pirates, Prostitutes and Pullers: Explorations in the Ethno-and Social History of Southeast Asia* (2008).

List of Figures

Fig. 1.1	The Indian Ocean World, showing locations covered by chapters in this book	4
Fig. 2.1	Major natural threats to Southeast Asia	36
Fig. 4.1	Map of India displaying study area (shaded). Place names current as of 2014	81
Fig. 4.2	Irrigation well near Deesa, Gujarat. From diary of M. Emma Walter, 19 March 1840	93
Fig. 5.1	Annual frequency of tropical cyclones within the PAR (1948–2000)	122
Fig. 5.2	Annual tropical cyclone El Niño/La Niña frequency, 1880–1898	124
Fig. 5.3	Frequency and intensity of Super Typhoons, 1600–2000. Based on an index compiled from PAGASA's records, Miguel Selga, *Charts of Remarkable Typhoons in the Philippines 1902–1934*, Miguel Selga, *Catalogue of Typhoons 1348–1934* (Manila: Bureau of Printing, 1935) and archival records found in Spain and the Philippines	125
Fig. 6.1	Madagascar, showing locations mentioned in the chapter	146
Fig. 7.1	The provinces of Vietnam during the reign of Tự Đức (prior to the territorial changes made due to French conquest after 1862)	171
Fig. 7.2	Incidence of disaster during Tự Đức's Reign based on data compiled from the *Veritable Records* and the *Annales of the Propagation of the Faith*	173
Fig. 8.1	The Philippines showing a detail of San Isidro and surrounding areas c.1887	205

Fig. 9.1 Ternate (*left*) and Sangihe Besar (*right*), Eastern Indonesia 228
Fig. 9.2 Java, showing the location of Mt Kelud 242
Fig. 10.1 Approximate paths of the cyclones of 1885 and 1890, with inset showing date growing region affected by 1890 cyclone 260
Fig. 10.2 Oman, showing places mentioned in chapter 262
Fig. 10.3 Muscat date exports, 1874–1905 266
Fig. 11.1 Broome and the Dampier Peninsula, showing the track of the March 1935 cyclone and the approximate position of the *Koolinda* (marked a 'K') when the cyclone was first reported 285
Fig. 11.2 Northwest Australia, showing places mentioned in the chapter and the locations of the severe cyclones of 1875, 1887, 1908 (April and November), 1910, and 1935 288
Fig. 11.3 Major losses through cyclones in the pearling industry, Northwest Australia, 1868–1935 (only losses of five or more lives and/or three or more vessels have been included; losses sustained by other industries, such as passenger or merchant shipping, are not counted) 290
Fig. 11.4 Monthly frequency of tropical storms, Northwest Australia, 1872–1923 292

List of Tables

Table 3.1	Households in the Coastal South, 2 BCE	61
Table 3.2	Taxpayers in Hai Duong and Nam Dinh, 1417–1819	69
Table 4.1	Details and codes for archival collections consulted in the UK, USA and India	82
Table 5.1	Number of remarkable typhoons per month, 1903–1918. Census of the Philippine Islands: Taken under the Direction of the Philippine Legislature in the Year 1918, (Manila: Philippine Islands Census Office, Bureau of Printing, 1920–1921), vol. 1, pp. 448–51	116
Table 5.2	Annual remarkable typhoon frequency, 1903–1918. Census of the Philippine Islands: Taken under the Direction of the Philippine Legislature in the Year 1918, (Manila: Philippine Islands Census Office, Bureau of Printing, 1920–1921), vol. 1, pp. 448–51	116
Table 5.3	The 25 most intense tropical cyclones in the Philippines (1951–2001), in order of wind speed	119
Table 5.4	Annual tropical cyclone numbers	123
Table 5.5	Number of typhoons affecting the Philippines from 1880 to 1898. *Report of the Philippine Commission to the President*, 1900, Vol. IV (Washington: Government Printing Office, 1901), p. 292	126
Table 10.1	Number of tropical storms > 34 knots and number of > 64 knots affecting the coastline of Oman and Yemen from 1890 to 1996	260
Table 10.2	Statement showing the number of date trees with their prices, destroyed by the cyclonic storm of June 5, 1890	265

List of Maps

Map 3.1	Dikes in the Red River delta, 1803	57
Map 3.2	Impact of sea intrusion at the current Red River delta	58
Map 3.3	Dikes of the Red River, 1926	59
Map 3.4	a. Historical evolution of the Red River delta, b. Sea level curve	63
Map 3.5	Sections of accretion and erosion in contemporary Nam Dinh coastlines	66
Map 5.1	Path of Super Typhoon Haiyan, November 2013	107
Map 5.2	Common typhoon paths in the Philippines	112
Map 5.3	Areas vulnerable to floods and typhoons, Philippines	114
Map 5.4	Areas vulnerable to storms surges, Philippines	127

CHAPTER 1

Bordering on Danger: An Introduction

Greg Bankoff and Joseph Christensen

Danger never lies far from the lives and minds of the people bordering the Indian Ocean. It is present in the air above, it bides its time in the ground below, and it lurks in the waters in between. Recent historiography has taken a turn towards a more transnational perspective as scholars attempt to come to terms with the difficulty of writing about the past in a globalised age. The term 'transnational', of course, is older, rooted in the sociological fabric of nineteenth-century America whose migrants had transcended Old World nationalisms and had created in its place a 'world federation in miniature'.[1] The notion of breaking with the nation-state as the preferred norm, of writing a more inclusive and wider history that seeks to cross geopolitical and cultural boundaries remains a signature of scholars who pursue this approach. This is particularly the case among historians of the Indian Ocean World (IOW), a vast arena spanning the Indian Ocean basin from the shores of East Africa to the South China Sea. A region, moreover, that has regained much of its former cohesion over the last few decades. Sugata Bose writes of 'the continuing relevance of the Indian Ocean as an inter-regional space in a time of intense global

G. Bankoff
Department of History, University of Hull, Hull, UK

J. Christensen
Asia Research Centre, Murdoch University, Perth, Australia

G. Bankoff, J. Christensen (eds.), *Natural Hazards and Peoples in the Indian Ocean World*, DOI 10.1057/978-1-349-94857-4_1

interconnections'.[2] A fatal reminder of this spatial and temporal unity took place on 26 December 2004 when giant tsunami waves triggered by a magnitude (Mw) 9 earthquake off the north-west coast of Sumatra struck communities around the Indian Ocean. Within seven hours, coastal populations as far distant as Sumatra and Somalia were left devastated as the common danger faced by all peoples of the IOW was 'demonstrated in the most tragic fashion by a great wall of water moving at the speed of a jet aircraft'.[3] The contributions in this collection respond to Bose's suggestion that disasters act to bind the peoples of the Indian Ocean together and link them to the world beyond. By placing natural hazards at the centre of Indian Ocean history, examining cross-cultural and trans-temporal themes in how communities coped with danger across time, the volume seeks to present a new, transnational understanding of the region's past.

Any attempt at discussing the historical unity of the IOW must begin with the work of Kirti N. Chaudhuri. While acknowledging a great debt to Fernand Braudel's recognition of the connection between the Mediterranean Sea and the peoples who live around its shores,[4] Chaudhuri argues that the countries of the Indian Ocean did not share a common destiny but that they did constitute 'a distinct sphere of influence' in terms of travel, the movement of peoples, economic exchange and climate, which created cohesion amongst the diverse religions, social systems and cultural traditions of the Irano-Arabic, Hindu, Indonesian and Chinese civilisations.[5] Both Braudel and Chaudhuri discuss their respective spaces as physical units and human units, the frontiers of which are not coterminous, which permit them to discuss hinterlands that were, at times, quite distant from the shore. As the title of his book suggests, *Trade and Civilization in the Indian Ocean: An Economic History from the Rise of Islam to 1750* (1985), Chaudhuri chose to focus on an historical study of long-distance trade in order to reveal the underlying sense of unity shared by the diverse civilisations and peoples that occupy the Indian Ocean rim. This trade was premised on climate, or more precisely on the monsoon cycle of alternating high- and low-pressure weather systems, which strictly governed regional food production and the sailing season and was thus indispensable to trans-oceanic economic activity across the basin. The monsoon, he writes, 'brought the whole area within the operation of a single global variable', and represented 'a cyclical component of time'.[6]

Chaudhuri's writings reveal a conscious effort to move beyond a historiographical framework focussed on the modern nations based on former European colonial empires of the Indian Ocean rim.[7] As such, he lays a path for other authors to follow. Kenneth McPherson's *The Indian Ocean: A History of People and the Sea* (1992) places the Ocean at the centre of a narrative that explores the common history of the societies and cultures linked through long-distance migration and trade regulated by the monsoon system. Michael Pearson, in *The Indian Ocean* (2003), also distinguishes a degree of unity and continuity revealed by the region's natural geography and includes an underlying 'deep structure' through topography, wind patterns and other physical phenomena.[8] Pearson went on to elaborate upon the commonalities of littoral societies, of finding around the ocean's shores a mixture of maritime and terrestrial influences where societies have 'more in common with other littoral societies than they do with their inland neighbours'.[9] Like Chaudhuri before them, both McPherson and Pearson acknowledge the influence of Braudel's monumental study of the Mediterranean and its central underlying theme of the role of climate and environment in regulating history across the *longue duree* of human civilisation. Their perspective is echoed more recently by Edward Alpers in his *The Indian Ocean in World History* (2014), where monsoons and ocean currents are described as the foremost of 'many deep continuities in the Indian Ocean'.[10]

It is perhaps to be expected that such histories would have a strong environmental bias in terms of both subject and method, that scholars who share a more transnational perspective would examine units that 'spill over and seep through national borders' and study 'the connections across national boundaries and the circulation of ideas, people and products these enable'.[11] While ecologies might be subnational or national in their focus, the environment per se and its principal manifestations such as climate, physical geography and hydrography would be recognised as inherently global or, at least, supranational in scale. In fact, one of the earliest attempts at a systematic transnational history, the concept of the Atlantic World, might seem to have just such 'an inbuilt geography'. David Armitage went on to elaborate a threefold typology of Atlantic history: *circum*-Atlantic history or the history of the ocean as an arena or zone of exchange, interchange, circulation and transmission; *trans-Atlantic* history or the history of meaningful comparisons between otherwise distinct histories; and *cis*-Atlantic history or the history of any particular place (nation, state, region, specific institution, etc.) in relation to the wider Atlantic world.[12] Some historians even talk about 'planetarity' and envisage human history in the context of planetary and cosmic evolution (Fig. 1.1).[13]

Yet the environment has rarely figured in transnational histories even though environmental histories of change that had little to do with national borders was one of the original examples offered by Ian Tyrell as a fruitful direction for further analysis.[14] Even if there has been an increasing interest in the notion of space and particularly the relationship between transnational spaces and politically bounded territories,[15] environmental historians, with few exceptions, have mainly continued to write histories that parallel the history of the nation-state, even if it is hard to see such boundaries in nature.[16] Histories of the IOW have long held the potential to challenge this wider neglect. On the one hand, the region has been often subdivided into subsystems based on 'natural' forces. Janet L. Abu-Lughod proposes circuits determined by the wind patterns of the separate monsoonal zones that limited any state's ability to influence events in an adjacent zone. She divides the pre-1500 Indian Ocean into three circuits: a westernmost circuit from the Red Sea–Arabian Peninsula–Persian Gulf to the south-west tip of India; a middle circuit,

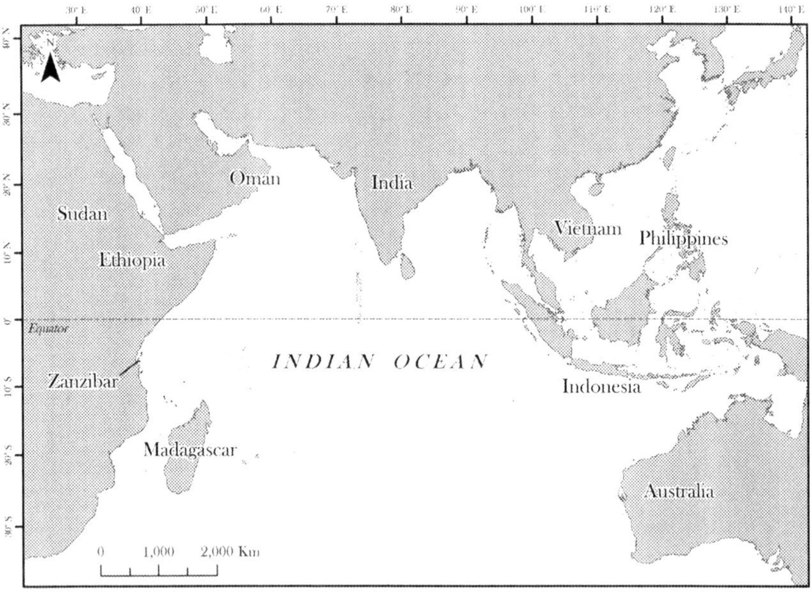

Fig. 1.1 The Indian Ocean World, showing locations covered by chapters in this book

from the south-eastern coast of India to the Straits of Malacca and Java; and an easternmost circuit, from the Straits of Malacca that includes the Indonesian and Philippine archipelagos to the great ports of south-east China.[17] Similarly, Eric Tagliacozzo identifies three principal 'ocean littorals' in his study of changes in Ocean-wide trade and production between the start of the seventeenth and the close of the nineteenth centuries; one in the eastern littoral of Southeast Asia, another in the northern littoral of the Indian subcontinent and a third in the western littoral of coastal East Africa.[18] Even Pearson, despite his insistence on the role of 'deep structures' underlying the Ocean's history, divides the region into separate systems, delimiting, for instance, an Afrasian Sea beginning at Sofala on the East African coast and extending right around the shoreline to the southern tip of India.[19]

Might there be a different basis for investigating the unity or otherwise of Indian Ocean history? If the wind patterns are central to determining the pace and rhythm of trade and cultural exchange across this ocean world and beyond, so the sea was also responsible for much of the risks that bound the people of these littoral societies together. On the water, craft of all kind, from fishing boat to merchant ship, faced the perils of wind and storm that shipwrecked many a mariner and sent many a vessel to the bottom. Off the water came hazards of a completely different magnitude and scale: tropical cyclones that caused havoc to town and country alike, destroying homes and ruining crops, and tidal surges that often obliterated all traces of human activity along the shoreline. The changing relationship of land and sea with the planet's rotation, the continental landmass to the north that warmed more quickly and the ocean to the south that cooled more slowly, not only generated the monsoons but also were partly responsible for the floods and droughts that all too frequently afflicted populations far into the interior. Danger, too, lay in the land itself, in the geophysical hazards that were more specific to certain sub-regional landmasses: the seismicity of its eastern edge (forming part of the Pacific Ring of Fire) and along its northern fringe (the Alpine–Himalayan belt), the volcanism of the Indonesian and Philippine archipelagos and its southern islands, Mauritius, Réunion and the Comoros, and finally the devastating tsunamis that infrequently ravage coastlines from shore to shore. Climate and geology are therefore responsible for many of the hazards that afflict the lands and peoples of the Indian Ocean so that the peoples and nations that border it also share a common historical experience of risk.

An Ocean of Risk

Hazards, then, bind the Indian Ocean together and link it to the world beyond. They also provide a basis for comparison and assessment of the peoples who live along its shores. How did different societies adapt to and manage risk? How have these social, cultural and economic adaptations over the centuries influenced culture, and how has culture influenced the nature of risk? Social scientists have long been intrigued by the idea that people who are frequently exposed to hazard adapt their behaviour to accommodate risk and that these patterns can become embedded in cultures over time. Interest is not so much in the one-in-one hundred-year 'big' events as with the smaller scale, more frequent hazards that can be regularly anticipated. The constant recurrence of such events, it is argued, can give rise to alternative normative structures that replace routine social norms with ones more in accord to a time of crisis: that is the very frequency of hazard can engender cultural adaptation or 'those adjustments, actual and potential, social, psychological and physical, which are used by residents of such areas in their efforts to cope with disasters which have struck or which tradition indicates may strike in the future'.[20] Adaptation to risk, however, is dependent on three crucial factors: that the hazard is repetitive, of a nature to allow forewarning, and that it inflicts significant damage to human and material resources.[21] The nature of risk in the Indian Ocean is such that it facilitates cultural adaptation but it is also, infrequently, of such a magnitude as to constitute a crisis or 'turning point' in the history of the people who have experienced it.[22] Both these kinds of hazards are present and have affected peoples and societies.

The dangers common to all peoples of the IOW are those that originate in the ocean itself in the form of climatic variability associated with the El Niño Southern Oscillation (ENSO) phenomenon. In popular usage, 'El Niño' refers to the recurrent large-scale ocean–atmosphere fluctuations that take place on an average three- to five-year cycle when variations in barometric pressure over the Indian Ocean are mirrored by variations in the Eastern Pacific. An El Niño event occurs when a pool of warm water moves eastward across the Pacific accompanied by high air surface pressure in the eastern Indian and western Pacific Oceans. The opposite La Niña phase takes place when warmer waters shift westward away from South America and lower than normal air pressure dominates over the western Pacific. In the Indian Ocean, the effects of ENSO variations are coupled by the Indian Ocean Dipole (IOD), which refers to variations in temperature gradients across equatorial waters between the western (Arabian Sea) and eastern Indian Ocean (eastern Timor Sea).[23] The ENSO phenomenon, as the most significant influence on global climate, has profound implications on the timing and volume of annual monsoonal rainfall across a wide section of the IOW, with El Niño phases responsible for periodic droughts in southern Africa, South and Southeast Asia and Australia, and the opposite La Niña phase for floods in these regions and drought in East Africa. Severe famines linked to El Niño droughts are known to have affected South and Southeast Asia in 1344–1345, 1631, 1685–1688, 1770 and 1792, while the severe fluctuation in global climate in the latter decades of the nineteenth century gave rise to famines that caused millions of deaths in India and China in 1876–1877 and 1896–1897, and which are counted amongst the worst famines in modern history.[24]

Danger comes off the sea, too, in the form of regular tropical cyclones and their frequently accompanying storm surges, as well as the much less common tsunamis. When the Indian Ocean cyclone season is merged with the Australian season and the annual north-west Pacific typhoons, the region accounts for over 70 % of the world's annual tropical cyclone activity. North of the equator tropical cyclones mainly form between May and November, and south of the equator from December to March, though destructive, out of season cyclones are not unknown. Early season cyclones mainly form in the warm waters of the Timor and Coral Seas and generally move in a south-south-westerly direction towards the Australian continent, while the few that originate over the Bay of Bengal move in straight lines west to strike India. These tropical cyclones often carry great amounts of precipitation. Late-season cyclones primarily develop north of 5° in the Arabian Sea and Bay of Bengal, are larger and have stronger winds that have the potential to cause enormous death and destruction as they make landfall on the low-lying but heavily populated coastlines of India, Bangladesh, Myanmar and less commonly the island of Sri Lanka. Some late-season cyclones will often travel great distances across the South Indian Ocean before swinging south-east to hit the Mascarene Islands, Madagascar and occasionally even mainland Africa along the Mozambican coast.[25] On its eastern outlier, the Philippines are regularly struck by tropical cyclones that originate over the warm waters of the Western North Pacific Ocean every year from April to December. The paths of some late-season typhoons may take them over 3000 miles of open, sun-drenched ocean before landfall allowing them to grow into 'heat-driven machines of enormous destructive potential'.[26] Data from the second half of the twentieth century suggests that on average approximately seven to nine typhoons make landfall over the Philippines each year.[27]

Data on the frequency and magnitude of tropical depressions is patchy before the twentieth century and is often more anecdotal with only the most severe events remembered. However, the record does extend back to the eighteenth century in some areas especially those subjected to European colonialism from an early date. Between 1737 and 1995, Indian coastlines were struck by no less than 640 tropical cyclones of varying size and intensity, an average of two per year, 200 of which were Category 3 or over events capable of inflicting extensive damage and high death tolls.[28] Similarly, a record exists for the Mascarene Islands of Mauritius, Réunion and Rodrigues from the seventeenth century showing that 89 cyclones made landfall there between 1656 and 2007, although the data are more reliable from the late eighteenth century onwards.[29] More recently, too, attempts have been made to use indigenous language sources to reconstruct the cyclone history of vulnerable subnational regions like Tamil Nadu.[30] In other areas, such as along the Australian coastline, archival records have been supplemented by palaeo-tropical cyclone evidence based on high-resolution isotope records. Isotope records have the advantage of identifying a broad range of tropical cyclone intensities and frequencies over long-term periods, centuries or even millennia. Their principal disadvantage, however, is that such studies focus on an examination of sedimentary deposits left by only the most severe cyclones accompanied by storm surges.[31] That is, they are principally concerned with the 'big' event rather than with smaller, more frequent ones that might lead to societal adaptation.

The record is complete enough, however, to show that tropical cyclones have historically been a principal source of death and destruction across the Indian Ocean. This is particularly the case in the Bay of Bengal, straddling the modern nations of India and Bangladesh, where tropical cyclones 'have routinely exacted near apocalyptic death tolls from the shallow farming and fishing settlements of the Ganges River Delta and Deccan plateaus'.[32] Low-lying with comparatively large normal tide swings ranging from three to five metres, large areas of the Bay of Bengal are flooded by relatively small increases in sea level. There are two typhoon seasons in the Bay of Bengal, one concentrated in May (pre-monsoon) and the other spanning October–November (post-monsoon).[33] Cyclones making landfall near the time of high tide have produced particularly devastating floods in the past. The Hooghly River Cyclone of October 1737 with its accompanying 10-

to 13-metre storm surge rolled across the heavily populated delta region around Calcutta drowning an estimated 300,000–350,000 people. It has long been considered one of the single worst cyclone disasters in history, though recent scholarship has raised serious doubts about the scale of the death toll.[34]

Tropical cyclones killing more than 10,000 people struck again in 1787, 1789, 1822, 1833 and 1839.[35] In October 1864, the Great Calcutta Cyclone, widely considered to be one of India's most vicious, drowned more than 50,000 people as it submerged large areas of the thriving port city and caused a further 30,000 deaths through subsequent outbreaks of cholera and dysentery.[36] Twelve years later, in October 1876, another tropical cyclone accompanied by a 13-metre storm surge tore through the Ganges Delta, almost annihilating the city of Bakergunj (modern-day Bangladesh) and claiming 100,000–200,000 lives.[37] The death toll of 40,000 during the little reported Bengal Cyclone of 16 October 1942 was undoubtedly exacerbated by military concerns for security during World War II and the failure or inability of authorities to issue adequate forewarning or organise evacuations.[38] Further killer cyclones followed in 1963 and 1965. The scale of these earlier disasters, however, pales in comparison to the tragedy of the disaster that unfolded in November 1970 when a cyclone made landfall at the mouth of the Ganges almost exactly at high tide. Between 1797 and 1991, Bangladesh has been hit by 60 severe cyclones, 32 of which were accompanied by storm surges.[39] However, the storm surge on this occasion is estimated to have been in excess of six metres. Coming ashore at night, while most people were indoors, it swept buildings, vegetation and animals before it, flooding almost a quarter of East Pakistan's landmass and taking the lives of between 300,000 and 500,000 people.[40] The failure of the Pakistani government to organise timely assistance and the rampant corruption evident in relief efforts is credited with contributing to East Pakistan's move towards independence and a vicious ten-month civil war that led to the formation of the sovereign state of Bangladesh in 1971.[41]

The people who live along the coast of the Bay of Bengal may represent one of the most acutely at risk populations in the Indian Ocean but societies all along its shores are periodically vulnerable to the danger of tropical cyclones. Recent research, too, indicates an upward trend in the magnitude if not the frequency of tropical cyclones in the last 30 years, mainly north of the equator as a consequence of global warming. This trend is due to storms having longer lifetimes and greater intensities and is highly correlated with sea surface temperatures.[42] Even regions that have hitherto only infrequently experienced tropical cyclones are being struck more often. For instance, since the 1990s, the number of tropical cyclones that have made landfall over Myanmar has increased both in number and intensity. Three mature tropical cyclones made landfall along the Myanmar coastline in almost consecutive years: 2004, 2006 and Cyclone Nargis in 2008.[43] According to official figures, 84,500 people died, a further 53,000 were reported missing and 2.4 million people were affected.[44]

In recent years, too, Oman has been struck by rare tropical cyclones. Tropical cyclones in the Arabian Sea seldom exceed tropical storm intensity but Cyclone Gonu in June 2007 caused extensive damage over the easternmost tip of the Sultanate: extensive wadi flooding, $4 billion in damages and at least 49 deaths. Gonu was followed three years later by Cyclone Phet, which passed over Sur in June 2010, resulting in 21 deaths and $700 million in damages. However, these were not the deadliest storms to hit Oman as a cyclone that struck in June 1890 reportedly caused 727 fatalities.[45] Further south, the African coastline is not except from the furry of cyclones. Cyclone Leon-Eline, the longest-lived Indian Ocean event on record, struck Madagascar in 2001 before crossing the Mozambican coast and dissipating over Namibia. Madagascar was hit again in quick succession by Cyclones Gloria and Hudah that caused the worst floods on Madagascar in a generation.[46] In contrast, the Western Australian coastline has shown a marked decrease in the frequency and destructiveness (in terms of fatalities) of tropical cyclone activity since 1960, although world-record wind gusts of 408 km/h were recorded by Cyclone Olivia in 1996. Projections suggest a continuing downward trend not only in Australia but also in the south-west Pacific and southern Indian regions in the twenty-first century.[47]

Typhoons and storm surges are not the only dangers that come off the sea. Tsunamis, of course, are not generated by cyclonic activity but are mainly caused by thrust-type subduction zone earthquakes which generate massive waves of water that cross vast ocean expanses before slowing down, increasing in height and crashing onto shallow-shelving, low-lying coastal regions. On landfall, waves vary in height between a few metres and tens of metres with their power accentuated by underwater landslides often associated with such earthquakes. The main tsunamigenic subduction zones in the Indian Ocean run along the Makran coast from Iran and Pakistan to the Indus Delta and then down the Kutch–Saurashtra coast, and in the Bay of Bengal from Bangladesh and southern Myanmar through the Andaman and Nicobar group of islands and then along the Sumatran coastline to Java and the islands of eastern Indonesia. Research on palaeo-tsunamis in the Indian Ocean based on inland deposits of marine and brackish fossils shows that these events are rare, occurring perhaps on a basin-wide scale only three times in the last two centuries in 1833, 1883 and 2004.[48]

Little is known about the 1833 tsunami except that it was generated by a Mw > 8.7 earthquake somewhere off the south central coast of Sumatra and that computer simulations suggest its effects were more directed towards Africa and Western Australia than towards Sri Lanka.[49] The 1883 tsunami is not associated with an earthquake but with the volcanic eruption of Krakatoa in Indonesia. The explosion was so powerful that part of the overlying land and sea bed collapsed into its underground magma chamber. A series of large tsunami waves were generated, some reaching a height of 35 metres above sea level. Tsunami waves were observed in the Indian Ocean, the Pacific, along the coasts of North and South America and even as far away as the English Channel. Evidence of an earlier, even larger eruption in the year 416 or 535 also exists in ancient Japanese scriptures.[50] The Indian Ocean Tsunami of 2004 was generated by a massive Mw 9.3 earthquake that released energy equivalent to 23,000 Hiroshima-type atomic bombs along the Andaman–Sumatran subduction zone. A tsunami struck the Indonesian coast less than one hour later with wave heights ranging between 15 and 27 metres that penetrated over a kilometre inland. Two hours later, 12-metre waves hit Sri Lanka and the Indian coastlines. Tsunami waves devastated ocean islands like the Maldives and reached as far as Somalia on the East African coast. All in all, nearly 300,000 people in 13 countries are estimated to have died, making it one of the deadliest such events known.[51]

Apart from these basin-wide tsunamis, smaller but more frequent events have affected the Bay of Bengal and the Arabian Sea. The Sumatran trench that runs parallel to that island's west coast is one of the most active plate tectonic margins of the world where the Indian–Australian plate gets subducted beneath Southeast Asia. Eighty per cent of the tsunamis of the Indian Ocean originate in this area. Major earthquakes in excess of Mw < 8.0 like those in 1833, 1861 and 2005 have a return period of 150–200 years. Smaller tsunamigenic earthquakes of Mw > 8.0 like those in 1843, 1907 and 1935 occur much more frequently on average every couple of decades.[52] The first recorded tsunami in the Bay of Bengal occurred in 1762 though there is now some dispute as to the scale of its effects along what is now the coastline of Bangladesh.[53] The first recorded tsunami in the Arabian Sea dates back to 326 BCE when the Macedonian fleet of Alexander the Great was destroyed on its return from India.[54] Limited historical records on the Makran subduction zone have resulted in an underestimation of the seismogenic potential for tsunamis along the Omani, Iranian, Pakistani and West Indian coastline. However, the Makran earthquake of 28 November 1945 generated tsunami waves up to 11.5 metres and killed more than 4000 people along the Iran–Pakistan border, penetrating even the usually protected Arabian Gulf.[55] More recent thermal modelling suggests that the Makran subduction zone may be capable of generating a very significant (Mw > 8.5) tsunamigenic earthquake.

Analysis of the records since 1701 entered on the World Data Centre (WDC) shows no conclusive trends in tsunami occurrence worldwide though the data exhibit a marked decadal periodicity (1701–1710, 1751–1760, 1861–1870, 1881–1890, 1891–1900 and 2001–2010) with higher fatalities coinciding with mega-events.[56] Tsunamis are also occasionally generated by the subduction zone system in the Philippines. In 1976, a strong earthquake struck the southern island of Mindanao producing a great tsunami that destroyed 700 km of coastline along the Moro Gulf in the North Celebes Sea. Waves up to 15 metres in height pulverised Padigan City and Cotabato City, killing approximately 7000 people and leaving over 93,000 homeless.[57]

The extent of destruction from earthquakes depends on several factors: its magnitude and intensity, that is, a measure of the amplitude or energy released during the event; its duration, with damage generally increasing over time; the local geology or the character of the underlying material; and the time of day, as more fatalities occur at night or in the early morning when people are indoors. Earthquakes are frequent along the Alpine–Himalayan arc and along that part of the Pacific Ring of Fire that includes Indonesia and the Philippines. In particular, Iran is located in an area of intense seismic activity and has repeatedly experienced highly destructive earthquakes of truly epic proportions: the Ardabil earthquake of 893 that caused as many as 150,000 deaths in an area stretching from Iran to Armenia, and the Khorasan earthquake of 1101 that reportedly killed 60,000 people. Some of the country's major cities have been periodically devastated by such events. Tabriz was initially destroyed by an earthquake in 858. The first well-documented event occurred in 1042 when perhaps as many as 40,000–50,000 people lost their lives. The next major earthquake occurred in 1721 with estimates of casualties ranging wildly between 8000 and 250,000. Six years later in 1727, the city was destroyed again with a reported death toll of 77,000. Other major earthquakes followed in 1755, when 40,000 died, and 1780, when whole villages in a damage zone extending 60 km disappeared into fissures. Some sources place the death toll as high as 50,000–60,000 people. More recent earthquakes have been equally as costly, most notably the Gilan earthquake of 1990 near the Caspian Sea that led to 50,000 deaths and left 200,000 people injured, and the Bam earthquake of 2003 that killed 43,000 people, left over 90,000 homeless, and flattened the latter's ancient walled citadel, which had been a major tourist destination.[58]

The Indian subcontinent, too, has been wracked by earthquakes. The Indian plate is currently penetrating into Asia at a rate of approximately 45 mm/year extending in an arc along the Himalayas. The record of earthquakes is patchy before 1800 despite a written tradition that extends beyond 1500 BCE. Little is known of earthquakes that occurred more than 500 years ago, and a more complete record only exists for the last two centuries. Early earthquakes are described in the Mahabharata (circa 1500 BCE) and several religious texts suggest a major Himalayan earthquake around the time of the Buddha's enlightenment (circa 538 BCE). Archaeological excavations in Sindh and Gujarat suggest earthquake damage to now abandoned Harappan cities. A single paragraph describes a massive earthquake in the Kathmandu Valley in 1255 and the arrival of Vasco de Gama's fleet in 1524 is said to have coincided with a violent tsumanigenic earthquake. A series of major earthquakes occurred in the Western Himalayas during the sixteenth century, another is known to have occurred in the Kathmandu Valley in 1668 and a severe earthquake rocked Bhutan and parts of Assam in 1713.[59] A massive earthquake with an epicentre to the east of Chittagong rocked the Bay of Bengal in 1762, causing liquefaction and compaction of riverbanks, submergence and subsidence, extensive flooding, the collapse of buildings and an unknown number of deaths.[60] Further costly earthquakes devastated the cities of Allahbund (Gujarat) in 1819 and Srinagar in northern India in 1885. Calcutta and Assam were struck by a huge Mw 8 earthquake in 1897 that offset telegraph poles by 3–3.5 metres and shifted one railway segment two metres laterally. The Brahmaputra River rose nearly eight metres in one area and reversed its flow, while landslides destroyed many of the hill towns in the area, most notably Shillong. Over 1500 people were killed and thousands were injured.[61]

The first major earthquake of the twentieth century happened at Kangra in April 1905 when an estimated 20,000 people died and over 100,000 buildings were destroyed.[62] Major earthquakes also occurred in successive years at Bihar near the Nepal border and Quetta in modern-day Pakistan in 1934 and 1935, respectively. The death toll in the latter case was particularly devastating, with approximately 35,000 people out of a total population of 40,000 meeting their death. A notable feature of the Bihar earthquake was that most victims were female; as the earthquake occurred during the day, many women were inside buildings engaged in domestic work.[63] Mounting death tolls characterise more recent earthquakes that continue to shake the modern nations of India, Pakistan and Bangladesh as population densities rise: the Bhuj earthquake of 2001 killed nearly 20,000 people, injured another 166,000 and left more than 600,000 homeless; and the Muzaffarabad earthquake in 2005 killed 86,000 and left 2.5 million people homeless. Many of the dead were children attending class in poorly constructed schools that collapsed.[64] Children fared better in the recent earthquake that devastated Nepal in 2015 as they were at home on a Saturday. The death toll, however, now estimated at 9000 is fast approaching the 10,000–12,000 fatalities of the even larger 1934 event that destroyed large parts of the capital, Kathmandu.[65]

While earthquakes are certainly not infrequent events in Indonesia and the Philippines, many of those that do occur are associated with the numerous active volcanoes that dominate the landscapes of these two archipelagic nation-states. The capital city of the Philippines, Manila, has been destroyed on numerous occasions since its founding by the Spanish in 1571, most notably in 1645, when the early colonial capital was almost completely levelled to the ground, and again in 1863 which destroyed many government buildings, hospitals and churches.[66] The capital of the Netherlands East Indies, Batavia (modern-day Jakarta) was also largely destroyed by an earthquake in 1699. More recently, major earthquakes have struck the modern cities of Baguio in the Central Cordillera of Luzon (1990) and the Bantul–Yogjakarta area of Java (2006) which destroyed more than 80 % of Bantul and killed over 6000 people. This earthquake, however, is considered to have been related to the renewed eruptive activity of nearby Mount Merapi.[67] Moreover, the media attention given to the Indian Ocean tsunami in 2004 often obscures the fact that the waves were generated by the fourth strongest earthquake recorded since measurements began in 1899.[68] As with the data on tsunamis, there is no discernible trend in the number of major earthquakes since 1701 though there appears to be periodic cycles when the occurrence of earthquakes greater than 7.5 magnitude peaked between 1721–1730, 1751–1760, 1771–1780, 1911–1920, 1971–1980 and 2001–2010. Four out of the top ten earthquakes with the highest number of fatalities on record have occurred since 2000.[69]

While volcanoes do exist elsewhere in the Indian Ocean basin—most notably in the Mascarene Islands, where Piton de la Fournaise on the island of Réunion is a highly active basaltic shield volcano; in the Comoros, where Mount Karthala has erupted more than 20 times since the start of the nineteenth century; and on Barren Island in the Andaman Sea, the only active volcano in the Indian subcontinent—most eruptions occur in Indonesia and the Philippines. In particular, Indonesia is an archipelago of volcanic islands beneath which the oceanic crust from the Australian plate is actively subducting into the Java Trench. Indonesia has more than 200 active volcanoes and was the site of the 1257–1259 Samalas eruption on Lombok that adversely affected climate and harvests around the world, and the 1883 eruption of Krakatoa, one of the most famous and violent volcanic events in recent history. The noise of the latter eruption, reputedly the largest sound in recorded history, was heard over 4800 km away on the island of Rodrigues where it was mistaken for the sound of ships' guns firing just over the horizon.[70] Lava rafts, 'some with human skeletons', even drifted as far as the east African coast.[71] The eruption of Tambora on the island of Sumbawa in 1815, a mega-colossal VEI (Volcanic Explosivity Index) 7 event is estimated to have killed more than 92,000 people, 80,000 of these as a result of starvation and disease. Much of the estimated 150 km^3 of solid material cast into the atmosphere remained airborne in the form of a fine ash and has been implicated in dramatic changes in climate and weather patterns in the northern hemisphere during the following year, the so-called year without a summer.[72] Among the many calderas formed by past eruptions throughout the archipelago is Toba on the island of Sumatra formed some 74,000 years ago as a result of an apocalyptic VEI 8 eruption, the largest such event in the last two million years. Many of Indonesia's volcanoes are very active, such as Merapi (Mountain of Fire), a stratovolcano that has erupted more than 68 occasions since 1548. One such eruption in 1006 is credited with the demise of the central Javanese kingdom of Mataram and its relocation to East Java.[73]

There are over 20 active volcanoes in the Philippines, too, that have erupted frequently in historical times. Mayon, in southern Luzon, is often described as having the world's most perfect cone and has erupted 47 times since 1616. Taal, in the centre of a caldera lake just south of Manila, has erupted at least 34 times since 1572, and, on occasion, has generated tsunamis that inundated lake-side communities and led to their relocation away from the shore.[74] The eruption of the Macaturin volcano around 1765 has been implicated in the movement of the Iranun people out of the Rio Grande valley in south-western Mindanao, where, over the course of the next century, they became feared maritime raiders and traders with settlements spread widely along the shores of the present-day Philippines, Indonesia and Malaysia.[75] Most recently, of course, the eruption of Mount Pinatubo in 1991 devastated the rice lands of central Luzon leaving more than one million people homeless and forcing the USA to abandon their strategically most important regional military post in East Asia, Clark Air Force Base. The scale of destruction and death was also prolonged by heavy rains associated with subsequent typhoons that unleashed pyroclastic flows and lahars onto surrounding farmlands and settlements for years to come.[76]

Bordering on Danger

As this brief catalogue of droughts, cyclones, storm surges, tsunamis, earthquakes and volcanoes so clearly shows: danger is never far away from the many different communities who live along the borders of the Indian Ocean. Over time, its peoples have had to learn to live with their consequences and adapt their cultures to meet the exigencies and challenges of living with these different forms of risk on a daily basis. The demands of living in a dangerous environment are explored in detail in the chapters that make up the remainder of this volume. For most regions, the historic impacts of 'natural' disasters remain poorly studied, even though hazards have been frequent and often potent events; as Anthony Reid observes in Chap. 2, in the case of Indonesian history, the interest of most scientists and historians in past disasters has been triggered comparatively recently by the 2004 Asian tsunami and the eruption of Mount Merapi in October 2010. Through his case study and the chapters that follow, a deeper understanding emerges of the complex links between populations, economies, climates and environments, where natural hazards loom as a constant and powerful force shaping the historical development of the IOW.

Reid's focus in this chapter is on the relationship between population history and the turbulent seismicity of the Indonesian archipelago.

The twentieth century demonstrates that a benign tropical climate and fertile volcanic soils have given rise to rapid demographic growth, particularly on Java and Sumatra. Yet historical and archaeological evidence shows that this has not always been the case. Both Dutch and Indonesian sources often tend to overlook volcanic eruptions and tsunamis. However, a close re-reading of these materials and the evidence of geological research expose periodic and devastating disasters that recast the archipelago's history as beset by a discontinuous pattern of development in which the relative quiescence of the more thoroughly documented twentieth century emerges as an exceptional period. Enhanced knowledge through cooperation between historians and scientists assists in planning for future seismic activity, as well as yielding new insights into the history, not only of the Malay world, but of the entire globe. As Reid points out, Southeast Asian volcanoes are now thought to be responsible for the northern hemisphere's 'year without summer' in 1816 and for inaugurating the seventeenth century's 'Little Ice Age'.

The next chapter, Li Tana's study of land reclamation and loss in Vietnam's Red River delta, also presents a long-term interaction of environment and society in a highly productive, yet hazard-prone region. Vietnam's Red River delta, 'one of the most altered landscapes on Earth', has been subject to reclamation and dike construction for more than two millennium. Her research pinpoints this large-scale reclamation to three crucial periods across two millennia, each marked by eventual abandonment and failure, a pattern she attributes to changes in the Red River's course and the inexorable influence of typhoon and flood on the despoliation of agricultural land and migration out of coastal areas. As she explains, such findings question established celebratory narratives of the Red River's reclamation by revealing 'the exposure and vulnerability of human projects in their ongoing battle against the sea'.

Both of these opening chapters reveal that it is not the immediate casualties of a particular event but rather the ensuing loss of agricultural productivity that often causes the greatest hardship and suffering. This is also a key insight of the next two chapters by George Adamson and James Warren. In Chap. 4, Adamson draws on scientific reconstructions of monsoonal variability to examine social responses to drought in western India between the late eighteenth and mid-nineteenth centuries, a comparatively neglected period in contrast to the more extensively studied droughts and famines of the later nineteenth century. The period between 1782 and 1857 marked the shift from indigenous to colonial governance in the Bombay Presidency, representing a shift towards a more laissez-faire approach to famine relief that contributed, in turn, to the breaking down of local adaptive responses in how agriculture and charity were administered. Warren's chapter on typhoons in the Philippines, on the other hand, points to long-term cycles in the frequency and intensity of cyclonic storms linked to the ENSO cycle that increased the likelihood of 'super typhoons' and for clusters of powerful typhoons towards the end of the eighteenth, nineteenth and twentieth centuries. The impact of these typhoons has also become more severe over time as agriculture has shifted from traditional forms of polyculture to raising cash crops for commodity export markets. Rural populations have been rendered more vulnerable to economic loss and starvation following the passage of large or repeat typhoons. As Warren suggests, the record of this historical experience will prove only too relevant to contemporary disaster mitigation and preparedness as the intensity if not the frequency of typhoons increases as a result of climate change in the twenty-first century. The cycles of severe drought and devastating typhoons explored by Adamson and Warren in their chapters suggest a fine balance between, on the one hand, the vital need for monsoonal rains, the regularity of extreme weather events and the well-developed indigenous systems of adaptation formed through repeated experience of hazard, and, on the other hand, the less frequent and severe events that can overcome adaptive strategies, particularly in contexts where indigenous practices of coping with risk have been undermined through the social and economic transformations wrought by European imperialism.

The history of bushfire in Madagascar in Chap. 6, by Christian Kull, explores the ambivalence of natural hazards as a central theme. Fire has been neglected in Madagascar's history, Kull argues, because it is 'multivalent, ambiguous, and flexible. It is both human and natural, both useful and problematic'. Kull builds on the concept of pyric phases by identifying six distinct periods when fire served, or is perceived to have served, different roles as an occasional hazard, a useful tool, a major transitional force, or as an agent of change. The use of fire by farmers and herders to manage the island's grasslands had become an established practise by the nineteenth century. Yet, in the period that followed French colonisation in 1896, the use of fire as a 'tool' increasingly came into conflict with attempts to restrict fire as a hazard to natural resource management and, in time, nature conservation. As Kull outlines, policy towards fire has necessitated a political compromise between the different pyric phases, and in the interests of the different economic and social groups who are touched, in one way or another, by the ambivalence of fire as both a useful and problematic influence in the Madagascan landscape.

The effectiveness of pre-colonial and colonial disaster preparedness and relief is explored in detail in the chapters by Greg Bankoff and Alicia Schrikker. Bankoff examines local responses to a typhoon and subsequent flood that affected the provincial town of San Isidro on the Philippines island of Luzon in September 1887. Although not a particularly significant flood by the standards of the Philippines, the 'storm over San Isidro' is remarkable for the detailed records of its passage and for the evidence these sources provide of a vibrant civil society that was responsible for organising an effective, grassroots response to the flood. This resilience, evident at the level of civic community in San Isidro, was shaped by a long history of coping with repeated storm and flood stretching back through the generations, and challenges prevailing assumptions about non-Western approaches to disaster risk mitigation. It also hints at the origins of the active civil society that is such a characteristic of Filipino society today. As Bankoff argues, the '[r]epeated experience of disasters can engender adaptations that become an integral part of local and even national culture'.

Cross-cultural interaction in disaster management is central to Schrikker's chapter, which explores indigenous and colonial approaches to response, recovery and reconstruction in relation to three crucial events: a volcanic eruption and earthquake on the island of Ternate in Eastern Indonesia in 1840; a volcanic eruption, earthquake and tidal wave on the island of Sangihe Besar, north of Menado in 1856; and the eruption of Gunung Kelud in 1919 in central Java. Her chapter problematises the Dutch colonial presence by examining how the extension of colonial rule led to greater intervention in disaster management, where different approaches and forms of knowledge came together by necessity in the face of hazard. Schrikker explores the evolution of a colonial humanitarian tradition, in which disaster victims were integrated into Dutch traditions of charity and humanitarianism, before showing how mitigation and management policy was often governed by a political agenda aimed at population control and avoiding mass migration after disasters. The Kelud disaster of 1919 is identified as a turning point in this broader history, marking the advent of a new prevention-oriented and scientifically based approach developed in response to local practises.

In Chap. 7, Kathryn Dyt looks at responses to 'bad weather' including storms, floods, typhoons and droughts and the resulting famines, epidemics and local uprisings that wreaked havoc on Vietnam's state infrastructure and human life in the second half of the nineteenth century. In accordance with the Confucian doctrine of the mandate of heaven, Vietnam's Emperor Tự Đức bore responsibility for these calamities, and, as Dyt explains, historical accounts have not only recycled this view but portray him as unresponsive to the suffering of his people. Contrary to this picture, her chapter shows, Tự Đức was not detached and inactive in the face of crisis but was deeply concerned about disasters and the impact of natural calamities. Dyt draws on Tự Đức's court records, folk songs and the accounts of French missionaries that she interprets through the prism of the socialist 'new history' of the twentieth century. The result is a more balanced picture of Tự Đức's reign and his response to natural hazards which suggests that the causes for disaster not only were ascribed to the emperor's wrongdoing but also were linked to a complex web of spiritual and social factors. These findings enable a wider re-evaluation of Tự Đức's reign and his place in Vietnam's nineteenth-century history. Dyt's chapter, like others in this book, highlights the value that the innovative use of new or untapped source materials holds for understanding the influence of natural hazards across the IOW.

The origins of vulnerability and thus the root causes of disasters are to be found 'in the political structures, economic systems and social orders of the societies in which they take place'.[77] This insight is particularly evident in the final two chapters by Matthew Hopper and Joseph Christensen. Hopper's chapter examines two cyclones that crossed the coast of Oman in 1885 and 1890 devastating the date plantations that supported a substantial global export trade in the late nineteenth century. These cyclones coincided with prolonged drought conditions, and associated famines and epizootic disease in East Africa, affecting population susceptible to enslavement at the same time as demand for their labour rose in Oman to replant devastated date plantations. As Hopper shows, the concurrence of environmental factors in East Africa and the Arabian Gulf acted synergistically to create an environment conducive to a growth in the Indian Ocean slave trade despite Western political and diplomatic efforts to end it. Christensen's chapter centres on a cyclone in March 1935 that devastated the local pearlshelling industry and became one of the most costly disasters recorded on Australia's western or Indian Ocean coast. The 1935 cyclone was also a milestone in the development of Australian disaster relief policy, marking the first occasion that the federal Commonwealth government granted financial aid to assist recovery from a natural catastrophe. The chapter shows that this important milestone in Australian disaster relief policy was linked to concerns over Japanese competition that threatened the viability of the pearlshelling industry and, in turn, the coastal communities it supported. In this way, official responses to the March 1935 disaster marked an extension of deep-seated anxieties in Australian society over the place of 'white' settlement in the nation's monsoonal north and the nation's position on the margins of the IOW during the first half of the twentieth century.

These two final chapters on Indian Ocean cyclones touch upon an important theme that emerges across the collection as a whole—that natural hazards do not pay heed to political boundaries, but rather extend beyond the borders of nation-states to regularly impact upon the lives and activities of geographically related but socially and culturally distinct populations. It is through such studies that hazard and risk emerge as unifying elements in the history of the IOW. There is a growing awareness that historical knowledge is essential, not only for appreciating the role of natural hazards in the past development of societies, but also for the development or improvement of risk reduction strategies in the present and future. Qualitative assessments of the impacts of past hazards on enhancing community resilience and the value of comparative historical studies in raising the awareness of cross-cultural and non-Western approaches to adaptation and mitigation is increasingly recognised by disaster risk reduction practitioners today.[78] This collection responds to the need for an enhanced understanding of the ways in which natural hazards have shaped the economic, social and political development of human societies, in areas that are highly susceptible to climatic shifts, extreme weather and seismic activity, and in a region that is of emerging global geopolitical and economic significance.

Notes

1. Bourne, Randolph, 'Trans-National America', *Atlantic Monthly*, 118 (July 1916), 86–97, available at: http://www.theatlantic.com/magazine/archive/1916/07/trans-national-america/304838/ (accessed 13 June 2014).
2. Bose, Sugata, *A Hundred Horizons: The Indian Ocean in the Age of Global Empire* (Cambridge: Harvard University Press, 2006): 1–2.
3. Bose, *A Hundred Horizon*, 1–2.
4. Braudel, Fernand, *La Méditerranée et Le Monde Méditerranéen a L' Époque De Philippe II*, 3 vols (Paris, 1949).

5. Chaudhuri, K.N., *Trade and Civilisation in the Indian Ocean: An Economic History from the Rise of Islam to 1750* (Cambridge: Cambridge University Press, 1985): 3, 21.
6. Chaudhuri, *Trade and Civilisation in the Indian Ocean*: 23, 127–128.
7. See Mukherjee, Rila, 'The Indian Ocean: Historians Writing History', *Asian Review of World Historians*, 1:2 (2013): 295–307; Vink, Markus P.M., 'Indian Ocean Studies and the "New Thalassology"', *Journal of Global History*, 2 (2007): 41–62; also Chaudhuri, K.N., *Asia Before Europe: Economy and Civilization of the Indian Ocean from the Rise of Islam to 1750* (Cambridge and New York: Cambridge University Press, 1990).
8. Pearson, Michael N., *The Indian Ocean* (London and New York: Routledge, 2003): 13.
9. Pearson, *The Indian Ocean*: 353.
10. Alpers, Edward A., *The Indian Ocean in World History* (New York: Oxford University Press, 2014): 146.
11. Seigel, Micol, 'Beyond Compare: Comparative Method After the Transnational Turn', *Radical History Review* 91 (2005): 63; Heilbron, Johan, Nicolas Guilhot and Laurent Jeanpierre, 'Toward a Transnational History of the Social Sciences', *Journal of the History of the Behavioural Sciences* 44, 2 (2008):147.
12. Armitage, David, 'Three Concepts of Atlantic History' in David Armitage and Michael Braddock (eds.) *The British Atlantic World, 1500–1800* (Basingstoke and New York: Palgrave Macmillan, 2009): 11–27.
13. Iriye, Akira, *Global and Transnational History: The Past, Present, and Future* (Basingstoke and New York: Palgrave Macmillan, 2013): 79–80.
14. Tyrrell, Ian, 'American Exceptionalism in an Age of International History', *American Historical Review* 96, 4 (1991): 1031–1055.
15. Müller, Michael G. and Cornelius Torp, 'Conceptualising Transnational Spaces in History', *European Review of History* 16, 5 (2009): 614.
16. White, Richard, 'The Nationalisation of Nature', *Journal of American History* 86, 3 (1999): 976. There are a few notable exceptions whose works could be considered as incipient transnational environmental histories even if their authors did not identify them as such: Lamb, Hubert H., *Climate, History and the Modern World* (London: Routledge, 1982); Crosby, Alfred W. *Ecological Imperialism: The Biological Expansion of Europe, 900–1900* (Cambridge: Cambridge University Press, 1986); Grove, Richard H., *Green Imperialism: Tropical Island Edens and the Origins of Environmentalism, 1600–1860* (Cambridge: Cambridge University Press, 1995).
17. Abu-Lughod, Janet, *Before European Hegemony: The World System A.D. 1250–1350* (New York and Oxford: Oxford University Press, 1989): 253–260.

18. Tagliacozzo, Eric, 'Trade, Production and Incorporation: The Indian Ocean in Flux, 1600–1900', *Itinerario* 26:1 (2002): 75–106.
19. Pearson, Michael N., *Port Cities and Intruders: The Swahili Coast, India, and Portugal in the Early Modern Era* (Baltimore and London: The Johns Hopkins University Press, 1998): 36–62.
20. Moore, Harry Estill, *And the Winds Blew* (Austin, Texas: Hogg Foundation for Mental Health, 1964): 195.
21. Wenger, Dennis E. and Jack M. Weller, *Disaster Subcultures: The Cultural Residues of Community Disasters, Preliminary Paper No. 9* (Columbus, Ohio: Disaster Research Centre, Ohio State University, 1973): 1–9.
22. Koselleck, Reinhart, 'Crisis', *Journal of the History of Ideas* 67, 2 (2006): 371, 399.
23. Bankoff, Greg, *Cultures of Disaster: Society and Natural Hazard in the Philippines* (London: Routledge Curzon, 2002): 123–127; Garden, Donald, *Droughts, Floods & Cyclones: El Niños that Shaped Our Colonial Past* (North Melbourne: Australian Scholarly Publishing, 2009): 1–11.
24. Fagan, Brian, *Floods, Famines, and Emperors: El Niño and the Fate of Civilizations* (New York, Basic Books, 1999), pp. 3–25; Davis, Mike, *Late Victorian Holocausts: El Niño Famines and the Making of the Third World* (London: Verso, 2001).
25. Longshore, David, *Encyclopaedia of Hurricanes, Typhoons and Cyclones* (New York: Facts on File, 1998): 82.
26. Longshore, *Encyclopaedia of Hurricanes, Typhoons and Cyclones*: 317.
27. Brown, Sophie, 'The Philippines Is the Most Storm-Exposed Country on Earth', *Time* 11 November (2013), available at: http://world.time.com/2013/11/11/the-philippines-is-the-most-storm-exposed-country-on-earth/ (accessed 3 December 2014).
28. Longshore, *Encyclopaedia of Hurricanes, Typhoons and Cyclones*: 186.
29. Garnier, Emmanuel and Jérémy Desarthe, 'Cyclones and Societies in the Mascarene Islands 17th-20th Centuries', *American Journal of Climate Change* 2 (2013): 4.
30. Kamala Devi, S. History of Cyclones on the Coromandel Coast with Special Reference to Tamil Nadu, 1800–1900, PhD dissertation, Bharathidasan University, 2013, available at: http://ietd.inflibnet.ac.in:8080/jspui/handle/10603/9665 (accessed 24 June 2014).
31. Nott, Jonathan, 'A 6000 Year Tropical Cyclone Record from Western Australia', *Quaternary Science Reviews* 30 (2011): 713–22.
32. Longshore, *Encyclopaedia of Hurricanes, Typhoons and Cyclones*: 186. For a recent discussion, see Amrith, Sunil S. *Crossing the Bay of Bengal: The Furies of Nature and the Fortunes of Migrants* (Cambridge, USA: Harvard University Press, 2013): 6–12.

33. Wang, Shih Yu, Brendan M. Buckley, Jin-Ho Yoon, Boniface Fosu, 'Intensification of Pre-Monsoon Tropical Cyclones in the Bay of Bengal and Its Impact on Myanmar', *Journal of Geophysical Research: Atmospheres* 11(2013): 4373–4384.
34. Bilham, Roger, 'Earthquakes in India and the Himalaya: Tectonics, Geodesy and History', *Annals of Geophysics* 47, 2/3 (2004): 839–858.
35. Emanuel, Kerry, *Divine Wind: The History and Science of Hurricanes* (New York: Oxford University Press, 2005a): 222.
36. Longshore, *Encyclopaedia of Hurricanes, Typhoons and Cyclones*: 187.
37. Emanuel, *Divine Wind: The History and Science of Hurricanes*: 222; Roy, Tirthankar, 'State, Society and Market in the Aftermath of Natural Disasters in Colonial India: A Preliminary Exploration', *Indian Economic and Social History Review* 45 (2008): 275.
38. Longshore, *Encyclopaedia of Hurricanes, Typhoons and Cyclones*: 188.
39. Khalil, Gazi MD, 'Cyclones and Storms Surges in Bangladesh: Some Mitigating Measures' *Natural Hazards* 6 (1992): 12.
40. Frank, Neill L. and S.A. Hussain, 'The Deadliest Tropical Cyclone in History?' *Bulletin of the American Meteorological Society* 52, 6 (1971): 438–444. Emanuel, *Divine Wind: The History and Science of Hurricanes*: 223–224.
41. Longshore, *Encyclopaedia of Hurricanes, Typhoons and Cyclones*: 30.
42. Emanuel, Kerry, 'Increasing Destructiveness of Tropical Cyclones over the Past 30 Years', *Nature* 436 (August 2005b): 686.
43. Wang et al. 'Intensification of Pre-monsoon Tropical Cyclones in the Bay of Bengal and Its Impact on Myanmar': 4373.
44. International Federation of Red Cross and Red Crescent Societies, *Myanmar—Cyclone Nargis Operation Rebuilding the Lives and Livelihoods of Communities Affected by Cyclone Nargis, Livelihoods Programme Report* (Yangon: Myanmar Red Cross Society, 2010), available at: http://myanmarredcrosssociety.org/publications/Rebuilding%20the%20lives%20and%20livelihoods%20of%20communities%20affected%20by%20Cyclone%20Nargis.pdf (accessed 25 June 2014).
45. Fritz, Hermann M., Christopher G. Blount, Fawzi B. Albusaidi, Ahmed Hamoud Mohammed Al-Harthy, 'Cyclone Gonu Storm Surge in Oman', *Estuarine, Coastal and Shelf Science* 86 (2010): 102–106.
46. For more information, see that annual cyclone reports of the RSMC La Réunion-Tropical Cyclone Centre, available at: http://www.meteo.fr/temps/domtom/La_Reunion/webcmrs9.0/anglais/archives/publications/index.html (accessed 18 June 2015).
47. Haig, Jordahna, Jonathan Nott and Gert-Jan Reichart, 'Australian Tropical Cyclone Activity Lower Than at Any Time over the Past 550–1500 Years' *Nature* 505 (January 2014): 667.

48. Kumar, K.A. and H. Achyuthan, 'A Record of Palaeo-Tsunami in the Indian Ocean', *Marine Geodesy* 29 (2006) 253–263.
49. Kumar and Achyuthan 'A Record of Palaeo-Tsunami in the Indian Ocean': 255.
50. Rastogi B.K., 'A Historical Account of the Earthquakes and Tsunamis in the Indian Ocean' in Tad S. Murty, U. Aswathanarayana and Niru Nirupama (eds.) *The Indian Ocean Tsunami* (London: Taylor & Francis, 2007): 12–13.
51. Gates, Alexander E. and David Ritchie, *Encyclopaedia of Earthquakes and Volcanoes, Third Edition* (New York: Checkmark Books, 2007): 24–25.
52. Rastogi, 'A Historical Account of the Earthquakes and Tsunamis in the Indian Ocean': 3–4.
53. Alam, Edris and Dale Dominey-Howes, 'An Analysis the AD1762 Earthquake and Tsunami in SE Bangladesh', *Natural Hazards* 70 (2014): 903–933.
54. Rastogi, 'A Historical Account of the Earthquakes and Tsunamis in the Indian Ocean': 13.
55. Kumar, Arun, 'Reclaim Islands and New Offshore Townships in the Arabian Gulf: Potential Natural Hazards', *Current Science* 96, 4 (2009): 482–483.
56. Dunbar, Paula, Kelly Stroker and Heather McCullouch, 'Do the 2010 Haiti and Chile Earthquakes and Tsunamis Indicate Increasing Trends?' *Geomatics, Natural Hazards and Risk* 1, 2 (2010): 110–111.
57. Gates and Ritchie, *Encyclopaedia of Earthquakes and Volcanoes*: 169–170.
58. Gates and Ritchie, *Encyclopaedia of Earthquakes and Volcanoes*: 13, 23–24, 97, 136–137, 252.
59. Bilham, 'Earthquakes in India and the Himalaya: Tectonics, Geodesy and History': 842–844.
60. Alam and Dominey-Howes, 'An Analysis the AD1762 Earthquake and Tsunami in SE Bangladesh': 903–933.
61. Gates and Ritchie, *Encyclopaedia of Earthquakes and Volcanoes*: 18–19; Roy, 'State, Society and Market in the Aftermath of Natural Disasters in Colonial India: A Preliminary Exploration': 275–276.
62. Gates and Ritchie, *Encyclopaedia of Earthquakes and Volcanoes*: 132.
63. Gates and Ritchie, *Encyclopaedia of Earthquakes and Volcanoes*: 30, 209; Roy, 'State, Society and Market in the Aftermath of Natural Disasters in Colonial India: A Preliminary Exploration': 278–289.
64. Gates and Ritchie, *Encyclopaedia of Earthquakes and Volcanoes*: 29, 171–172.
65. Vervaeck, Armand and James Daniell, 'Deadly Nepal Gorkha/Kodari Earthquakes', *Earthquake-Report.com*, 26 May 2015,available at: http://earthquake-report.com/2015/04/25/massive-earthquake-nepal-on-

april-25-2015/ (accessed 18 June 2015). Brahma Shumsher Jung Bahadur Rana, *The Great Earthquake in Nepal 1934 A.D.* (Kathmandu: Ratna Pustak Bhandar, 2013).

66. Greg Bankoff, 'A Tale of Two Cities: The Pyro-seismic Morphology of Nineteenth Century Manila' in Greg Bankoff, Uwe Luebken and Jordan Sand (eds.) *Flammable Cities: Urban Fire and the Making of the Modern World* (Madison: University of Wisconsin Press, 2012), 170–189.
67. Gates and Ritchie, *Encyclopaedia of Earthquakes and Volcanoes*: 26.
68. Gates and Ritchie, *Encyclopaedia of Earthquakes and Volcanoes*: 24.
69. Dunbar et al., 'Do the 2010 Haiti and Chile Earthquakes and Tsunamis Indicate Increasing Trends?': 109–110.
70. Winchester, Simon, *Krakatoa: The Day the World Exploded: August 27, 1883* (London: Penguin Books, 2004).
71. Alpers, *The Indian Ocean in World History*: 137.
72. Gates and Ritchie, *Encyclopaedia of Earthquakes and Volcanoes*: 252–253.
73. Soegiharto, Handewi, 'Merapi and the Demise of the Mataram Kingdom', *The Jakarta Post*, 13 June 2006, available at: http://www.thejakartapost.com/news/2006/06/13/merapi-and-demise-mataram-kingdom.html (accessed 8 July 2014).
74. Hargrove, Thomas R., *The Mysteries of Taal: A Philippine Volcano and Lake, Her Sea Life and Lost Towns* (Manila: Bookmark Publishing, 1991).
75. Warren, James Francis, 'Volcanoes, refugees and raiders: the 1765 Macaturin eruption and the rise of the Iranun', paper presented at the conference *Dimensions of the Indian Ocean World Past: Sources and Opportunities for Interdisciplinary Work in Indian Ocean World History, 9th–19th Centuries* (Western Australian Maritime Museum, Fremantle, 12–14 November 2012).
76. Rantucci, Giovanni, *Geological Disasters in the Philippines: The July 1990 Earthquake and the June 1991 Eruption of Mount Pinatubo* (Rome: Dipartimento per L'Informazione e L'Editoria, 1994).
77. Bankoff, Greg, 'The Historical Geography of Natural Disaster', in Bankoff, Greg, G. Frerks and D. Hilhorst (eds.) *Mapping Vulnerability: Disasters, Development and People* (London: Earthscan, 2004): 35.
78. Bankoff, Greg 'Historical Concepts of Disasters and Risk' in B. Wisner, J.C. Gaillard & I. Kelman (eds.) *Handbook of Natural Hazards and Disaster Risk Reduction*, (Routledge, 2011): 37–47.

Bibliography

Abu-Lughod, Janet. 1989. *Before European hegemony: The world system A.D.1250–1350*. New York and Oxford: Oxford University Press.

Alam, Edris, and Dale Dominey-Howes. 2014. An analysis the AD1762 earthquake and tsunami in SE Bangladesh. *Natural Hazards* 70: 903–933.

Allen, James de V., 'A proposal for Indian Ocean studies' in C. Mehaud (ed.) *Historical relations across the Indian Ocean world*, (Paris: UNESCO 1980), 137–151.

Alpers, Edward A. 2014. *The Indian Ocean in world history*. New York: Oxford University Press.

Amrith, Sunil S. 2013. *Crossing the Bay of Bengal: The furies of nature and the fortunes of migrants*. Cambridge: Harvard University Press.

Armitage, David. 2009. Three concepts of Atlantic history. In *The British Atlantic world, 1500–1800*, eds. David Armitage, and Michael Braddock, 11–27. Basingstoke and New York: Palgrave Macmillan.

Bankoff, Greg. 2002. *Cultures of disaster: Society and natural hazard in the Philippines*. London: Routledge Curzon.

———. 2004. The historical geography of natural disaster. In *Mapping vulnerability: Disasters, development and people*, eds. Bankoff Greg, G. Frerks, and D. Hilhorst, 25-36. London: Earthscan.

———. 2011. Historical concepts of disasters and risk. In *Handbook of natural hazards and disaster risk reduction*, eds. B. Wisner, J.C. Gaillard, and I. Kelman, 37–47. London: Routledge.

———. 2012. A tale of two cities: The pyro-seismic morphology of nineteenth century Manila. In *Flammable cities: Urban fire and the making of the modern world*, eds. Greg Bankoff, Uwe Luebken, and Jordan Sand, 170–189. Madison: University of Wisconsin Press.

Bilham, Roger. 1994. The 1737 Calcutta earthquake and cyclone evaluated. *Bulletin of the Seismological Society of America* 84(5): 1650–1657.

———. 2004. Earthquakes in India and the Himalaya: Tectonics, geodesy and history. *Annals of Geophysics* 47(2/3): 839–858.

Bose, Sugata. 2006. *A hundred horizons: The Indian Ocean in the age of global empire*. Cambridge: Harvard University Press.

Bourne, Randolph. 1916. Trans-national America. *Atlantic Monthly*, 118: 86–97. Available at: http://www.theatlantic.com/magazine/archive/1916/07/trans-national-america/304838/. Accessed 13 June 2014.

Braudel, Fernand. 1949. *La Méditerranée et Le Monde Méditerranéen a L' Époque De Philippe II*, 3 vols. Paris: Armand Colin.

Brown, Sophie. 2013. The Philippines is the most storm-exposed country on earth. *Time* 11 November. Available at: http://world.time.com/2013/11/11/the-philippines-is-the-most-storm-exposed-country-on-earth/. Accessed 3 Dec 2014.

Chaudhuri, K.N. 1985. *Trade and civilisation in the Indian Ocean: An economic history from the rise of Islam to 1750*. Cambridge: Cambridge University Press.

——— 1990. *Asia before Europe: economy and civilization of the Indian Ocean from the rise of Islam to 1750.* Cambridge and New York: Cambridge University Press.

Crosby, Alfred W. 1986. *Ecological imperialism: The biological expansion of Europe, 900–1900.* Cambridge: Cambridge University Press.

Davis, Mike. 2001. *Late Victorian Holocausts: El Niño famines and the making of the third world.* London: Verso.

Dunbar, Paula, Kelly Stroker, and Heather McCullouch. 2010. Do the 2010 Haiti and Chile earthquakes and tsunamis indicate increasing trends? *Geomatics, Natural Hazards and Risk* 1(2): 95–114.

Emanuel, Kerry. 2005a. *Divine wind: The history and science of hurricanes.* New York: Oxford University Press.

———. 2005b. Increasing destructiveness of tropical cyclones over the past 30 years. *Nature* 436: 686–688.

Fagan, Brian. 1999. *Floods, famines, and emperors: El Niño and the fate of civilizations.* New York: Basic Books.

Frank, Neill L., and S.A. Hussain. 1971. The deadliest tropical cyclone in history? *Bulletin of the American Meteorological Society* 52(6): 438–444.

Fritz, Hermann M., Christopher G. Blount, Fawzi B. Albusaidi, and Ahmed Hamoud Mohammed Al-Harthy. 2010. Cyclone Gonu storm surge in Oman. *Estuarine, Coastal and Shelf Science* 86: 102–106.

Garden, Donald. 2009. *Droughts, floods & cyclones: El Niños that shaped our colonial past.* North Melbourne: Australian Scholarly Publishing.

Garnier, Emmanuel, and Jérémy Desarthe. 2013. Cyclones and societies in the Mascarene Islands 17th-20th centuries. *American Journal of Climate Change* 2: 1–13.

Gates, Alexander E., and David Ritchie. 2007. *Encyclopaedia of earthquakes and volcanoes*, 3rd edn. New York: Checkmark Books.

Grove, Richard H. 1995. *Green imperialism: Tropical island Edens and the origins of environmentalism, 1600–1860.* Cambridge: Cambridge University Press.

Haig, Jordahna, Jonathan Nott, and Gert-Jan Reichart. 2014. Australian tropical cyclone activity lower than at any time over the past 550–1500 years. *Nature* 505: 667–673.

Hargrove, Thomas R. 1991. *The mysteries of taal: A Philippine volcano and lake, her sea life and lost towns.* Manila: Bookmark Publishing.

Heilbron, Johan, Nicolas Guilhot, and Laurent Jeanpierre. 2008. Toward a transnational history of the social sciences. *Journal of the History of the Behavioural Sciences* 44(2): 146–160.

International Federation of Red Cross and Red Crescent Societies. 2010. *Myanmar – Cyclone Nargis operation rebuilding the lives and livelihoods of communities affected by Cyclone Nargis, livelihoods programme report.* Yangoon: Myanmar Red Cross Society. Available at: http://myanmarredcrosssociety.

org/publications/Rebuilding%20the%20lives%20and%20livelihoods%20of%20communities%20affected%20by%20Cyclone%20Nargis.pdf. Accessed 25 June 3014.

Iriye, Akira. 2013. *Global and transnational history: The past, present, and future.* Basingstoke and New York: Palgrave Macmillan.

Kamala Devi, S. 2013. History of cyclones on the Coromandel coast with special reference to Tamil Nadu, 1800–1900. PhD dissertation, Bharathidasan University. Available at: http://ietd.inflibnet.ac.in:8080/jspui/handle/10603/9665. Accessed 24 June 2014.

Khalil, Gazi M.D. 1992. Cyclones and storms surges in Bangladesh: Some mitigating measures. *Natural Hazards* 6: 11–24.

Koselleck, Reinhart. 2006. Crisis. *Journal of the History of Ideas* 67(2): 357–400.

Kumar, Arun. 2009. Reclaim islands and new offshore townships in the Arabian Gulf: Potential natural hazards. *Current Science* 96(4): 480–485.

Kumar, K.A., and H. Achyuthan. 2006. A record of palaeo-tsunami in the Indian Ocean. *Marine Geodesy* 29: 253–263.

Lamb, Hubert H. 1982. *Climate, history and the modern world.* London: Routledge.

Longshore, David. 1998. *Encyclopaedia of hurricanes, typhoons and cyclones.* New York: Facts on File.

Moore, Harry Estill. 1964. *And the winds blew.* Austin: Hogg Foundation for Mental Health.

Mukherjee, Rila. 2013. The Indian Ocean: Historians writing history. *Asian Review of World Historians* 1(2): 295–307.

Müller, Michael G., and Cornelius Torp. 2009. Conceptualising transnational spaces in history. *European Review of History* 16(5): 609–617.

Nott, Jonathan. 2011. A 6000 year tropical cyclone record from Western Australia. *Quaternary Science Reviews* 30: 713–722.

Pearson, Michael N. 1998. *Port cities and intruders: The Swahili Coast, India, and Portugal in the early modern era.* Baltimore and London: The Johns Hopkins University Press.

——— 2003. *The Indian Ocean.* London and New York: Routledge.

——— 2006. Littoral society: The concept and the problems. *Journal of World History* 17(4): 353–373.

Rantucci, Giovanni. 1994. *Geological disasters in the Philippines: The July 1990 earthquake and the June 1991 eruption of Mount Pinatubo.* Rome: Dipartimento per L'Informazione e L'Editoria.

Rastogi, B.K. 2007. A historical account of the earthquakes and tsunamis in the Indian Ocean. In *The Indian Ocean tsunami*, eds. Tad S. Murty, U. Aswathanarayana, and Niru Nirupama, 3–18. London: Taylor & Francis.

Roy, Tirthankar. 2008. State, society and market in the aftermath of natural disasters in colonial India: A preliminary exploration. *Indian Economic and Social History Review* 45: 261–294.

Seigel, Micol. 2005. Beyond compare: Comparative method after the transnational turn. *Radical History Review* 91: 62–90.

Smith, Gemma L., Lisa C. McNeill, Kelin Wang, Jiangheng He, and Timothy J. Henstock. 2013. Thermal structure and megathrust seismogenic potential of the Makran subduction zone. *Geophysical Research Letters* 40: 1528–1533.

Soegiharto, Handewi. 2006. Merapi and the demise of the Mataram Kingdom. *The Jakarta Post*, 13 June. Available at: http://www.thejakartapost.com/news/2006/06/13/merapi-and-demise-mataram-kingdom.html. Accessed 8 July 2014.

Tagliacozzo, Eric. 2002. Trade, production and incorporation: The Indian Ocean in flux, 1600–1900. *Itinerario* 26(1): 75–106.

Tyrrell, Ian. 1991. American exceptionalism in an age of international history. *American Historical Review* 96(4): 1031–1055.

Vervaeck, Armand, and James Daniell. 2015. Deadly Nepal Gorkha/Kodari earthquakes. *Earthquake-Report.com*, 26 May. Available at: http://earthquake-report.com/2015/04/25/massive-earthquake-nepal-on-april-25-2015/. Accessed 18 June 2015.

Vink, Markus P.M. 2007. Indian Ocean studies and the 'new thalassology'. *Journal of Global History* 2: 41–62.

Wang, Shih Yu, Brendan M. Buckley, Jin-Ho Yoon, and Boniface Fosu. 2013. Intensification of pre-monsoon tropical cyclones in the Bay of Bengal and its impact on Myanmar. *Journal of Geophysical Research: Atmospheres* 11: 4373–4384.

Warren, James Francis. 2012. Volcanoes, refugees and raiders: The 1765 Macaturin eruption and the rise of the Iranun, paper presented at the conference *Dimensions of the Indian Ocean World Past: Sources and Opportunities for Interdisciplinary Work in Indian Ocean World History, 9th–19th Centuries*, Western Australian Maritime Museum, Fremantle, 12–14 November 2012).

Wenger, Dennis E., and Jack M. Weller. 1973. *Disaster subcultures: The cultural residues of community disasters, preliminary paper no. 9.* Columbus: Disaster Research Centre, Ohio State University.

White, Richard. 1999. The nationalisation of nature. *Journal of American History* 86(3): 976–986.

Winchester, Simon. 2004. *Krakatoa: The day the world exploded: August 27, 1883.* London: Penguin Books.

CHAPTER 2

Revisiting Southeast Asian History with Geology: Some Demographic Consequences of a Dangerous Environment

Anthony Reid

The longer the time-depth considered, the more human history is dependent on the beneficence of the planet we inhabit. Historians have tended to write as if humans were masters of their own destiny, not only out of ignorance of climatic and other changes, but also because of a bias towards the political. Historians of Indonesia, in particular, were understandably preoccupied with the political upheavals of the past century, also a relatively mild one in terms of earthquakes, tsunamis, volcanic eruptions and climatic aberrations. Since the sequence of disasters between the Aceh tsunami of December 2004 and the Merapi eruption of October 2010, however, it has become clear that these events must have had antecedents at least as destructive.

The new geological research strengthens a growing sense of Indonesian history as one unusually exposed to the vagaries of nature. In periods of relative quiescence on the ring of fire, such as the twentieth century, a benign climate and fertile soils can produce rapid population growth and

A. Reid
College of Asia and the Pacific, Australian National University, Canberra, Australia

G. Bankoff, J. Christensen (eds.), *Natural Hazards and Peoples in the Indian Ocean World*, DOI 10.1057/978-1-349-94857-4_2

development. But rather than forming a constant, this pattern appears to have been interrupted by periodic disasters. The scientific exploration of this pattern on the ground, though still in its infancy, offers considerable hope for a better understanding. Cooperation between scholars in different disciplines, and the pooling of data from around the Indian Ocean, offers a way forward in understanding the long-term pattern of the region.

It is now known that ruptures in the earth's crust occur with necessary regularity and that the bigger events every few centuries have enormous impacts not only on Indonesia but also on the planet as a whole. Politically focused histories look woefully inadequate to explain this pattern. The scientists use the data of historians when they can, though in Southeast Asia we have not been particularly helpful to them. A comprehensive record began to be available for the whole planet only once the seismograph was invented around 1900, and long after that descriptions of the effects of seismic movements on human societies remained under-reported in Indonesia. Before 1900, our understanding is dependent firstly on the written records of observers, though the ring of fire around the Pacific where the periodic pattern should be clearest is, except for Japan, poorly provided with historical records before 1600. Increasingly therefore we turn to scientific investigation of the traces left by volcanic eruptions, earthquakes and tsunamis on the ground.

The geologists have done a little better than the historians at recording data for Southeast Asia, but this region remains badly under-reported in comparison with other parts of the ring of fire. The US National Geophysical Data Center records only four tsunamis in Sumatra before 1800, because they damaged the Dutch pepper-posts, and none in Java where the endangered south coast was then of no interest to the Dutch. By contrast, 29 Indonesian tsunamis have been recorded in the 22 years since 1990, including nine massive ones each killing over 100 people, two of them in Java. The Philippines was no better recorded, with 14 tsunamis noted in 1600–1800, overwhelmingly of the minor kind found in the areas of Spanish interest on the west coast of Luzon. Only after 1890 are there reports of serious tsunamis in the most affected areas of the east coast and the Sulu Sea. In the 112 years since 1900, some 56 tsunamis have been reported, yet only after 1960 do these appear to show realistically the devastating effects of tectonic movements on the subduction plate on the eastern and southern flanks of the southern Philippines. The three biggest tsunami disasters are all those recorded in the last 50 years: 355 killed in a volcano-generated tsunami of 1965, 4376 from a tsunami generated by an 8.1 earthquake in the Moro Gulf in 1976 and 81 killed in Mindoro in 1994 by a tsunami generated by a 7.1 earthquake.[1]

The Smithsonian Institute volcanism database shows that 40 % of the world's major eruptions (of VEI 4 or more—Volcanic Explosivity Index) since 1982 have been in Southeast Asia or Papua New Guinea. The further we go back in time, however, the less well represented is Southeast Asia in the record. Before 1500, we are dependent chiefly on scientific work on the ground for the knowledge of major eruptions, which has produced a Smithsonian Institute list of over 500 probable major eruptions dated by various systems between 10,000 BCE and 1500 CE, mostly in northern sections of the Pacific ring of fire, and in Iceland and New Zealand. Of these, only 19 eruptions, or 3.7 % of the global total, have been located in Southeast Asia including New Guinea. Almost everything remains to be done in Indonesia, where only five very questionable entries before 1500 are in the Smithsonian list.[2] No destructive (VEI 4+) Southeast Asian eruptions before 1500 have been documented to the more exacting standards of the National Geophysical Data Center, although 35 are acknowledged elsewhere.[3]

The problems of this lack of research not only for Southeast Asia but for the world have been again demonstrated by the example of the global disaster of 1257–59 CE, evident from sulphur deposits in the polar ice caps, and recently discovered to have been responsible for thousands of mass burials in the Spitalfields cemetery of London, as well as other crop failures resulting from the most dramatic drop in global temperature for the last thousand years.[4] A tropical location must be assumed for any such eruption shown to affect the whole planet, and it looks finally as though it has been found in Lombok, the island east of Bali. The huge Samalas caldera adjacent to the remaining Mount Rinjani was already thought to have been formed in a thirteenth-century explosion, but the glass geochemistry of the pumice deposits from the eruption on Lombok has now been found to match the shards found in the coring of polar ice caps, as other possible culprits do not.[5]

Southeast Asian sources themselves were surprisingly little concerned to record natural disasters, except insofar as volcanic eruptions served as portents for human events. They were relatively little affected by earthquakes because of the light and flexible construction of their buildings, and by tsunamis because earlier generations had learned no doubt painfully how devastating they could be. The tsunami-prone coasts of Nias, Mentawai, western Sumatra and southern Java were relatively deserted before 1800, and only a careful re-reading of sources begins to show that past disasters played a role in this choice. The people of Nias, probably the most vulnerable to tsunamis of all complex Indonesian societies, spurned their coasts completely before sea-based Dutch infrastructure arrived in the second half of the nineteenth century. They fished only in freshwater rivers, and built their villages on hilltops where possible.[6] One of their origin myths shows the creator god sending his sons to people the earth. The first was sent to the south, which began to submerge beneath the sea because of his weight, so he sent the second to the north. When that too began to submerge with the centre rising upward he sent the last two sons to the centre, and balance was restored.[7] Mentawai islanders also built their villages away from the coast. Their mythology had the lower tier of the cosmos, the underworld, as the abode of the earthquake god, who had to be propitiated with the erection of each house.[8]

Southeast Asia Takes Shape in a Quiet Century

Until the current round of disasters in Sumatra and Java since 2004, neither historians nor scientists had given the past record the attention it deserved. After the Krakatau eruption of 1883, there had been more than a century of relative geological calm, with a death toll that by hindsight appears uncannily low. In this quiet time (geologically), between 1885 and 2000, the Indonesian population grew from about 25 million to 205 million, and its urban (predominately coastal) population from little over a million to 90 million. The Philippines grew even more dramatically, from about 6 to 76 million, of which 45 million urban, including dramatic growth in Bikul and Samar, the most exposed to both typhoons and tsunamis. Two large modern states were created, amidst much bloodshed and suffering. Their infrastructures were built with little consideration of the geological dangers, with urban centres at Banda Aceh, Padang, Bengkulu, Cilacap and Ambon all directly facing some of the world's most dangerous subduction zones. In the cities, moreover, building was predominately of poor-quality brick construction, so that the new series of earthquakes since 2000 has produced much greater carnage than previously. Padang's 1833 earthquake was much bigger than that of 2009 (7.6 magnitude), and accompanied by a major tsunami, but the population of the West Sumatran littoral was then less than 80,000 compared with over two million today. Whereas the 1833 toll was a few hundred, that of 2009 accounted for over 1000 dead with 135,000 homes destroyed and 1.25 million people affected.

The relatively low number of casualties from Indonesia's natural disasters in the twentieth century, and the very high numbers of casualties of political conflict, mostly state-sponsored, make it unsurprising that historians should have focused on political events. The deaths attributed to the eruption of either Tambora (1815) or Krakatau (1883) alone exceed all the 32,000 recorded deaths from earthquakes, tsunamis and volcanoes in the twentieth century.[9] The minimal death toll (excluding missing persons) of the 2004 tsunami in Sumatra, over 170,000, alone exceeds that twentieth-century figure more than fivefold. In another paper,[10] I showed that the Indonesian deaths directly attributed to political conflict in the twentieth century (mostly in the period 1940–80) outnumbered those caused by geological spasms by about 30 to 1. The recent reversal has been dramatic, not only because of the extraordinary death toll of the 2004 tsunami. Interethnic violence following the 1998 fall of Suharto and Indonesia's transition to democracy declined sharply in 2002,[11] since when the death toll of tectonic disasters has far outweighed that from political violence even without including the massive 2004 tsunami. The biggest of the other natural disasters were earthquakes in Jogjakarta (2006, killing 5749) and Padang (2009, killing 1117), and tsunamis in Nias (2005, killing 1313) and Pangandaran, West Java (2006, killing 802). Steven Pinker has recently shown that violent deaths from human conflict have in fact declined dramatically in most places as a proportion of total deaths,[12] even though the figures look very high in the mid-twentieth century as the world system (notably including Indonesia) readjusted from empires to nation-states. Since 1980, Southeast Asia has joined the global downward trend for violent conflict deaths.

The point has already been made in relation to Indonesia's most-studied volcano, Gunung Merapi near Yogyakarta, that the relatively 'benign' twentieth century should not be a guide to the twenty-first. A geological study of the last two centuries concluded that:

> a major difference in eruption style exists between the twentieth and nineteenth centuries, although the periodicity between larger events seems about the same. ... In the 1800s, however, explosive eruptions of relatively large size occurred (to VEI 4), and some associated "fountain-collapse" *nuées ardentes* were larger and farther reaching than any produced in the twentieth century. ... The nineteenth century activity is consistent with the long-term pattern of one relatively large event every one or two centuries... Merapi could soon be due for another large event and its occurrence with

> only modest (or inadequately appreciated) precursors could lead to a disaster unprecedented in Merapi's history because the area around the volcano is now much more densely populated.[13]

Others have noted more spectacular events within the framework of thousands of years, including the suggestion that the major collapses of the whole volcanic cone and building of another one could be roughly dated to incidents around 9630, 3400 and 1900 years before the present, with a partial collapse around 1130 and frequent explosive eruptions capable of burying key sites on the Mataram plain (Fig. 2.1).[14]

Volcanic Disruptions to Population

There are puzzles about the population history of the tectonically threatened island arc from Sumatra to Luzon, and especially its volcanic jewel, Java. The whole region had population densities well below the rest of humid Asia in the early 1800s when population began to be assessed accurately, even though humans had flourished in this region for longer than in most

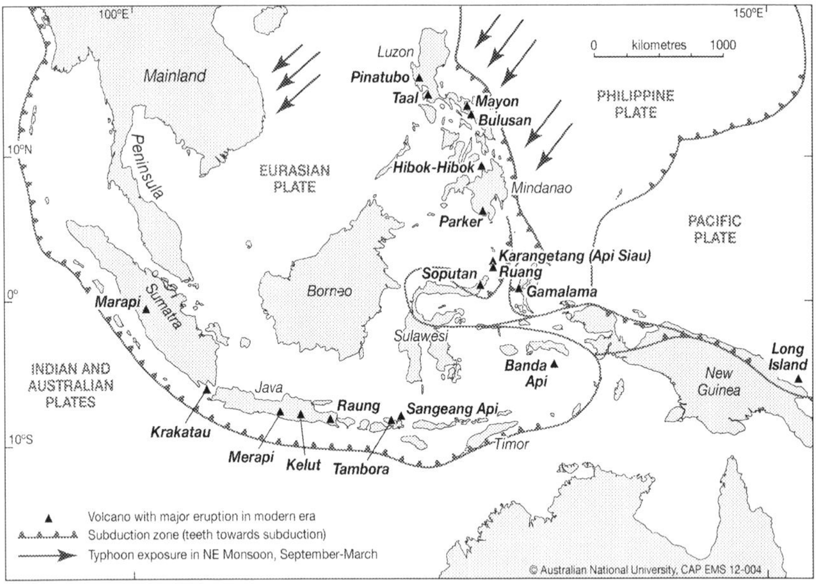

Fig. 2.1 Major natural threats to Southeast Asia

of Asia, surviving the ice ages as was not possible further north. Its environmental conditions were moreover congenial to human life, and its volcanic soils and humid climate were favourable to agriculture, which had been practised there for at least 5000 years. This was demonstrated by the astonishing growth of population of the two main volcanic rice baskets in the 'mild' century described above, and to some extent the whole period between 1835 and 2000. Having been still relatively underpopulated by Asian standards in the 1830s, Java and Luzon soared until they reached the highest rural population densities in the world. In general, demographers have either struggled to show the earliest estimates must have been much too low, or explained extremely low growth rates prior to the colonial peace by a pattern of disease and small-scale warfare (as indeed did I).[15] We need also to look much more carefully at a discontinuous pattern of periodic disasters. Volcanic eruptions alternately enabled intense agricultural production through the rich volcanic soils, and destroyed it by ash deposits and the blocking out of sunlight. But so little of the destructive power was displayed in the relatively calm period after 1835 that even the historians neglected this likelihood.

Although Java was far better known to the Dutch than the other Indonesian islands in the eighteenth and nineteenth centuries, too little is still known about tectonic disasters even there before Krakatau (1883) captured the world's attention. When relatively detailed population surveys began to be made by Raffles around 1813 and the returning Dutch from 1820, the low population of two areas stands out: the Sundanese area of Parahyiangan (West Java) and the eastern salient of Java. Boomgaard shows population densities for 1820 of ten per square kilometre (km) for the whole of Parahyiangan (covering fully half of West Java), nine for Krawang to its north and only four for Banyuwangi, the easternmost of Java's residencies, against an overall density for Java of 43.[16] The Tasikmalaya area of West Java was devastated in 1822 by the lahar flow from a big eruption of Mount Galunggung (VEI 5), recorded as having directly killed 4011 people.[17] These poorly populated areas are for the same reason among the most fertile in Java, with among the highest productivity per hectare in Java. Their populations increased particularly dramatically during the geologically quieter times after 1830, even though the 1883 Krakatau eruption again depopulated the south-western corner, making possible the survival of the Javan rhino in today's Ujong Kulon National Park. Today Sundanese speaking West Java (minus Jakarta and Banten) is Indonesia's most populous province with the densest population (except for the city–provinces) of 1235 per square km.

West Java's discontinuity is particularly striking, as it was the site of the earliest inscriptions in Java, evidence for some kind of polity named Tarumanagara on the coast as early as the fifth century. Javanese legends acknowledge the western area as the source of ancient legitimacy associated with the Baron Sakundar myth as well as that of Ratu Kidul (of whom more below).[18] It appears to have flourished in the ninth to eleventh centuries, with its Sunda kingdom initially having relations with the first Mataram in Central Java. According to the Carita *Parahiyangan*, a manuscript probably of the late seventeenth century found in Cirebon, the rule of Sri Jayabhupati, given as 1030–42/43 (*Saka* 952–64), appears to have been particularly successful. In 1042/1043 (*Saka* year 964), however, 'clouds of ash overshadowed the kingdom of Sunda', the king mysteriously departed and the glorious times were over.[19] The earliest European accounts in the eighteenth century judged it a sparsely peopled place of 'simple and uneducated mountaineers'.[20] Raffles noted formerly cultivated areas of Parahyangan abandoned in his time, though predictably blaming Dutch oppression rather than geology for it.[21]

In the easternmost salient of Java (the *Oosthoek* to the Dutch), the almost total depopulation of Banyuwangi by 1820 is usually attributed to the prolonged warfare over the area between Hindu Balinese and Muslim Mataram, with the Vereenigde Oostindische Compagnie (VOC) adding a nasty kind of scorched earth policy when they intervened after taking over Mataram's claims in 1734. Raffles again blamed the 'desolating system' of Dutch control for the collapse of Banyuwangi's population from 80,000 in 1750 to 8000 in 1811.[22] More careful modern research blames warfare, flight, starvation and epidemic disease for the collapse of the population of Blambangan to less than 10 % in the 1770s of what it had been in the 1750s, and a continuing deterioration thereafter up to the 1800s.[23] But it is likely that the eruption of Tambora in 1815, and of the Oosthoek's own unusually active volcanoes, also played a significant role in perpetuating this desolation.[24] Mount Ijen possesses the largest caldera in Java at 20 km, and the world's largest acidic crater lake, which confers both benefits of sulphur deposits and great dangers of spillage in eruptions, which occurred in 1797 and 1817. Nearby Mt Raung has erupted 43 times since 1880, but its earlier eruptions were more severe, causing numerous deaths in 1638, 1730, and January 1817.

The Tambora eruption of April 1815 has recently attracted more attention than other Southeast Asian disasters, since it was identified as the cause of the notorious 'year without summer' in Europe and America in

1816. Besides leaving a substantial ash deposit in the ice caps of both polar regions, it lowered temperatures in the northern hemisphere and caused disastrous crop failures and freakish weather. Its effects in Southeast Asia itself were certainly greater, and vital for my theme, but predictably less well studied. Although we have no reliable population figures so early, it is reasonable to assume that the damage to crops, and therefore livelihoods, from the ash clouds was far greater in the immediate fallout zone of the volcano than on the other side of the world. Those directly killed by the explosion of gases and lava flows were mainly on the Tambora Peninsula formed by the mountain on the island of Sumbawa, where the explosion killed virtually everybody—around 11,000 people. The Tambora language, a word list of which had been collected by Raffles before the eruption, was wiped out, eliminating what is now understood to have been by far the most westerly survival of a Papuan-type (non-Austronesian) language.[25]

Some 60,000 more people were estimated to have died of hunger and disease in the remainder of Sumbawa Island and in Lombok and Bali to its west, as agriculture was destroyed by ash deposits and lack of sunlight. Harvests in Bali (the best-documented) were drastically affected for the next four years. The effect was worsened by a severe earthquake in Bali in November of the same year which caused the crater lake of Mount Pangilinan to burst its banks, destroying 17 villages in north Bali and killing an estimated 10,000 people. The devastated population was then assailed by infestations of rats which consumed much of the little food left. A Dutch visitor to Bali in 1818 counted 34 corpses lying beside the 25 km track between Badung and Gianyar, having presumably failed to survive the desperate walk in search of food. The same Dutch observer pointed out 'The lords are dirt-poor, while the people suffer great deprivations and often go hungry'. Only after 10–15 years of misery did the ecological curse turn again to a blessing as the nutrients in the ash were absorbed to fertilise Bali's soils.[26]

Estimating Longer-Term Effects

Whether or not people realised it at the time, therefore, the most numerous fatalities and disruptions to civilisation in earlier mega-eruptions would have arisen not from the dramatic immediate victims of gasses and lahar, but from the failure of crops as a result of the ash and lack of sunlight over a period of several years. In attempting to understand the longer-term effects of this extremely dangerous environment, a first step is to recognise that densely settled wet-rice (*sawah*) farmers were the most vulnerable because of their reliance on a single crop delicately adjusted to climate and environment. Their vulnerability to disasters helps explain Southeast Asia's remarkable human and biological diversity, particularly evident in the most exposed arc of tectonic subduction around the region's southern and eastern rim.

The most remarkable human survival in the region is the diminutive 'hobbit', *Homo floresiensis*, discovered in 2003 to have survived as recently as 12,000 years ago, and thus long coexisting with *Homo sapiens*. The find was in Flores, easterly neighbour of Sumbawa and itself one long volcanic spine including eight active volcanoes still causing significant damage in the twentieth century. The dominant Flores population today itself has more visible genetic links with older Australo-Melanesian inhabitants (pre-Austronesian and closer to Australian and New Guinea populations) than elsewhere, and has vivid memories of different 'wild men' surviving up to a few generations ago.[27] Since the sea crossing to Flores from other islands is not difficult, the likelihood is that Austronesian agriculturalists repeatedly colonised the island but were checked from taking over as fully as elsewhere by the effects of major eruptions (and tsunamis on the coast).

The Philippine islands was also a place of low, though presumably very long-standing, Austronesian agricultural population at the Spanish arrival in the sixteenth century, and of robust survival by darker Australo-Melanesian hunter-gatherers, estimated to be still 10 % of the population in 1600. The Spanish called these people *negros* or *negritos*, and named as Negros the large island in the Visayas where they still dominated.[28] Negros is also home to the Central Philippines' most active volcano, Kanlaon (26 eruptions of VEI 1 or 2 since 1919). Nothing is yet known of the history of this volcano before Negros was taken over by sugar cultivation in the nineteenth century, but the minimal place of agriculture on its rich soils in earlier times suggests an eruption strong enough to have destroyed or deterred earlier agriculturalists. After sugar in Negros destroyed their habitat, the strongest Negrito survivals were the Aetas, on slopes exposed to Mount Pinatubo's mega-eruptions, the last of which before 1991 were around 1450 CE and 1000 BCE.[29] The Aetas demonstrated their exemplary flexibility and mobility in locating food sources at the time of Pinatubo's 1991 eruption (VEI 6), which destroyed a quarter of a million homes and livelihoods in Luzon. After typhoons also, it had been noted, poor agriculturalists sought to marry into Aeta families for survival.[30] It seems likely, therefore, that just as rapid expansion of agriculture and population has threatened the survival of hunter-gatherers in the last 200 years, similar expansions threatened them in the past only to be checked by natural disaster.

The most celebrated presumed victim in Southeast Asia of destructive volcanoes is the early Mataram civilisation of Central Java, which built the magnificent temples of Borobudur (Buddhist) and Prambanan (Saivite) among many others in the period 600–900 CE. Since at least the 1950s, it has been conventional in Java to point to an eruption of Merapi in about 1000 CE as the cause, based on hypotheses of Dutch geologists such as van Bemmelin. Jan Christie has summarised more recent progress on both geological and historical fronts, to propose a number of major eruptions that contributed to the absence of any dated inscriptions after 928 CE. There are three distinct layers of volcanic ash in the Borobudur area, but the thickest of up to half a metre clearly put an end to occupation of the ninth and early tenth centuries. To the south of Merapi around modern Yogyakarta, perhaps the densest area of settlement then as now, several layers of lahar buried temples in quick succession. Recently, excavated temples such as Sambisari and Kedulan were under seven metres of lahar. Christie shows in the ninth and tenth centuries the centre of activity shifted to East Java, and surviving inscriptions of this period were increasingly concerned to propitiate local ancestor spirits associated with the volcanoes rather than Indic deities. She proposes therefore a number of successive eruptions with the worst in or about 928 CE.[31] Other disasters, however, may have played a role in causing this fertile Central Java region to fall silent for six centuries. A bigger eruption from a more distant volcano, less given to regular small eruptions than Merapi, may have caused the crops to fail for a year or more, while massive earthquakes and tsunamis on the south coast cannot be ruled out (see below).

The earliest Javanese chronicle with reliable dates for events such as volcanoes is the East Javanese *Pararaton*, probably compiled in the late fifteenth or sixteenth century. It clearly regarded eruptions as important and listed nine of them between 1311 and 1481, the last of which ends the chronicle.[32] For the remoter past, speculation based on Javanese legends must serve until there is more systematic work on the ground. There is a tradition that Sumatra and Java were one island until a vast eruption, perhaps in the site of Krakatau, separated them early in the Common Era. Some have argued that Krakatau was the likeliest source of the ash that caused extreme weather events and a marked cooling of the planet in 535–36 CE, which has now been established on scientific as well as historical grounds.[33]

Climatologists are also looking for a volcanic cause for the global cooling of 1258, mentioned above, and of the 1590s, inaugurating the seventeenth-century 'little ice age'.[34] The probabilities are indeed high that these and other climatic disturbances were caused by Southeast Asian volcanoes, but much more work is required on the ground before we can locate them.

Tsunamis of Early Modern Sumatra and Java

The world's worst known tsunami unexpectedly struck Sumatra on 26 December 2004, after a 9.2 megathrust earthquake, causing 168,000 deaths there and another 60,000 around the Indian Ocean. The dangers of Sumatra's subduction zones had already been pointed out in an important 1987 paper, which drew on historical records to show an 1861 earthquake to have been the most recent rupture of the northern section of the Sumatra subduction plate, and an 1833 quake as having the same role in its south-central section.[35] An earlier major earthquake and tsunami that destroyed much of Padang in 1797 may also have resulted from a significant rupture. After the 2004 disaster, research was naturally more intensive to establish a longer record from scientific evidence. Sand deposits on the tsunami-prone coasts of southern Aceh and southern Thailand revealed a tsunami on the scale of 2004 in the fourteenth century, and another in the ninth or tenth.[36] More specific dates have emerged from U–Th (uranium–thorium) disequilibrium dating of the death of coral in the major uplifts that accompany tectonic ruptures. These showed big uplifts of more than a metre in 1347, 1381, 1607 and 1613 at the Pagai Islands (southern Mentawai chain), and 1394 and 1450 at Simeulue.[37] There is no doubt that mega-earthquakes and tsunamis do affect this Sumatra coast relatively regularly, but the matching of scientific and historic data requires much further work.

Little has been done hitherto to document major tectonic events prior to 1797 from the fragmentary historical records on Sumatra. Dutch records report that their post at Padang was heavily damaged by earthquakes in March 1691, February 1697, November 1756 (swallowing 'whole villages' in the interior) and 1770 (with a tsunami).[38] British records from their post at Bengkulu, further south, add a tsunami in 1714.[39] Marsden also noted that the 1770 earthquake had a major impact at Manna, south of Bengkulu, creating a coastal plain seven miles long and dropping the hill on which the British resident's house stood by fifteen feet.[40] For Aceh and Nias, the sites most devastatingly affected by the 2004–05 earthquakes and tsunamis, we have not hitherto been able to go back beyond 1800.

The diaries of Dutch factors in Aceh in the period 1636–60, explored by Takeshi Ito, reveal another major tsunami of a scale probably not seen again until 2004. On 5 January 1660, 'in less than three hours beginning at nine in the evening the water carried over the whole land, even streaming over the walls of the forts along the beach as well as over the lofts or dwellings of most of the inhabitants, of whose houses more than a hundred were carried away by its force, and more than a thousand souls were drowned as well as innumerable cattle'.[41] The Pidië coast (around modern Sigli) was also badly affected, with 14 vessels wrecked and many men drowned. The Dutch report characteristically paid more attention to the loss of VOC textile stocks, as well as the damage to its rivals in the market, than to the effect on the government of the Queen, who was lucky to survive though losing many of her servants and slaves. The 1660 date coincides with evidence from Siberut (northern Mentawai) of a substantial coral uplift in that year.[42] This dramatic event undoubtedly set Aceh's city and port back for many years. Indeed, its population and its importance in Asian trade would never be quite the same, though this trauma has never before been recognised as among the reasons for the setback. The VOC post was withdrawn, and we hear little about the city over the next decade.

The 1987 paper that pointed to the danger of the Sumatra subduction zone judged that by contrast Java appeared to be aseismic, with few tectonic earthquake or tsunami events on record.[43] This view, partly responsible for the inadequate research into the record of Java's southern coast, needs serious revision since the recent run of disasters. These include deadly tsunamis sparked by moderate (7.8 and 7.7, respectively) deep earthquakes to the south of Java. The eastern section of this coast was hit in 1994, with 223 deaths and a thousand houses destroyed, and the western section around Pangandaran in 2006, with over 800 deaths. These events were also recorded on Australia's sparsely inhabited northwest coast, which provides a second avenue of potential research into the physical evidence and dating of earlier major tsunamis. For the record before 1900, most of the existing geological literature relies on Dutch reportage, although only after 1840, when a Dutch official was placed at the Southwest Java port of Cilacap, was there any Dutch presence at all on the endangered south coast. As we have seen elsewhere, pre-twentieth-century Dutch sources seldom recorded natural disasters unless their own infrastructure, personnel or vital interests were affected. The Australian north coast is little better researched, though one research team has cited geophysical and mythic evidence for a truly mammoth tsunami in the Kimberley region at some time in the seventeenth century.[44]

Javanese sources must be carefully re-examined, even when they are inadequately dated and generally opaque. They do make clear that the supernatural power of the southern seas was even more important a theme of palace literature and ritual since the seventeenth century than volcanoes, these being the twin poles, north and south, around which the cosmology of the Mataram kingdom and its successors revolved. The Queen of the South Seas (*Ratu Kidul*) is believed to have given the conquering founders of the Mataram dynasty, the shadowy and possibly mythic Senopati (believed to have died in 1601) and the mighty Sultan Agung (reigned 1613–46) their power to rule by her mystic union with them, though ordinary mortals who provoked her would be carried to a watery grave. The most chronologically reliable of the early Javanese chronicles, the *Babad ing Sangkala*, does provide the most explicit evidence for a tsunami with words translated by Merle Ricklefs as 'In Mataram, they moved [the court] to Karta, indeed, when *disappearing, all was turned into sea*'.[45] The italicised words, repeated three times in slightly different form for emphasis, are a chronogram (a date in words) for the Javanese year 1540, which began in February 1618 CE, at the beginning of Sultan Agung's campaigns to conquer Java.[46]

The scholars who worked on these texts before 2004 were looking not for tsunamis, but for the Babad's usefulness in dating political events, so the possibility remained unremarked. Other Javanese chronicles, written down considerably later than the *Babad ing Sangkala*, appear not to describe this disaster as such, but they may have mythologised it into the myth of the Queen of the South Seas (*Ratu Kidul*). The apparent confusions in the great chronicle of Java, the *Babad Tanah Jawi*, between the events of Senopati's and Sultan Agung's reigns make it not unlikely that a single disastrous tsunami could have been mythologised into both the beginnings of Sultan Agung's reign and the raging sea event that provided a subsequent legitimation for the violent beginnings of the dynasty's rise some 40 years earlier. Several manuscripts regard the Javanese date 1540 as powerfully and ritually linked to Senopati as well as Ratu Kidul, even if the date must be in Agung's reign according to accepted modern chronologies. At least two cite it as the year of Senopati's accession (and therefore of the encounter with Ratu Kidul which empowers Javanese kingship).[47] An explanation from the 1930s Surakarta court of its annual ritual at Parangtritis also dates the beginnings of the ritual from the moment 'Senopati' established the kingdom of Mataram in this same Javanese year 1540, or 1618–19 CE.[48] Although chronologically puzzling, since modern scholarship believes Senopati died in 1601, the occurrence of this date in several key documents may contain a more important truth about when the world changed on the south coast of Java. It also strengthens C.C. Berg's idea[49] that some features of Senopati's legendary life were in fact taken from that of Sultan Agung.

Together this evidence adds to the likelihood that a major tsunami occurred in 1618, most likely embracing Parangtritis and the mouths of Mataram's two great rivers, Kali Progo and Kali Opak, in the central (Mataram) section of Java's south coast where the Ratu Kidul rituals are centred. If so, Sultan Agung's supporters were able to convince themselves and their rivals that the flood was a sign of the ruler's supernatural power, not of his impotence before nature. Most important is the probability that the whole of Java's south coast is subject to devastating tsunamis, and ought to be prepared for another one.

Conclusion

The evidence of Southeast Asian historical sources, in other words, is of great value to the geologists in their quest to determine the past record of events and their human consequences, which in turn will determine the assessment of future risks. The mega-earthquakes and resultant tsunamis that release pressure on the subduction of the India–Australia plate beneath the Sunda Shelf are now known to be part of a necessary cycle of events, the regular recurrence of which is understood even though exact timing is not. They are, however, the least well reported in the historical records, so that every fragment of evidence needs to be weighed carefully. The islands off the west coast of Sumatra—Simeulue, Nias, Siberut and the Pagai group in particular—appear to be the only regions known so far where tsunamis occur with sufficient frequency to be remembered by inhabitants, making the oral histories and mythologies of these relatively isolated communities particularly important for research.

Volcanic eruptions were more spectacular portents to pre-modern inhabitants, and reports of the regular eruptions of the stratovolcanoes of Java, Bali, north Sulawesi and Maluku occur much more frequently in the historical records. The crater-forming explosive mega-eruptions capable of forming local devastation and global climate change are a different matter. Since 1800, there have been five such Southeast Asian eruptions listed as VEI 5 or above by volcanologists: Tambora, Sumbawa (7) in 1815; Galunggung, west Java (5) in 1822; Krakatau (6) in 1883; Agung, Bali (5) in 1963; and Pinatubo, Luzon (6) in 1991. While this record would suggest a similar frequency before 1800, too little is yet known to trace them, making the recent identification of Samalas/Rinjani in 1257 as a VEI 7, bigger than Tambora, a very helpful standout. Most of these mega-eruptions occurred at sites not remembered for previous activity. The reasons why explosive mega-eruptions occur when and where they do are still not well understood, and risk prediction in any particular location is extremely difficult. Their destructive impact on human society and agriculture is likely to have been so great that there were no recorders left to describe them, as seems the case for Lombok–Bali in 1257 and for Central Java at the end of the Mataram temple-building civilisation around 930. Nevertheless, scientists and historians can together make a huge difference to our understanding of the Southeast Asian past, and therefore likely future.

Notes

1. NOAA National Geophysical Data Center, register of tsunami events, http://www.ngdc.noaa.gov/hazard/hazards.shtml, accessed 6 July 2011.
2. Smithsonian Institute Global Volcanism Program, 'Large Holocene Eruptions', http://www.volcano.si.edu/world/largeeruptions.cfm, accessed 7 June 2012.
3. NOAA National Geophysical Data Center, Significant Volcanic Eruption Database, http://www.ngdc.noaa.gov/nndc/servlet/ShowDatasets?dataset, accessed 9 November 2012.
4. Carly Hilts, 'London's Volcanic Winter', *Current Archeology* 6 August 2012, http://www.archaeology.co.uk/articles/features/londons-volcanic-winter.htm. Julien Emile-Geay et al., 'Volcanoes and ENSO over the Past Millennium', *Journal of Climate* 21 (2008), pp. 3134–48.
5. Franck Lavigne et al. 'Source of the Great A.D. 1257 Mystery Eruption Unveiled, Samalas Volcano, Rinjani Volcanic Complex, Indonesia', *Proceedings of the National Academy of Science*, approved 4 September 2013, www.pnas.org/doi/10.1073/pnas.1307520110. This discovery was first brought to global attention in Alexandra Witze, 'Thirteenth Century Volcano Mystery May Be Solved', *Science News* 182, 14 July 2012.
6. Petra Gruber, 'Adaptation and Earthquake Resistance of Traditional Nias Architecture, http://www.nirn.org/pdf/01_070614_gruber.pdf, accessed 21 July 2012.

7. Johannes Hämmerle, *Asal Usul Masyarakat Nias: Suatu Interpretasi* (Gunungsitoli: Penerbit Yayasan Pusaka Nias, 1999), p. 44.
8. Reimer Schefold, *Lia: Das große Ritual auf den Mentawai-Inseln (Indonesien)* (Frankfurt: Dietrich Reimer Verlag, 1988), pp. 73–5; Edwin Loeb, *Sumatra: Its History and People* (Vienna: Institut für Völkerkunde der Universität Wien, 1935), pp. 134, 161.
9. See Anthony Reid, 'History and Seismology in the Ring of Fire: Punctuating the Indonesian Past', in *Environment, Trade and Society in Southeast Asia: A Longue Durée Perspective* ed. David Henley and Henk Schulte-Nordholt (Leiden: Brill/KITLV, 2015), pp. 62–77.
10. See Note 9.
11. Gerry van Klinken, *Communal Violence and Democratization in Indonesia: Small Town Wars* (London: Routledge, 2007), pp. 4–5.
12. Steven Pinker, *The Better Angels of Our Nature: The Decline of Violence in History and Its Causes* (London: Allen Lane, 2011).
13. B. Voight, E.K. Constantine, S. Siswowidjoyo and R. Torley 'Historical Eruptions of Merapi Volcano, Central Java, Indonesia, 1768–1998', *Journal of Volcanology and Geothermal Research* 100 (2000) 69–138.
14. C.G. Newhall et al., '10,000 Years of Explosive Eruptions of Merapi Volcano, Central Java: Archaeological and Modern Implications', *Journal of Volcanology and Geothermal Research* 100 (2000) 9–50.
15. Previously explained in other terms, see Anthony Reid, 'Low Population Growth and Its Causes in Pre-colonial Southeast Asia', in *Death and Disease in Southeast Asia: Explorations in Social, Medical, and Demographic History*, ed. Norman Owen (Singapore, OUP for ASAA Southeast Asia Publications Series, 1987), pp. 33–47.
16. Peter Boomgaard, *Children of the Colonial State: Population Growth and Economic Development in Java, 1795–1880* (Amsterdam: Free University Press, 1989), p. 75.
17. Sutikno Bronto, 'Volcanic Geology of Galunggung, West Java, Indonesia', PhD Dissertation, University of Canterbury, New Zealand, 1989. Smithsonian Volcanism Program, 'Galunggang Summary', online. The Galunggung caldera is estimated to have been formed in a massive eruption in about 2200 BCE.
18. See M.C. Ricklefs, *Jogjakarta Under Sultan Mangkubumi 1749–1792: A History of the Division of Java* (London: SOAS, 1974), especially pp. 373–5; Theodore Pigeaud, *Literature of Java* Vol. II (Leiden: Bibliotheca Universitatis Leidensis, 1968), p. 249 (LOr 4935).
19. I have only been able to consult the Indonesian summary in Etti R.S. Abdurrachman and Edi S. Ekajati, *Carita Parahiyangan, karya Pangeran Wangsakerta: Ringkasan, Konteks Sejarah, Isi Naskah dan Peta*

(Jakarta: Yayasan Pembangunan Jawa Barat, 1991), pp. 23–8. The quote is at p. 28.

20. T.S. Raffles, *The History of Java* (1817, reprinted Kuala Lumpur: Oxford University Press, 1965), Vol. I, p. 358.
21. Raffles 1817, I: 64.
22. Ibid.
23. Sri Margana, *Java's Last Frontier: The Struggle for Hegemony of Blambangan c.1763–1813* (Leiden University published dissertation, 2007), pp. 207–16.
24. Ann Kumar, 'Javanese Historiography in and of the "Colonial Period": A Case Study', in *Perceptions of the Past in Southeast Asia*, ed. Anthony Reid and David Marr (Singapore: Heinemann, 1979), p. 191.
25. Mark Donahue, 'The Papuan Language of Tambora', *Oceanic Linguistics*, 46, 2 (December 2007).
26. Alfons van der Kraan, 'Trade, Rajas and Bandars in South Bali', in *The Rise and Fall of Revenue Farming: Business Elites and the Emergence of the Modern State in Southeast Asia*, ed. John Butcher and Howard Dick (Basingstoke: Macmillan, 1993), pp. 107–10 (the quotation is at p. 109); also Helen Creese, 'New Kingdoms, Old Concerns: Balinese Identities in the Eighteenth and Nineteenth Centuries', in *The Last Stand of Asian Autonomies: Responses to Modernity in the Diverse States of Southeast Asia and Korea, 1750–1900*, ed. Anthony Reid (Basingstoke: Macmillan, 1997), p. 356.
27. Gregory Forth, 'Flores after Floresiensis: Implications of Local Reaction to Recent Palaeoanthropological Discoveries on an Eastern Indonesian Island', *Bijdragen tot de Taal-, Land-, en Volkenkunde* 162 (2006), pp. 336–49.
28. The sugar industry quickly transformed the formerly forested Negros after 1860. In 1889, Blumentritt cited a Spanish source that the Negritos were still numerous in the forests around the mountain, but 'weak and sickly, and decreasing in number'. Diaz Arenas had estimated their population at 3475 in 1850, before the sugar invasion, but this number declined quickly thereafter to be only a handful, having lost their languages, by the 1950s. Deforestation in the twentieth century removed their remaining habitat. Rudolf Rahman and Marcellino Maceda, 'Notes on the Negritos of Northern Negros', *Anthropos* 50 (1955), pp. 810–36.
29. Smithsonian Global Vocanism Program: Pinatubo.
30. S. Seitz, 'Coping Strategies in an Ethnic Minority Group: The Aeta of Mount Pinatubo', *Disasters* 22 (1998), pp. 76–90.
31. Jan Wisseman Christie, 'Under the Volcano: Stabilizing the Early Javanese State in an Unstable Environment', in *The Longue Durée in Southeast Asian*

History ed. David Henley and Henk Schulte-Nordholt (Leiden: Brill/KITLV, forthcoming).

32. I. Gusti Putu Phalgunadi, *The Pararaton: A Study of the Southeast Asian Chronicle* (New Delhi: Sundeep Prakashan, 1996).
33. David Keys, *Catastrophe: A Quest for the Origins of the Modern World* (New York: Ballantine Books, 2000).
34. Sam White, *The Climate of Rebellion in the Early Modern Ottoman Empire* (*Studies in Environment and History*) (Cambridge: Cambridge University Press, 2011).
35. K.R. Newcomb and W.R. McCann, 'Seismic History and Seismotectonics of the Sunda Arc', *Journal of Geophysical Research* 92, B1 (10 January 1987).
36. Katrin Monecke, Willi Finger, David Klarer, Widjo Kongko, Brian McAdoo, Andrew Moore and Sam Sudrajat, 'A 1,000-year sediment record of tsunami recurrence in northern Sumatra', *Nature* 455, 1232–4 (30 October 2008); Stein Bondevik, 'Earth Science: The Sands of Tsunami Time', *Nature* 455, 1183–4 (30 October 2008).
37. Kerry Sieh, 'The Sunda Megathrust: Past, Present and Future', in Patrick Daly, Michael Feener and Anthony Reid, eds., *From the Ground Up: Conflict Resolution and Reconstruction in Post-tsunami Aceh* (Singapore: ISEAS, 2012), pp. 1–22.
38. W.Ph. Coolhaas (ed.), *Generale missiven van gouverneurs-generaal en raden aan Heren XVII der Verenigde Oostindische Compagnie*, Vol. 5 (The Hague: Instituut voor Nederlandse Geschiedenis), pp. 413, 844; Newcomb and McCann 1987, p. 424.
39. Alan Harfield , *Bencoolen: A History of the Honourable East India Company's Garrison on the West Coast of Sumatra 1685–1825* (Barton-on-Sea, Hampshire, privately printed by A & J Partnership, 1995), p. 61.
40. William Marsden, *History of Sumatra*, pp. 30–1.
41. Jan van Groenewegen to Batavia, 28 January 1660, VOC 1233 (KA 1123), f. 502r (my translation). I gratefully acknowledge the labours of Takeshi Ito in transcribing this material, now being prepared for publication by Brill.
42. Corespondence with Kerry Sieh and Belle Philibosian, November–December 2012. They point out, however, that other evidence does not so far suggest that an event in Siberut would have such major effects as far away as Banda Aceh. Clearly, this is important new evidence in that direction, however.
43. Newcombe and McCann 1987: 43–4.
44. E. Bryant, G. Walsh and D. Abbott, 'Cosmogenic Mega-Tsunami in the Australia Region: Are They Supported by Aboriginal and Maori Legends?' (2007), http://ro.uow.edu.au/cgi/viewcontent.cgi?article=1035&context=scipapers.

45. Ricklefs, M.C., *Modern Javanese Historical Tradition: A Study of an Original Kartasura Chronicle and Related Materials.* London: SOAS, 1978), pp. 32–3.
46. Anthony Reid, 'Historical Evidence for Past Tsunamis in the Java Subduction Zone', *Asia Research Institute Working Paper* No. 178, Singapore (February 2012). www.ari.nus.edu.sg/docs/wps/wps12_178.pdf.
47. Ricklefs 1978: 166n.55. One of the texts he refers to is the LOr7699 described in Pigeaud, *Literature of Java* (1968) II: 459, as including two versions of dated chronicles, *Babad Sangkala.*
48. 'Heilige pleffen Kasunan Surakarta', 15 March 1932, in Leiden University Library Cod. Or. 11 December 2012.
49. See Ricklefs 1978, pp. 160–1, 164–5, and 204–6, and the literature there cited.

Bibliography

Abdurrachman, Etti R.S., and Edi S. Ekajati. 1991. *Carita Parahiyangan, karya Pangeran Wangsakerta: Ringkasan, Konteks Sejarah, Isi Naskah dan Peta.* Jakarta: Yayasan Pembangunan Jawa Barat.

Bondevik, Stein. 2008. Earth science: The sands of tsunami time. *Nature* 455: 1183–1184.

Boomgaard, Peter. 1989. *Children of the colonial state: Population growth and economic development in Java, 1795–1880.* Amsterdam: Free University Press.

Bronto, Sutikno. 1989. Volcanic geology of Galunggung, West Java, Indonesia. PhD Dissertation, University of Canterbury, New Zealand, Smithsonian Volcanism Program.

Bryant, Edward A., G. Walsh, and D. Abbott. 2007. Cosmogenic mega-tsunami in the Australia region: Are they supported by Aboriginal and Mairo legends? In *Myth and Geology*, eds. L. Piccardi and W.B. Masse, 203–214. Geological Society, London. Special Publications, 273. http://ro.uow.edu.au/cgi/viewcontent.cgi?article=1035&context=scipapers.

Christie, Jan Wisseman. forthcoming. Under the volcano: Stabilizing the early Javanese state in an unstable environment. In *The longue Durée in Southeast Asian history*, eds. David Henley and Henk Schulte-Nordholt. Leiden: Brill/KITLV.

Coolhaas W.Ph. (ed.), *Generale missiven van gouverneurs-generaal en raden aan Heren XVII der Verenigde Oostindische Compagnie*, 5 The Hague: Instituut voor Nederlandse Geschiedenis.

Creese, Helen. 1997. New kingdoms, old concerns: Balinese identities in the eighteenth and nineteenth centuries. In *The last stand of Asian autonomies: Responses to modernity in the diverse states of Southeast Asia and Korea, 1750–1900*, ed. Anthony Reid, 345–366. Basingstoke: Macmillan.

Donahue, Mark. 2007. The Papuan language of Tambora. *Oceanic Linguistics* 46(2): 520–537.

Emile-Geay, Julien, Richard Seager, Mark A. Cane, Edward R. Cook, and Gerald H. Haug. 2008. Volcanoes and ENSO over the past millennium. *Journal of Climate*, 21: 3134–3148. doi: http://dx.doi.org/10.1175/2007JCLI1884.1.

Forth, Gregory. 2006. Flores after floresiensis: Implications of local reaction to recent palaeoanthropological discoveries on an eastern Indonesian island. *Bijdragen tot de Taal-, Land-, en Volkenkunde* 162(2–3): 332–345.

Gruber, Petra. Adaptation and earthquake resistance of traditional Nias architecture. Department of Design and Building Construction Vienna University of Technology, Austria. http://www.nirn.org/pdf/01_070614_gruber.pdf, accessed 21 July 2012.

Hämmerle, Johannes. 1999. *Asal Usul Masyarakat Nias: Suatu Interpretasi*. Gunungsitoli: Penerbit Yayuasan Pusaka Nias.

Harfield, Alan. 1995. *Bencoolen: A history of the honourable East India Company's garrison on the west coast of Sumatra 1685–1825*. Barton-on-Sea: privately printed by A & J Partnership.

'Heilige pleffen Kasunan Surakarta', 15 March 1932, in Leiden University Library Cod. Or. 11 December 2012.

Hilts, Carly. 2012.. London's volcanic winter. *Current Archeology* 6 August [270]. http://www.archaeology.co.uk/articles/features/londons-volcanic-winter.htm.

I Gusti Putu Phalgunadi. 1996. *The pararaton: A study of the Southeast Asian chronicle*. New Delhi: Sundeep Prakashan.

Jan van Groenewegen to Batavia, 28 January 1660, VOC 1233 (KA 1123), f. 502r.

Keys, David. 2000. *Catastrophe: A quest for the origins of the modern world*. New York: Ballantine Books.

Kumar, Ann. 1979. Javanese historiography in and of the "colonial period": A case study. In *Perceptions of the past in Southeast Asia*, eds. Anthony Reid, and David Marr. Singapore: Heinemann.

Lavigne F., Degeai, J.-P., Komorowski, J.-C., Guillet, S., Robert, V., Lahitte, P., Oppenheimer, C., Stoffel, M., Vidal, C.M., Surono, Pratomo I., Wassmer, P., Hajdas, I., Hadmoko, D.S., De Belizal, E. 2013. Source of the great AD 1257 mystery eruption unveiled, Samalas volcano, Rinjani volcanic complex, Indonesia. *Proceeding of the National Academy of Sciences of the United States of America (PNAS)* 110: 42. DOI: 10.1073/pnas.1307520110.

Loeb, Edwin. 1935. *Sumatra: Its history and people*. Vienna: Institut für Völkerkunde der Universität Wien.

Monecke, Katrin, Willi Finger, David Klarer, Widjo Kongko, Brian McAdoo, Andrew Moore, and Sam Sudrajat. 2008. A 1,000-year sediment record of tsunami recurrence in northern Sumatra. *Nature* 455: 1232–1234.

Newcomb, K.R., and W.R. McCann. 1987. Seismic history and seismotectonics of the Sunda Arc. *Journal of Geophysical Research* 92(B1): 421–439.

Newhall, C.G., et al. 2000. 10,000 years explosive eruptions of Merapi volcano, Central Java: Archaeological and modern implications. *Journal of Volcanology and Geothermal Research* 100: 9–50.

Pigeaud, Theodore. 1968. *Literature of Java*, vol II. Leiden: Bibliotheca Universitatis Leidensis.

Pinker, Steven. 2011. *The better angels of our nature: The decline of violence in history and its causes*. London: Allen Lane.

Raffles, T.S. 1965. *The history of Java* (1817, reprinted Kuala Lumpur: Oxford University Press.

Rahman, Rudolf, and Marcellino Maceda. 1955. Notes on the Negritos of northern Negros. *Anthropos* 50: 810–836.

Reid, Anthony. 1987. Low population growth and its causes in pre-colonial Southeast Asia. In *Death and disease in Southeast Asia: Explorations in social, medical, and demographic history*, ed. Norman Owen. Singapore: OUP for ASAA Southeast Asia Publications Series.

Reid, Anthony. 2012. Historical evidence for past tsunamis in the Java subduction zone. Asia Research Institute Working Paper No. 178, Singapore, February. www.ari.nus.edu.sg/docs/wps/wps12_178.pdf.

Reid, Anthony. 2016. 'History and Seismology in the Ring of Fire: Punctuating the Indonesian Past,' in Environment, Trade and Society in Southeast Asia: A Longue Durée Perspective ed. David Henley and Henk Schulte-Nordholt (Leiden: Brill/KITLV, 2015), pp. 62–77.

Ricklefs, M.C. 1974. *Jogjakarta under Sultan Mangkubumi 1749–1792: A history of the division of Java*. London: SOAS.

Ricklefs, M.C. 1978. *Modern Javanese historical tradition: A study of an original Kartasura chronicle and related materials*. London: SOAS.

Schefold, Reimer. 1988. *Lia: Das große Ritual auf den Mentawai-Inseln (Indonesien)*. Frankfurt: Dietrich Reimer Verlag.

Seitz, S. 1998. Coping strategies in an ethnic minority group: The Aeta of Mount Pinatubo. *Disasters: The Journal of Disaster Studies, Policy and Management* 22(1): 76–90.

Sieh, Kerry. 2012. The Sunda Megathrust: Past, present and future. In *From the ground up: Conflict resolution and reconstruction in post-tsunami Aceh*, eds. Patrick Daly, Michael Feener, and Anthony Reid, 1–22. Singapore: ISEAS.

Smithsonian Institution. National museum of natural history. Global Vocanism Program: Pinatubo. http://www.volcano.si.edu/volcano.cfm?vn=273083.

Sri Margana. 2007. Java's last frontier: The struggle for hegemony of Blambangan c.1763–1813. Leiden University Published Dissertation.

van der Kraan, Alfons. 1993. Trade, rajas and bandars in South Bali. In *The rise and fall of revenue farming: Business elites and the emergence of the modern state in Southeast Asia*, eds. John Butcher, and Howard Dick. Basingstoke: Macmillan.

van Klinken, Gerry. 2007. *Communal violence and democratization in Indonesia: Small town wars*. London: Routledge.

Voight, B., E.K. Constantine, S. Siswowidjoyo, and R. Torley. 2000. Historical eruptions of Merapi volcano, Central Java, Indonesia, 1768–1998. *Journal of Volcanology and Geothermal Research* 100: 69–138.

White, Sam. 2011. *The climate of rebellion in the early modern Ottoman Empire*, Studies in environment and history. New York: Cambridge University Press, Cambridge.

Witze, Alexandra. 2012. Thirteenth century volcano mystery may be solved. *Science News* 182, 14 July. DOI: 10.1002/scin.5591820113.

CHAPTER 3

'The Sea Becomes Mulberry Fields and Mulberry Fields Become the Sea': Dikes in the Eastern Red River Delta, c.200 BCE to the Twenty-First Century CE

Li Tana

It is well known that the Red River delta is essentially land reclaimed from the sea. Because this territorial expansion is well recorded, the story of this delta is told as a more or less continuous series of human victories over nature. The ever increasing number and length of dikes spreading across the delta over time seems to provide testimony for such a claim. In 1803, there were but 124 kilometer (km) of sea and river dikes in the Red River delta.[1] By 1945, however, this figure had grown to 3000 km, while by 1996 it had nearly tripled to 8000 km.[2] As with the Mekong delta, it thus seems that human activities have altered every aspect of this land.[3] But how historically accurate is this conventional view? Unlike the comparatively recently formed Mekong delta, human intervention in the ancient Red River delta has a much longer history, and the apparently triumphant onward march of dikes over the last two centuries needs to be understood

L. Tana
College of Asia and the Pacific, Australian National University, Canberra, Australia

G. Bankoff, J. Christensen (eds.), *Natural Hazards and Peoples in the Indian Ocean World*, DOI 10.1057/978-1-349-94857-4_3

within this far older historical context. Surviving records suggests that countless advances and setbacks occurred in this millennially old arena, so much so that the story of the Red River delta resembles that of Magu, a legendary Chinese fairy who, during her long life, three times saw the sea become mulberry fields and mulberry fields become the sea. When we take a very much wider view, we do not see a narrative of human triumphs. Instead, with the exception of the mountains—the only relatively stable feature in this land—every other element changes repeatedly over time: river courses shift under the joint forces of nature and human actions; coastlines expand or erode; and human settlements, and their irrigation systems, are repeatedly set up only to be later abandoned.[4]

This chapter is a preliminary attempt to look beyond the simple triumphalism of the existing narrative of the Red River delta's past to explore its untold story, to reveal the exposure and vulnerability of human projects in their ongoing battles against the sea. Rather than a tale of continuous growth and the accumulation of land area and productivity, we find, interleafed with those success stories, discontinuity, reduction, abandonment and the flight of people from their own establishments in the face of natural disasters. When viewed in the longue durée, the dikes of the Red River delta in fact tell a story of how fragile, temporary and fluid human efforts have actually been in this ancient delta area.

Three maps below introduce our investigation. Map 3.1 shows the dikes in the delta in 1803; Map 3.2 indicates the impact of sea intrusion into the Red River delta. As the second map shows, the eastern delta (centered on Hai Duong, as opposed to the western flood plain centered on Nam Dinh) is most heavily affected by saltwater intrusion. One would expect, therefore, that this area should have had numerous dikes designed to prevent the sea from pushing saltwater inland. But curiously, as the first map reveals, no dikes whatsoever existed in the eastern delta in 1803. As a nineteenth-century manuscript collection of dike reports to the Nguyen emperors reveals, the Hai Duong area, the heart of the eastern delta, had never had dikes.[5] The *Kham dinh Dai Nam hoi dien su le* [Official compendium of institutions and usages of Imperial Vietnam] confirms this. From a total of 72 km of river dikes, and 52 km of sea dikes, there were no river dikes recorded in Hai Duong, and only 366 meters of sea dikes out of a total of 52 km.[6] This means that the vast majority of sea dikes were located in the Western flood plain. Indeed, it was not until 1839 that Hai Duong finally built 21 km of sea dikes.[7] Yet this is not the end of the story: 90 years later, those 21 km of dikes had disappeared, according to the French map of 1926 (Map 3.3).[8]

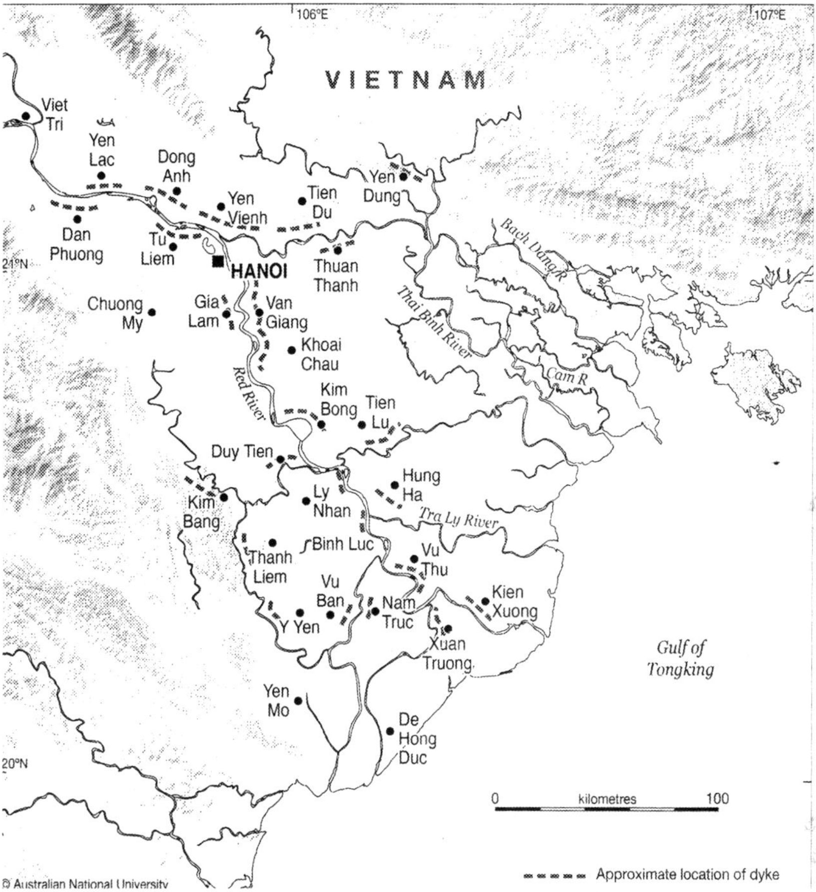

Map 3.1 Dikes in the Red River delta, 1803 (Source: *Kham dinh Dai Nam hoi dien su le*, vol. 13 (Hue: Nha xuat ban Thuan Hoa, 1993): 229–237)

When carrying out his fieldwork in the 1920s Red River delta, Pierre Gourou observed a string of underpopulated cantons in the eastern delta. 'The explanation is simple', he reported: 'in the north, the eastern divisions of the province of Bac Ninh, Que Duong and Lang Tai, are made up of low lying lands inundated in the summer since, up until recently, the Song Cau, the Canal de Rapides (Song Duong) and the Thai Binh were not diked in this region and only the fifth month crop (*lua chiem*) could be grown'.[9] A Vietnamese source confirms this situation. It says that before 1930 in the

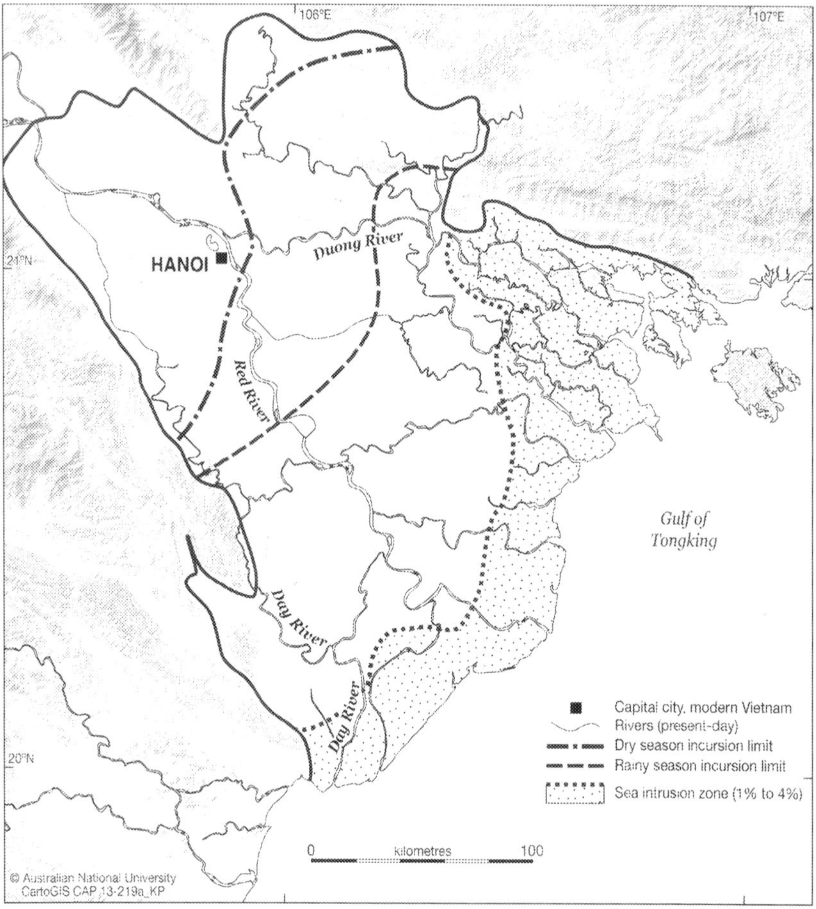

Map 3.2 Impact of sea intrusion at the current Red River delta (Source: Vu Tu Lap et al., *Van hoa va cu dan Dong bang Song Hong* [Culture and peoples in the Red River Delta] (Hanoi: Khoa hoc xa hoi, 1991), p. 18)

area between the Thai Binh river and the Kinh Thay river only an alluvial dike (*de boi*) or natural levee existed. It was only two meters higher than the surrounding paddy land, even though the maximum tidal range along the coast of the delta is approximately four meters.[10]

How do we explain this phenomenon? Why was dike building apparently started so late and prove so unsuccessful in an area that obviously needed it the most?

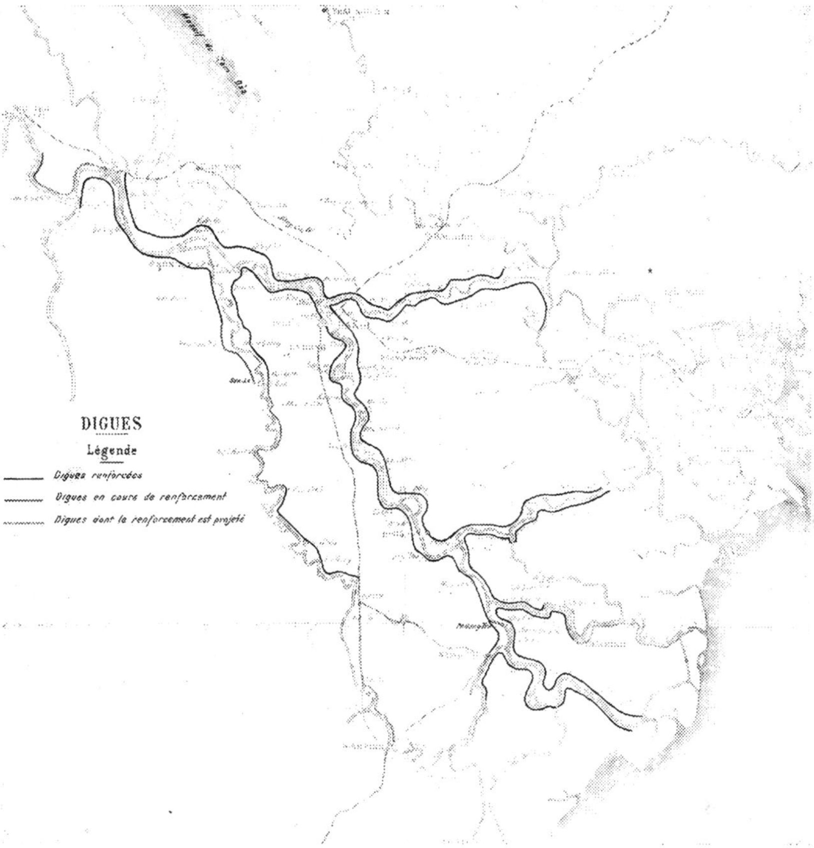

Map 3.3 Dikes of the Red River, 1926 (Source: Tuan Pham Anh and Kelly Shannon, 'Water Management in Vietnam: Indigenous Knowledge and International Practices –The Case of the Red River Delta', paper for 11th N-AERUS Conference 2010, Brussels, 28–30 October 2010, http://www.n-aerus.net/web/sat/workshops/2010/pdf/PAPER_pham_t.pdf, p. 3)

The Earlier Dikes in the Eastern Delta

The historical record shows that the dike situation here during the early nineteenth century was not typical of all periods. The fourth-century CE Chinese source recorded that before the Chinese occupation of second century BCE, in today's Red River delta 'there had been fields called Lac, which changed according to the rise and fall of the tides'.[11] French geographer Gourou identified these Lac fields as the eastern delta: 'This means that the Chinese observer who noticed the fact, was struck by a peculiarity which still subsists in Tonkin, for in the region of Hai Duong, the tides freely rise and fall in the rice fields'. Gourou seemed to have regarded this as dikes in the eastern delta when he commented that 'but this does not mean that there were no dikes elsewhere'. Gourou argued that without small-scale dike building, designed to reclaim small portions of low land, the delta would never have been colonized by the Viet people. From this he concluded that dikes must have existed well before the eleventh century.[12]

This would make the Red River delta one of the earliest regions of Asia to have dikes, however elementary they may have been. The earliest record about river dikes in the Red River delta referred to the second century CE. It said that there was a dike in the Phong Khe district. The legend has it that because the water head here drops down to a hundred *xun* (1 *xun* = 8 *chi* = 2.4 meters), fish which were able to swim up against the river would become dragons. The dike thus was called the 'dike at the Dragon Gate'.[13] The Phong Khe district included today's Co Loa and the land north of the Duong River. [14]This area is quite far from the present main stream of the Red River, but it is logical to assume that the earliest river dike would have been built along the most powerful river of the delta—the Red River—to protect the land and the residence nearby. At one point of time, it seemed, the Red River used to run its course at the northeast of today's Red River on the eastern delta.

This eastern delta region was also the area where the fifth month rice that is popularly known as Cham rice was initially bred, according to the Vietnamese agronomist Dao The Tuan. The introduction of this Cham rice to China is thought to have occurred in the eleventh century and led to its first population boom. Because it is called Cham rice, people believed it derived from Champa, but Dao The Tuan has found no species of 'Cham rice' in central Vietnam, including not in the former Champa principality to the north of Hue (Binh Tri Thien), the very region in which 'Cham rice' was supposedly bred. In fact, as Dao The Tuan points out, the word 'Cham' came from the Vietnamese word 'Chăm', meaning 'swampy areas'. 'Chăm' was the name in demotic Vietnamese (Chu Nom) of Mo Trach, a famous village in Hai Duong. [15] This rice is extremely hardy; it withstands drought and cold as well as saline–alkali soil, and requires less fertilizer. Equally important, since it ripens in lunar May, the harvest escapes the typhoon season between July and August, thereby guaranteeing minimum yields. Local climatic evidence supports his point, since cold winters and the intrusion of saltwater are both characteristic features of the eastern delta but quite absent from the old Cham area of modern Binh Tri Thien. Cham rice (*lua chiem*) could have been bred to suit the cultivation requirements of the eastern delta, so well does it match them. This helps us to understand another record of the second century CE, which states that 'in Jiaozhi rice is ripe again in winter. Farmers plant rice twice a year'.[16] Only the quick ripening Cham rice could make double cropping possible.

The discussion above also sheds light on a puzzling Chinese record which indicates that in the second century BCE, Jiaozhi households were four times as many households the size of those of Guangzhou (prefecture of Nanhai in Table 3.1, below),[17] and the average size of households in Jiaozhi was larger than in some parts of contemporaneous northern China, at the time the most developed area in Han China.[18]

Remarkably, the practice of double cropping was never recorded again in Jiaozhi later than the second century CE. Thereafter, most of the Red River delta does not seem to have practiced double cropping until the nineteenth century. Coupled with its apparent abandonment was the decline of Jiaozhi's population after the third century. This seems to indicate the possibilities of scattered population, lost memories and techniques, including the disappearance of the dikes.

It would be the thirteenth century before records of the building of sea dikes appeared again. A fifteenth-century history noted that: 'The royal families tended to make their slaves build dikes and weirs on the coast to stop the saltwater. In two to three years the land would be ready to cultivate. Those slaves would marry each other and live on these lands.

Table 3.1 Households in the Coastal South, 2 BCE

Prefecture	*Household*
Nanhai	19,613
Jiaozhi	92,440
Jiuzhen	35,743
Hepu	15,398
Cangwu	57,510

Source: Hanshu [History of the Former Han dynasty], Wenyuange Edition, *juan*. 28

[This was how] many of the estates were founded'.[19] Technically, this process involved blocking the sea with seawalls, desalinating the enclosed lands by connecting them to a river system, and creating polders in which to cultivate. The coastal areas now being opened and cultivated were in the western delta, and it was at this time that today's coastal area of Thai Binh and Nam Dinh began to be populated.[20]

Sea dikes were also reported in the fifteenth-century Hai Duong area. The Viet Chronicle (*Dai Viet su ky toan thu*) recorded that in 1404 the King (Ho Han Thuong) had promoted Tran Quoc Kiet, an officer of the Tan Hung prefecture, 'because he organised building sea dikes to protect the fields of his prefecture from saltwater invasion'.[21] The Tan Hung prefecture included both the Hai Duong and Quang Ninh areas of the eastern delta.[22] Again, in 1460, we find this record: 'Typhoon struck, the sea rose extremely high in Nam Sach, Hiep Son, Thai Binh and Kien Xuong, dikes were broken and [saltwater] inundated rice fields, many people at the coast died of starvation'.[23] The first two districts mentioned here were in Hai Duong, so clearly dikes existed there in the fifteenth century as they were breached by the typhoon, causing flooding and famine. In the sixteenth century, the Mac dynasty, whose family roots lay in Hai Duong, also built dikes in today's Hai Phong and Quang Ninh areas.[24] None of these dikes remained by the early nineteenth century, as Map 3.1 shows. There are ruins of old dikes which local people called *De nha Mac* ('dikes of the Mac dynasty'), but no real dikes were left.[25]

As far as the scattered records reveal, sea dikes had been built at least three times (second century BCE, through the Tran and sixteenth century CE) in the Hai Duong area. On each occasion, their construction was recorded as a new phenomenon and none of them had survived by the early nineteenth century. The Nguyen built 21 km of sea dikes in Hai Duong in 1839, but these sea dikes were unknown to the French map by 1926 and no reference to them appeared in Gourou's influential study of the early 1930s.

These are the few sketchy references to sea dikes mentioned in chronicle sources for the last 2000 years. Many more dikes must have been constructed and decayed in between these dates in the eastern delta. The question arises: Why were dikes being repeatedly built, sometimes being recorded historically and mostly not, only for them all to disappear? The next section argues a hypothesis that seeks to address this intriguing issue.

Hypothesis: The Shifting Red River and the Erosion of the Eastern Coast

Map 3.4 charts the historic evolution of the Red River delta. A striking feature of this map is the rapid expansion of the eastern delta before the tenth century CE and its complete cessation after it. As the map indicates, this delta area, virtually non-existent before the first century CE, swelled to double the size of the western delta in a matter of ten centuries. Such a speed of accretion needed a strong discharge from the upper delta and there is only one river in the region that has such a huge capacity—the Red River. Most of the maps of the Red River delta to date, however, show a linear evolution which showcases the expansion of the delta. Yet when this expansion is broken into different stages, as is shown in Map 3.4, it strongly hints at an untold story of reduction over the last ten centuries.

Undoubtedly, many factors would have been involved in causing the reductions that occurred in the eastern delta during that millennium, among them intensive human intervention in the upper and mid-Red River delta. The modern Hanoi, Bac Ninh and Hai Duong regions were the locations in which the most frequent human actions were concentrated before the fourteenth century. A clear sign of this is shown by the

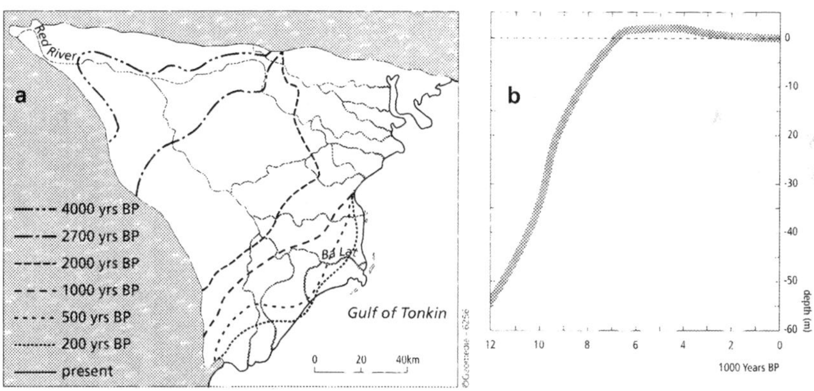

Map 3.4 a. Historical evolution of the Red River delta, b. Sea level curve (Source: Maren, D.S. van, 'Morphodynamics of a cyclic prograding delta: the Red River', PhD thesis, Nederlandse Geografische Studies, Utrecht University (2004), Chap. 2

channel fragmentation of the river system from Hai Duong down to the sea. A Vietnamese expert in hydrological history observes:

> The river network of Thai Binh becomes more complex from Pha Lai (Bac Ninh) downstream. [At the upper area] we can still differentiate the Kinh Thay river from the Thai Binh river, but when one comes to the section of Hai Duong to the sea, the river network becomes so mixed up and confusing that no one can tell which river originated from where and goes to where.[26]

Channel fragmentation is a significant symptom of stress which is particularly associated with dams, canals, withdrawals and diversions of the free flow of river systems.[27] The degree of fragmentation in the lower eastern Red River delta noted above would have taken centuries to develop. Similar cases are found in the second-century BCE Yellow River delta, and the eighteenth-century Pearl River delta. Both result from human activities, specifically diverting river flow and reclaiming floodplains along rivers.[28]

First, the construction of the Dinh Nhi dike in 1248 presents the most direct evidence of river diversion. According to an 1852 imperial report, the Red River used to run in two courses, one to the east to Hai Duong and the other to west to Nam Dinh. In the words of an author:

> There had been thousands of years from the Hung kings down to the Former Le and Ly kings and in all these years water damage had been unheard of. It was not until the Tran dynasty [twelfth–thirteenth centuries] that the Dinh Nhi dike was built. It was said that the Tran king was so deluded by belief in *fengshui* that he ordered the eastern tributary be filled in and made all the rivers flow toward the Nam Dinh and Vi Hoang rivers.[29]

This generalized account almost sounds too simplistic and improbable to believe except that it does reflect one important historical reality that the building of the Dinh Nhi dike damaged the interests of the eastern delta. The major river of the delta—the Red river, was directed to the southwestern coast, where the Tran's base and the second capital, Thien Truong, was located.[30] This also confirms that the 1248 diking was a strong central government act but one that integrated local efforts which may have been carried out in small scales. The construction of the Dinh Nhi dike was therefore not simply one event and a single project. It makes much more sense to treat it as a series of big and small, communal and private projects that were undertaken to protect different local interests at different localities, with most, if not all, being located in the middle Red River delta. As one later scholar pointed out, other than the official project of the Dinh Nhi, 'ordinary people also carried out their own projects of building dams and weirs thus all the old waterways—rivers, streams and canals were silted'.[31]

Second, clear evidence for reclaiming of floodplains in the Red River delta appears in the huge difference in distances between dikes. According to a nineteenth -century Vietnamese officer, the distance between the dikes at the left hand and right hand sides of the upper Red River area in Son Tay was as great as 1500 meters, while comparable distances in the section from Hanoi and Bac Ninh down to the lower delta were one- to two-third less, at between 500 and 1000 meters.[32] The basic reason for this difference was because the people in the lower delta competed with each other to occupy and reclaim the fertile floodplain between the dikes to such an extent that they rebuild the dikes for more spaces for land. Spaces between dikes became narrower and narrower. These people were rendered easy targets in any flood.

All these alterations to the river must have influenced the directions and volumes of water flow. For this reason and possibly others, some significant changes seemed to have happened to the major course of the Red River between the thirteenth and fifteenth centuries. Put simply, when the main stream shifted its course from the eastern delta, water discharge to this area was greatly reduced. This reduced the amount of sedimentation, which in turn reduced the earlier speed of the natural expansion of the eastern delta, effectively bringing it to a standstill, as shown on Map 3.4 above. Renewed freshwater gave way to saltwater intrusions into the eastern delta; and the damaged soil, which lacked oxygen, was even more vulnerable to damage by typhoons. The end result was the serious and continuous erosion of the eastern delta coast.

While we cannot chart this historical evolution directly from extant sources, what is happening in the current Red River delta does shed light on the hypothesis outlined above of what happened a few hundred years ago. In the Red River delta, as a recent report by Ngo Ngoc Cat et al. points out, 'in the coastal sections far from river mouths [there] occurs erosion, while in river mouth areas [there] occurs sedimentation with high rate as in Ba Lat and Day river mouths'.[33] This story of reduction will be examined below through a brief case study of the Hai Hau district, at the western delta (Map 3.5).

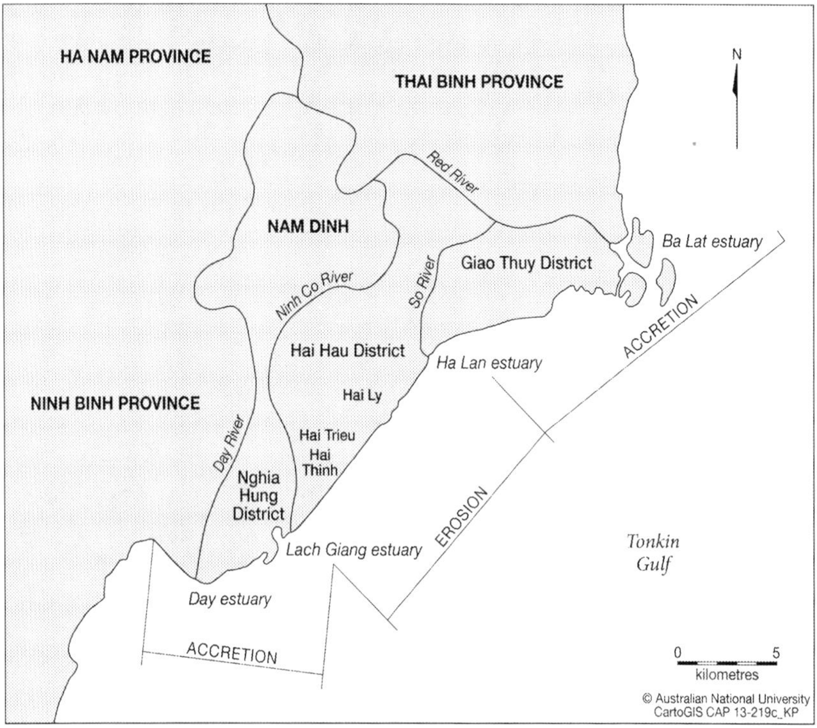

Map 3.5 Sections of accretion and erosion in contemporary Nam Dinh coastlines (Source: Cong V. Mai, Marcel J.F. Stive, and Pieter H.A.J.M. Van Gelder, 'Coastal Protection Strategies for the Red River Delta', *Journal of Coastal Research*, 25:1(2009):108)

The Disappearing Coast: The Case of Hai Hau

Modern Hai Hau in the western delta is suffering from erosion for the same reason that the eastern delta suffered it a few hundred years ago. Sometime between the twelfth and fifteenth centuries, the Red River apparently changed its course and began to empty at Hai Hau, meaning that Hai Hau was where the eighteenth-century main stream of the Red River met the sea. The main channel of the Red River must have terminated here from at least the fifteenth century, since the well-known Hong Duc dike (*De Hong Duc*) of 1471 was said to be in Hai Hau.[34] This places it about 40 km from the current main gate of the Red River, at Ba Lat. Local tradition maintains that the Red River in this section used to be huge, while the stream at Ba Lat was narrow, with some people saying that it was but a canal. This situation apparently persisted until 1787. When the delta was struck by a huge flood that year, the Red River broke into the Ba Lat waterway from Hai Hau and from then on the Red River emptied into the sea at Ba Lat.[35]

A more recent factor might have even more directly led to the erosion of Hai Hau. The Hoa Binh hydropower dam upstream of the Red River's main branch, the Da River, was built in 1979–1994. This dam project regulated 70 % of the flow discharge in the Red River system and has halved the sediment supply to the coastal areas.[36]

When the main stream changed its course and the old waterway was left with much less flow than previously, the coastal sections that no longer received sedimentation from the river began to erode. Geologists term this 'structural erosion'. It must have been happening earlier in Hai Hau, although data is available only from 1905. The data we do have indicate a clear pattern of retreat of the Hai Hau coastline, of about 29 meters per year.[37] As Cong et al. (2009) point out, 'the main reason for the heavy erosion along the Hai Hau coastline is changes in the distribution of the natural fluvial sediment discharge to the whole system. This could have been a result of either a natural development or man-made river training works, which changed the Red River flow direction'.[38]

The places where erosion occurred then became the most vulnerable to typhoons. In the Hung Yen area (part of the former Hai Duong province), from 1806 to 1882, the dikes were breached 21 times.[39] Dike breaching often occurred after severe typhoons, due to erosion and erosion-induced weakening. As a consequence, the coastline of the Red River here has shifted 3000 meters inland in the last 100 years, an average of 30 meters per year. The total area of land loss is approximately 18,000 hectares (nearly as large as the current area of the Hai Hau district). In September 2005, typhoon Damrey, one of the most powerful storms of the past 50 years, breached more than 25 km of sea dikes in Hai Hau and Hau Loc districts. Extensive sea flooding of the protected regions behind the dikes and flash floods created by the typhoon submerged more than 130,000 hectares of rice fields.[40]

The foregoing examples help explain why there were no surviving dikes in the Hai Duong and Quang Ninh area in 1803 and also in the early twentieth century: no dikes could have survived if the coastline was shifting inland continually. A contemporary report on the coastal area of Van Ly in Hai Hau states that the average annual coastline retreat has resulted in one destroyed dike line every ten years,[41] while another report estimates the rate of dike failure at about one every seven years, instead of twenty years as the expected return period.[42] This illustrates the point Gourou made in the last century: 'The newly claimed land and the sea walls are particularly vulnerable to typhoons, weakened by torrential rains the sea would easily take back the larger part of the lands occupied by men'.[43] Vietnamese geographer Le Ba Thao described the sense of destruction and loss with a vivid picture of the Hai Hau area. In his words: 'The erosions could create such incredible sites in the lower delta area that on the coast between Cua Lach and Van Ly, one can see some sluices standing alone: they used to be a part of certain dikes that no longer exist'.[44]

As might be expected, the maintenance of dikes puts tremendous pressure on local people. It is reported that men living in the erosion-prone coastal districts in Vietnam are on average obliged to work for free about ten days per year, building and maintaining the dikes in their district. If they refuse, or for any reason are not able to work, they are required to pay for a hired worker to take their place. In Van Ly village in Hai Hau, a place suffering from severe erosion, the men are obliged to work around 40 days a year for free.[45] There is every reason to expect that pressure for dike maintenance on local people in centuries past would have been similar, if not worse. At times of harvests failure or civil strife, when people were already hungry, the pressure might have become unbearable, leading to the flight of local populations. Dikes that were no longer maintained would be breached and eventually collapse. Those who remained in the watery desert would have little choice but to relocate.

These repeated losing battles between man and nature on the coast were summed up perfectly by a thirteenth-century Chinese officer: 'People provide themselves with the money to build dikes, which cause the mudflats to turn into farmland. [But] if they fail to repair them in good time, the dike walls collapse or leak, and the farmland reverted once more to being mudflats'.[46] Another description is even more succinct: 'People successively built [dikes], only [to see them] successively broken and breached, and have had in succession to take flight'.[47] There is evidence that this pattern occurred historically in the eastern delta, and this chapter ends with some comments on population flight in the Hai Duong area in the last 600 years.

Outmigration from the Eastern Delta

From scattered sources, it is possible to track some of the human consequences of the pattern of dike construction and demise outlined above. As the table below shows, in 1417, 27 % of Vietnamese villages were located in the eastern delta, with the 357 villages in the coastal areas around Hai Duong comprising 11.2 %. This meant that the eastern delta was second only to the capital area in terms of population density. In comparison, today's most populous part of Nam Dinh (Nam Truc, Truc Ninh, Giao Thuy, My Loc and Thuong Nguyen) contained merely 84 registered villages.[48] Four centuries later, however, we find that taxpayer numbers in the Nam Dinh area had grown fourfold, while in the Hai Duong area there numbers had shrunk by half (Table 3.2).

These figures are very eloquent and call to mind an observation by Chinese scholars regarding the eighteenth-century Yangzi River delta, where expansion in one area often led to constriction in another rather than producing net population growth.[49] This process of suppression arose from a combination of ecological and social elements. The above table on the two delta economic and population centers at different times conforms to the pattern observed in the Yangzi River delta. It does not seem too much to suggest that what we see here is an indication of the centuries-long process of outmigration from Hai Duong to the area of modern Nam Dinh.

One major reason for outmigration was the impact of frequent harvest failure. One example occurred in 1594, when it was recorded that the har-

Table 3.2 Taxpayers in Hai Duong and Nam Dinh, 1417–1819

	Hai Duong	*Nam Dinh*
	Estimated taxpayers	
1417	39,270	9240
	Registered taxpayers	
1819	23,000	38,700

For 1417, Ngan-nan tcheyuan (Hanoi: Imprimeried'Extrême-Orient, 1932), p. 61; For 1819, see Dao The Tuan and Le Duc Thinh, 'Su phat trien cua he thong nong nghiep Dong bang Song Hong', in P. Papin & O. Tessier eds. *The Village in Questions* (Hanoi: Ecole française d'extrême-Orient and Trung tam khoa hoc xa Hoi va nhan van quoc gia, 2002), p. 187

vest in districts in the Hai Duong area was 'extremely poor: people are so hungry that they eat others. A third of the population has died of starvation'.[50] And again in the mid-eighteenth century, we read that:

> From the years of Vinh Huu (1735–1739), the whole area was in chaos, especially in Hai Duong where people planted nothing and all the stored rice had been eaten. Conditions in the Son Nam (Nam Dinh) area were slightly better, and thus the roads were congested with starving people trying to go there. The price of rice was so high that 100 cash was not enough to buy one meal. People had to eat wild herbs, snakes, or even rats. The land was strewn with bodies of those who had starved to death. Only one out of ten people survived this famine. Although Hai Duong used to be the most densely populated area, now in some villages only three to five families remained.[51]

This long-term current of outmigration from the Hai Duong area was also captured in some family genealogies and the biographies of village tutelary genies, as well as being recorded by Gourou. For example, the genie (*than tich*) of Hy Ha village in Quynh Coi, Thai Binh, recorded that the people had moved there from Gia Loc, Hai Duong.[52] Migration also happened among families that were relatively better off economically. One of the descendants of Vu Hon (a Chinese governor from before the tenth century) reportedly migrated from Mo Trach in Hai Duong to Nam Truc in Nam Dinh.[53]

From the fifteenth century onward, the western delta centered on Nam Dinh became the land reclamation focus of the Red River delta. This change of focus is quite clear historically, as one Vietnamese scholar in Thai Binh province recently observed. After examining Nguyen Trai's 1430s geography, *Du dia Chi*, Doan Ngoc Han noted that it only mentioned one river and one canal in the western delta, while records between the late fifteenth and early nineteenth centuries in the Thai Binh area recorded an additional ten rivers, three dikes and one canal.[54] While the land of Thai Binh was not even mentioned by Nguyen Trai, today it contains hundreds of rivers and canals that crisscross the province, along with hundreds of kilometers of dikes, and thousands of villages.[55] When we focus on the western delta it does seem that human beings have won their battle with nature, but the experiences of contemporary Hai Hau as discussed above, and of Hai Duong in the past, cast a threatening shadow over this success story in the long run.

Conclusion

The Red River delta is arguably one of most altered landscapes in the world and the most important feature of this reshaped land is its dike systems. Tremendous human effort has been invested in these innumerable projects, and countless instances of construction, destruction and reconstruction have been taking place on this land for the last 2000 years. As this chapter has discussed, a dynamic interplay between nature and human society has shaped life in the Red River delta for millennia, with varying results for the human actors at different times. Two distinct parts of the delta—the Hai Duong area in the eastern delta and the Hai Hau district in the western delta—have been examined as illustrations of repeated processes whose operations can be hypothesized to have occurred far more often than the scattered sources can confirm. While the story of the eastern delta extends over 2000 years, that of Hai Hau in the western delta only stands about 600 years. In both cases, we see traces of significant human intervention and its consequences. The economically most vibrant and active eastern delta, the earliest part of the delta developed with both sea and river dike building, fell victim to the changed river course and the coastal erosion. After the Red River moved its main channel to the south and the Hong Duc sea dike was built in the late fifteenth century, the erosion of the western coastal area began, driven by the endless cycle of random but inescapable typhoons. This long, slow process was accelerated with the building of Hoa Binh dam in the later twentieth century, a hydrological intervention whose impact is most noticeably shown with the destruction of Hai Hau. In both cases, human activities seem to have made a major contribution to disrupting the balance between the deposition of river-borne sediments and ebb tides in the Gulf of Tongking. The historic battle between human beings and nature over recapturing the land gained from each other in the evolving Red River delta is far from over.

Notes

1. This included 72 km of river dikes and 52 km of sea dikes. *Kham dinh Dai Nam hoi dien su le* [Official compendium of institutions and usages of Imperial Vietnam] vol. 13 (Hue: Nha xuat ban Thuan Hoa, 1993), pp. 229–237.
2. For figures of 1945 and 1996, see Tuan Pham Anh and Kelly Shannon, 'Water Management in Vietnam: Indigenous Knowledge and International

Practices – The Case of the Red River Delta', paper for 11th N-AERUS Conference 2010, Brussels, 28–30 October 2010, http://www.n-aerus.net/web/sat/workshops/2010/pdf/PAPER_pham_t.pdf. The figure of 1996 includes 5000 km of river dikes and 3000 km of sea dikes.

3. D. Biggs, 'Between the River and Tides: A Hydraulic History of the Mekong Delta, 1820–1975', PhD thesis, University of Washington, 2004, p. 2.
4. Pierre-Étienne Will, 'Clear Waters Versus Muddy Waters: The Zheng-bai Irrigation System of Shaanxi Province in the Late Imperial Period', in Mark Elvin and Liu Ts'ui-jung, *Sediment of Time: Environment and Society in Chinese History* (Cambridge: Cambridge University Press, 1998), p. 283.
5. For example, see *De chinh tap* [Collections of the Nguyen officers' reports on the dikes], manuscript kept in the Han-Nom Institute, Hanoi, shelf no. A.615, vol. 4.
6. *Kham dinh Dai Nam hoi dien su le*, vol. 13 (Hue: Nha xuat ban Thuan Hoa, 1993), pp. 229–237.
7. *Kham dinh Dai Nam hoi dien su le*, vol. 13, p. 238. Whitmore discusses that hydraulic system was badly neglected because of the social, political and economic instability of the eighteenth century. See J. Whitmore, 'Literati Culture and Integration in Dai Viet, c.1430–1840', *Modern Asian Studies* 31:3 (1997): 684.
8. 'Travaux Publics au Tonkin_ 125/1729', cited from Tuan Pham Anh and Kelly Shannon, 'Water Management in Vietnam, Indigenous Knowledge and International Practices: The Case of the Red River Delta', p. 3. http://www.n-aerus.net/web/sat/workshops/2010/pdf/PAPER_pham_t.pdf.
9. Pierre Gourou, *The Peasants of the Tonkin Delta: A Study of Human Geography* (New Haven: Human Relations Area Files, 1955), vol. 1, pp. 176–177.
10. S. Andrew Enticknap Smith, 'Water First: A Political History of Hydraulics in Vietnam's Red River Delta', PhD thesis, ANU, 2002, pp. 60–61.
11. 郦道元 Li Daoyuan, *Shuijingzhu* [Commentary to the rivei classic]. juan 17 (北魏郦道元《水经注·叶榆河》quoted 《交州外域记》:'交趾昔未有郡县之时,土地有雒田。其田从潮水上下,民垦食其田,因名为雒民.'). http://zh.wikisource.org/zh/%E6%B0%B4%E7%B6%93%E6%B3%A8, accessed 3 December 2014.
12. Pierre Gourou, *The Peasants of the Tonkin Delta: A Study of Human Geography* (New Haven: Human Relations Area Files, 1955), vol. 1, p. 83.
13. Fan Ye, *Houhanshu* [History of the Later Han dynasty]: junguozhi, 5 (《后汉书.郡国志五》'〔交趾郡〕封溪建武十九年置'刘昭注引晋刘欣期《交州记》: '有堤防龙门,水深百寻,大鱼登此门化成龙,不得过,曝鳃点额,血流此水,恒如丹池.').

14. Dao duy Anh 1964:37.
15. Le Duc Thinh and Dao The Tuan, 'Nong nghiep o Mo Trach xua va nay', in P. Papin & O. Tessier eds. *The Village in Questions* (Hanoi: École française d'Extrême-Orient/&Trung tam Khoa hoc Xa Hoi va Nhan van Quoc Gia, 2002), p. 523.
16. 异物志辑佚校注, '交州异物志. In: Yang Fu, *Jiaozhou Yiwuzhi* [Record of Rarities of Jiaozhou], collected in Yiwuzhi jishi jiaozhu [Collections of the Fragments of Yiwuzhi] 异物志辑佚校注 (Guangzhou: Guangdong renminchubanshe, 2000), p. 15.
17. *Hanshu*, j.28B.
18. 葛剑雄Ge Jianxiong (1986) 西汉历史地理 [Historical Geography of the Western Han Dynasty] (Beijing: Renmin chubanshe), p. 46.
19. *Dai Viet sit ky toan thu* [Chronicles of the Great Viet; hereafter *Toan Thu*] *Toan Thu*, p. 4 73.
20. See Nishimura Masanari and Nishino Noriko, 'Nghien cuu khao co hoc ve hinh thanh lang xa o dong bang song Hong: Truong hop lang Bach Coc va khu lan can' [Archaeologist Findings on the Formation of Villages in the Red River Delta: Case Studies of Bach Coc and Adjacent Areas], in *Thong tin Bach Coc*, So Dac biet (Tokyo: Hoi Nghien cuu land xa Viet Nam, July 2006), p. 22.
21. *Dai Viet sit ky toan thu* (Tokyo: Tokyo University, 1984), p. 484 (甲申春二月,漢蒼以新興府路安撫使兼東都判陳國傑,能築隄障鹹水,便民耕種,賞爵一資。).
22. Dao DuyAnh, *Dat nuoc Viet Nam qua cac doi*, pp. 108–111.
23. *Toan Thu*, p. 669. (1460: '飓风.南策峡山太平建昌濒海等府海水大涨,堤防决裂,禾谷浸淹,濒海之民多饿死').
24. Dinh Khac Thuan, *Lich su nha Mac qua thu tich va van bia* [A History of the Mac Dynasty via Books and Inscriptions] (Hanoi: Nha xuat ban Khoa hoc xa hoi, 2001), p. 199.
25. Ibid.
26. Do Duc Hong, *Van de tri thuy o Dong Bang Bac Bo duoi thoi Nguyen, the ky XIX* [Hydrologic Issues of the Northern Vietnam During the Nguyen Dynasty, Nineteenth Century] (Hanoi: Nha xuat ban khoa hoc xa hoi, 1997), p. 60.
27. Mats Dynesius & Christer Nilsso, 'Fragmentation and Flow Regulation of River Systems in the Northern Third of the World', *Sciences*, new series, 266:5186 (4 November 1994): 753.
28. *Zhongguoshuilishigao* [A Draft History of Chinese Hydrology] (Beijing: Shuilidianlichubanshe, 1989), vol. 1, p. 176; vol. 2, p. 291.
29. Report of the Minister of Bo Cong to Tu Duc, 12 Oct (Lunar) 1852, in *De Chinh Tap*, n.p.
30. I am grateful to John Whitmore for this insightful suggestion.
31. *De chinh tap* [Collections of the Nguyen Officers' Reports on the Dikes], manuscript kept in the Han-Nom Institute, Hanoi, shelf no. A.615, n.p.

32. *De chinh tap* [Collections of the Nguyen Officers' Reports on the Dikes].
33. Ngo Ngoc Cat et al. 'Status of Coastal Erosion of Vietnam and Proposed Measures for Protection' (2006): 11.
34. Nguyen Hai Ke, 'De Hong Duc va cong cuoc khan hoang vung ven bien namSong Hong thoi Le so' [The Hong Duc Dike and the Opening of the Coastal Area in the Early Le Period], *Nghien cuu Lich su* [Historical Studies], no. 5 (1985): 36.
35. Phan Dai Doan, 'May net ve cong cuoc khai hoang thanh lap Hai Tong Hoanh Thu, Ninh Nhat' [History of the Cultivation and Founding of the Two Cantons of Hoanh Thu and Ninh Nhat], *Nghien cuu Lich su* [Historical Studies], no. 3 (1982): 25.
36. Cong V. Mai, Marcel J.F. Stive, and Pieter H.A.J.M. Van Gelder, 'Coastal Protection Strategies for the Red River Delta': 108.
37. Martin Häglund and PärSvensson, 'Coastal Erosion at Hai Hau Beach, in the Red River Delta, Vietnam' , Master of Science thesis in Coastal Engineering, Department of Water Resources Engineering, Lund Institute of Technology, Lund University, Sweden, 2002, p. 40.
38. Cong V. Mai, Marcel J.F. Stive, and Pieter H.A.J.M. Van Gelder, 'Coastal Protection Strategies for the Red River Delta': 108.
39. Le Duc Thinh and Dao The Tuan, 'Nong nghiep o Mo Trach xua va nay', p. 528.
40. Cong V. Mai, Marcel J.F. Stive, and Pieter H.A.J.M. Van Gelder, 'Coastal Protection Strategies for the Red River Delta': 106, 109.
41. M. Häglund and P. Svensson, 'Coastal Erosion at Hai Hau Beach, in the Red River Delta, Vietnam', p. 39.
42. Cong V. Mai, Marcel J.F. Stive, and Pieter H.A.J.M. Van Gelder, 'Coastal Protection Strategies for the Red River Delta': 113.
43. Gourou, vol. 1, pp. 255–256.
44. Le Ba Thao, *Thien nhan Viet Nam* [Nature of Vietnam] (Hanoi: Nhaxua ban khoa hoc ky thuat, 1990), p. 156.
45. M. Häglund and P. Svensson, 'Coastal Erosion at Hai Hau Beach, in the Red River Delta, Vietnam', p. 39.
46. Elvin and Su, 'Action at a Distance: The Influence of the Yellow River on Hangzhou Bay Since AD 1000' , in Mark Elvin and Liu Ts'ui-jung, *Sediment of Time: Environment and Society in Chinese History* (Cambridge: Cambridge University Press, 1998), p. 376.
47. Ye Mengzhu, *Yueshibian* [On the world I have seen], juan 1. ([清]叶梦珠,《阅世编》 '其如水势汹涌,非土木所能捍卫,随筑随溃,随决随避,迄今塘距石塘旧地不下数里,潮退,一望微茫,如在烟云之外').
48. I use a rough way of 110 households per village, to estimate the taxpayers of the Hai Duong and Nam Dinh areas in 1417, to compare with the number of registered taxpayer in the same areas in 1819. This was because

village numbers are often available and more constant, and more reliable than taxpayer numbers in the pre-modern Vietnam. See Li Tana, *Nguyen Cochinchina* (Ithaca: Cornell University SEAP, 1998), Appendix 1.

49. A. Osborne, 'Economy and Ecology of the Lower Yangzi', in Elvin and Liu, *Jijian suozhi*, p. 221.
50. *Toan Thu*, p. 902.
51. *Kham dinh Viet su thong giam CuongMuc* [The Imperially Ordered Mirror and Commentary on the History of the Viet], (Taibei: Taibei zhongyang tushuguan, 1969), vol. 7, p. 3523.
52. P. Gourou, *Peasants of the Tonkin delta*, vol. 1, p. 123.
53. P. Gourou, *Peasants of the Tonkin delta*, vol. 1, pp. 124–125. Migrants from Hai Duong to Hanoi in the 36 streets, recorded in Nguyen Vinh Phuc & Tran Huy Ba, *Duong pho Ha Noi* [Hanoi Streets] (Hanoi: Nha xuat ban Ha Noi, 1979): Hang Bac, silver smith and money changers, from Binh Giang, Hai Duong, migrated at the end of the fifteenth century, p. 135; Hang Dao, silk dyeing and trade from Binh Giang, p. 172; Hang Dong, copper trade, from Khoai Chau, migrated at the end of the Le, p. 176; Hang Giay shoe makers, from Tu Ky, Hai Hung. They founded the village Hai Tuong ('shoe making craftsmen'), p. 184; Hang Huong, joss stick, from Kim Dong, Hai Duong, p. 189; Hang Trong, drum making and trade, from Yen My, Hai Hung), p. 211.
54. Doan Ngoc Han, 'Cong cuoc tri thuy o vung Hung Ha, Quynh Phu thoi Ly Tran' [Irrigation Projects in the Hung Ha and Quynh Phu Areas in the Ly and Tran Period], in Historical Studies Group, *Hồi Thảo khoa học ngươi dân Thái Bình trong lịch sư*, bô phân lịch sư dân tôc, ban nghiên cưu lịch sư [Proceedings of the Conference on People in Thai Binh's History] (Thái Bình: Đảng tỉnh Thái Bình, 1986), pp. 67–68.
55. Doan Ngoc Han, 'Cong cuoc tri thuy o vung Hung Ha, Quynh Phu thoi Ly Tran', p. 68.

Bibliography

Biggs, D. 2004. Between the river and tides: A hydraulic history of the Mekong Delta, 1820–1975. PhD thesis, University of Washington.

Cong V. Mai, Stive, Marcel J.F., and Pieter H.A.J.M. Van Gelder, (2009) 'Coastal protection strategies for the Red River delta', *Journal of Coastal Research*, 25:1:105–116.

Dai Viet sit ky toan thu [Chronicles of the Great Viet]. 1984. Tokyo: Tokyo University.

Dao, Duy Anh. 1964. *Dat nuoc Viet Nam qua cac doi* [Vietnamese territory throughout the history]. Hanoi: Khoa Hoc.

De chinh tap [Collections of the Nguyen officers' reports on the dikes], 1820–1880 manuscript kept in the Han-Nom Institute, Hanoi, shelf no. A.615.

Dinh, Khac Thuan. 2001. *Lich su nha Mac qua thu tich va van bia* [A history of the Mac dynasty via books and inscriptions]. Hanoi: Nha xuat ban Khoa hoc xa hoi.

Do, Duc Hong. 1997. *Van de tri thuy o Dong Bang Bac Bo duoi thoi Nguyen, the ky XIX* [Hydrologic issues of the northern Vietnam during the Nguyen dynasty, nineteenth century]. Hanoi: Nha xuat ban khoa hoc xa hoi.

Doan, Ngoc Han. 1986. Cong cuoc tri thuy o vung Hung Ha, Quynh Phu thoi Ly Tran [irrigation projects in the Hung Ha and Quynh Phu areas in the Ly and Tran period]. In *Hoi thao khoa hoc nguoi dan Thai Binh trong lich su* [Proceedings of the conference on peopling in Thai Binh's history], 63–79. Thái Bình: Đảng tỉnh Thái Bình.

Dynesius, Mats, and Nilsso, Christer. 1994. Fragmentation and flow regulation of river systems in the northern third of the world. *Sciences*, new series, 266(5186):753–762.

Elvin, Markand, and Su Ninghu. 1998. Action at a distance: The influence of the Yellow River on Hangzhou Bay since AD 1000. In *Sediment of time: Environment and society in Chinese history*, eds. Mark Elvin, and Liu Ts'ui-jung, 344–410. Cambridge: Cambridge University Press.

范晔 Fan Ye. *Houhanshu* [History of the Later Han dynasty]: junguozhi, 5.

葛剑雄Ge Jianxiong. 1986. 西汉历史地理 [Historical geography of the Western Han dynasty]. Beijing: Renmin chubanshe.

Gourou, Pierre. 1955. *The peasants of the Tonkin delta: A study of human geography*, vol 1. New Haven: Human Relations Area Files.

Häglund, Martin and Pär Svensson. 2002. Coastal erosion at Hai Hau beach, in the Red River delta, Vietnam. Master of Science thesis in Coastal Engineering, Department of Water Resources Engineering, Lund Institute of Technology, Lund University, Sweden.

汉书 *Hanshu* [History of the former Han dynasty], Wenyuange Edition, *juan*. 28.

Kham dinh Dai Nam hoi dien su le [Official compendium of institutions and usages of Imperial Vietnam]. 1993. Reprinted Hue: Nha xuat ban Thuan Hoa.

Kham dinh Viet su thong giam CuongMuc [The imperially ordered mirror and commentary on the history of the Viet]. 1969. Reprinted Taibei: Taibei zhongyang tushuguan.

Le Ba Thao. 1990. *Thien nhan Viet Nam* [Nature of Vietnam]. Hanoi: Nha xuat ban khoa hoc ky thuat.

Le Duc Thinh, and Dao The Tuan. 2002. Nong nghiep o Mo Trach xua va nay. In *The village in questions*, eds. P. Papin, and O. Tessier, 505–549. Hanoi: École française d'Extrême-Orient/&Trung tam Khoa hoc Xa Hoi va Nhan van Quoc Gia.

郦道元 Li Daoyuan (5 CE), 水经注 [Commentary to the river classic]. http://zh.wikisource.org/zh/%E6%B0%B4%E7%B6%93%E6%B3%A8, accessed 3 Dec 2014.

Nishimura Masanari, and Nishino Noriko. 2006. Nghien cuu khao co hoc ve hinh thanh lang xa o dong bang song Hong: Truong hop lang Bach Coc va khu lan can [Archaeologist findings on the formation of villages in the Red River delta: Case studies of Bach Coc and adjacent areas]. In *Thong tin Bach Coc*, So Dac biet [The Bach Coc newsletter, special edition]. Tokyo: Hoi Nghien cuu land xa Viet Nam.

Ngo Ngoc Cat, Pham Huy Tien, Do Dinh Sam, and Nguyen Ngoc Binh (2006) Status of coastal erosion of Vietnam and proposed measures for protection. http://www.fao.org/forestry/11286-08d0cd86b-c02ef85da8f5b6249401b52f.pdf, accessed 30 Aug 2014.

Nguyen Hai Ke. 1985. De Hong Duc va cong cuoc khan hoang vung ven bien nam Song Hong thoi Le so [The Hong Duc dike and the opening of the coastal area in the early Le period]. *Nghien cuu Lich su* [Historical studies] (5): 35–42.

Phan Dai Doan. 1982. May net ve cong cuoc khai hoang thanh lap Hai Tong Hoanh Thu, Ninh Nhat' [History of the cultivation and founding of the two cantons of Hoanh Thu and Ninh Nhat. *Nghien cuu Lich su* [Journal of Historical Studies] 3: 24–31.

Osborne, A. 1998. Economy and ecology of the lower Yangzi. In *Sediment of time: Environment and society in Chinese history*, eds. Mark Elvin, and Liu Ts'ui-jung, 203–232. Cambridge: Cambridge University Press.

Shiba, Yoshinobu. 2012. 宋代江南经济史研究 [Studies on econimic history of the Jiangnan area in the Song period]. Nanjing: Jiangsu renmin chubanshe.

Smith, S. Andrew Enticknap. 2002. Water first: A political history of hydraulics in Vietnam's Red River delta. PhD thesis, Australian National University.

Pham, Tuan Anh, and Shannon, Kelly. 2010. Water management in Vietnam: Indigenous knowledge and international practices – The case of the Red River delta. paper for 11th N-AERUS Conference 2010, Brussels. http://www.n-aerus.net/, accessed 28 Aug 2014.

杨孚Yang Fu (1 CE). 2000. *Jiaozhou Yiwuzhi* [Record of Rarities of Jiaozhou], collected in Yiwuzhi jishi jiaozhu [Collections of the fragments of Yiwuzhi] 异物志辑佚校注. Guangzhou: Guangdong renminchubanshe.

Van Maren, D. S. 2004. Morphodynamics of a cyclic prograding delta: The Red River, Vietnam. PhD thesis, Utrecht Universit.

Vu Tu Lap et al. 1991. *Van hoa va cu dan Dong bang Song Hong* [Culture and peoples in the Red River delta]. Hanoi: Khoa hoc xa hoi.

Will, Pierre-Étienne. 1998. Clear waters versus muddy waters: The Zheng-bai irrigation system of Shaanxi province in the late imperial period. In *Sediment of time: Environment and society in Chinese history*, eds. Mark Elvin, and Liu Ts'ui-jung, 283–343. Cambridge: Cambridge University Press.

Ye Mengzhu, *Yueshibian* [On the world I have seen], juan 1.

张芳 Zhang, Fang. 隋唐宋元海塘的修筑和塘工技术的演进 [Building of sea dikes and the progression of techniques in sea dikes building between the Sui and Yuan dynasties] In中国古代灌溉工程技术史 [A history of irrigation engineering in ancient China], 中国经济史论坛. http://economy.guoxue.com, accessed 28 Aug 2014.

中国水利史稿. 1989. *Zhongguoshuilishigao* [A draft history of Chinese hydrology]. Beijing: Shuilidianlichubanshe, 2 vols.

CHAPTER 4

'The Most Horrible of Evils': Social Responses to Drought and Famine in the Bombay Presidency, 1782–1857

George Adamson

Introduction

The monsoon climatology characteristic of the Indian Ocean World renders the region particularly sensitive to drought and floods.[1] Small variations in the quantity of monsoon rainfall or length of the rainy season can have catastrophic effects on agriculture, with cascading impacts across the economy.[2] Disentangling the adaption of societies to climate-related natural hazards stress in the past, and examining the ways in which vulnerability is differentiated across society, is vital for effective preparation against climatic hazards in the future.[3] However, the history of drought in the Indian Ocean World—India in particular—remains understudied, given the region's size and its relative vulnerability.[4] This is less true of the period between 1860 and the Bengal Famine of 1943, which have received attention through analyses of material contained within the Indian Famine

G. Adamson
Department of Geography, King's College London, London, UK

G. Bankoff, J. Christensen (eds.), *Natural Hazards and Peoples in the Indian Ocean World*, DOI 10.1057/978-1-349-94857-4_4

Reports.[5] Both the periods after South Asian independence and before 1860, however, require substantial further analysis.[6]

One reason for the paucity of studies into social responses to climate variability in the past has been the lack of reliable meteorological information. In India, the national instrumental monitoring network was established in 1871. Before this date, meaningful analysis of social responses to climate is difficult. Periods in the historical record that apparently indicate famine may be related to warfare or political tensions, rather than meteorological drought. Food price shocks may have occurred because of trade disruption rather than harvest shortage, or by shortages in regions geographically distant from the area under study. The ability to understand the role of drought in Indian history is changing, however, through new developments in climate reconstruction, particularly using tree-ring analysis and documentary sources.[7] These detailed, often annually or sub-annually resolved reconstructions of meteorological conditions allow the role of climate to be determined with a far higher degree of confidence.

This chapter focuses on social vulnerability to climatic fluctuations—particularly drought—in the Bombay Presidency (Fig. 4.1) from 1782 to 1857. The purpose of the chapter is to analyse in detail the socioeconomic effects of a series of droughts identified within a new semi-quantitative reconstruction of monsoon rainfall developed by Adamson and Nash (2014). The authors of this study used references to climate within documentary material and newly uncovered instrumental rainfall observations to generate a five-category reconstruction of monsoon intensity, adopting Indian Meteorological Department terminology (namely: 'deficient monsoon', 'below normal', 'normal', 'above normal' and 'excess'). As well as identifying periods of climatic stress, this reconstruction indicated that the overall climatic regime over western India before and after 1860 had not changed significantly, meaning that the climatic fluctuations experienced by the societies living during this period were broadly similar to that of today. This renders this period ideal for analysing the response of societies to climatic stress.

The period chosen spans the era of East India Company (EIC) hegemony in western India, from the end of the First Anglo-Maratha War (1782) through to the collapse of the EIC following the 1857 Indian Uprising. This period witnessed substantial social and political shifts, including the wholescale change of governance from indigenous to colonial. The period also witnessed a number of major droughts, including three particularly severe famines in 1790, 1803, and 1812. Using the archives of the India

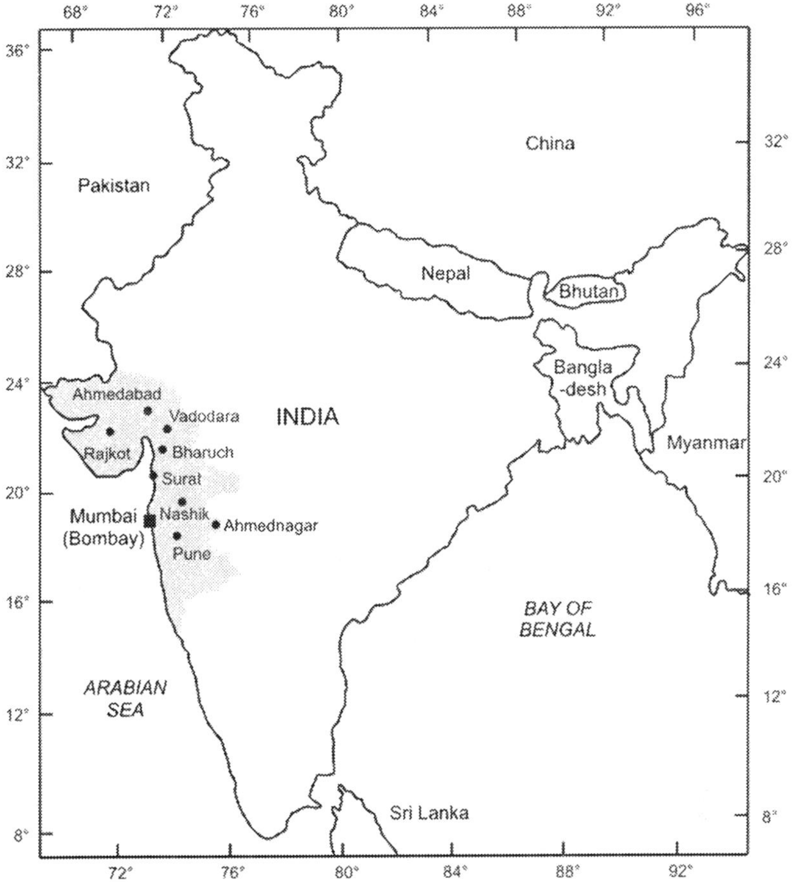

Fig. 4.1 Map of India displaying study area (shaded). Place names current as of 2014

Office Records, English-language newspapers and government archives (Table 4.1), this chapter aims to detail for the first time the impact of these droughts in detail, and analyse adaptation options adopted across a variety of scales. It will address the differing levels of vulnerability between social strata within the region and attempt to identify factors that led to increased vulnerability or decreased the ability for individuals to adapt to drought. This will then provide a framework upon which to map the influence of drought

Table 4.1 Details and codes for archival collections consulted in the UK, USA and India

Name of collection	*Archive location*	*Archive code*
India Office Records	British Library, St Pancras, London, UK	BL followed by catalogue number
India Office Private Papers	British Library, St Pancras, London, UK	BL followed by catalogue number
British Library Newspaper Collections	British Library, Colindale, London, UK	BLNC followed by catalogue number
Church Missionary Society	University of Birmingham Library, UK	UBL followed by catalogue number
American Board of Commissioners for Foreign Missions	Houghton Library, Harvard, USA	HLH followed by catalogue number
Scottish Missionary Societies	National Library of Scotland, Edinburgh, UK	NLS followed by catalogue number
National Library of Scotland Archives	National Library of Scotland, Edinburgh, UK	NLS followed by catalogue number
Archives of the Government of Maharashtra	Department of Archives, Government of Maharashtra, Mumbai, India	DAGM followed by catalogue number
Royal Society Archives	Library of the History of Science, Royal Society, London, UK	RSA followed by catalogue number
Aberdeen Medico-Chirurgical Society	University of Aberdeen Library, UK (available online)	UAL followed by catalogue number

on the early colonial history of the Bombay Presidency. It is hoped that this will allow us to better understand social responses to climate stress in a broader sense.

A Note on the Political Economy of the Bombay Presidency, 1782–1857

The period 1782–1857 saw substantial political change in western India. In 1782, the region was governed predominantly by the Maratha confederacy, with the majority under the control of the *Peshwa* of Poona (Pune) and the north of the region—present-day southern Gujarat—governed by the *Gaekwad* dynasty of Baroda (Vadodara). The British EIC-controlled minor territories, comprising only the islands of Bombay and neighbouring Salsette. However, Maratha hegemony fractured during the first years of the nineteenth century, leading to EIC ascendancy and eventually dominance. The first major territorial shift occurred in 1802–03. During this period, the Maratha chieftain Yashwant Rao Holkar, based at Indore, attacked the territories of the *Peshwa*. This schism within the Maratha confederacy resulted in the signing of the Treaty of Bassein in 1802 between the *Peshwa* and the EIC, in which the EIC were granted territories in southern Gujarat in exchange for military support. In addition to territorial acquisitions, the EIC also placed Residents in several key Maratha cities, including the *Peshwa's* capital at Poona. Frustrated with EIC interference in the governance of Maratha territories, the *Peshwa* attacked the residency of the British Resident in Poona in 1817, leading to the short-lived Third Anglo-Maratha War in which the *Peshwa* was defeated. Upon this, all of the *Peshwa's* territories fell under EIC control, and the *Gaekwad* and other princely territories became EIC Protectorates (de facto colonies).

Despite the change of rulers, the rural economy of western India did not change significantly upon the move from Maratha to EIC control. The EIC maintained their hegemony through a system of regional Collectors and subordinate Indian *Mamlutdars*, who were largely retained upon the transfer from Maratha to EIC rule. The amount of Revenue collected was largely left to the discretion of the Collectors, and varied throughout the region.[8] Peasants commonly had ownership over land (the *ryotwadi* system). However, chronic indebtedness was a prevailing feature,[9] with many peasants having insufficient capital to fund them through an entire year. Rural capital was maintained through local moneylenders or *sahukars* (generally of the *Baniya* caste), who loaned seed, clothing or farming tools.[10] Interest rates were often high, with up to 300 % reported[11]; in some cases crops were bought on loan before the harvest, at a rate at least 25 % below post-harvest prices, and at the worst case all of the annual crop would be removed by the *sahukars* as payment for loans previously accrued.[12]

Sahukars owned much of the stores of grain, and were therefore responsible for providing 'support' to cultivators in times of scarcity, as either charity or further loans. *Sahukars* were responsible for the flow of grain from the village level to larger, speculative markets, through economic relationships with town-based merchants who were in turn credited by city banks.[13] *Sahukars* therefore controlled the supply of grain within villages and at town bazaars, and held a virtual monopoly, which could be exploited during times of droughts, as will be seen.

Droughts in Western India, 1782–1857

The Indian Meteorological Department defines a 'deficient' monsoon as one where rainfall is below 90 % of a long-period average. Using this definition, ten 'widespread' droughts occurred between 1781 and 1860, that is, periods where deficient rainfall occurred over much or all of western India.[14] These were 1790–92, 1803, 1812, 1823–24, 1833, 1838, 1845, 1847–48, 1850 and 1855. With the exception of 1847–48, 1850, and 1855, significant socioeconomic impacts were reported during each of these periods. Each of these will be addressed in turn.

'Paternalistic' Governance and Warfare: 1790–92 and 1803

Little information about the famine of 1790–92 has been preserved within English-language archives. This is unfortunate, as other studies have suggested that it was global in nature and potentially devastating.[15] Information sent from the Public Department of the Bombay Government to the Court of Directors in December 1790 mentioned 'the greatest apprehensions of a famine from the almost total failure of the Crops of Grain.'[16] A letter from Bharuch in southern Gujarat also reports a total crop failure 'for want of rain'.[17] The predominant effect of this drought is one that was noted regularly in reports for the entire study period: heavy and sustained increases in the price of grain.[18]

The response of the Maratha Government at Poona was to buy grain from 'the Nizam's country' and distribute it to the poor 'at 6 to 7 seers to the Rupee around Poona and in the Concan'.[19] By 1792–93, this extended to the provision of charitable donations of food.[20] The EIC Government at Bombay adopted a similar approach, employing a Custom Master to import 150,000 bags of grain into Bombay.[21] These responses apparently represented a continuation of earlier Maratha drought responses, a 'paternalistic' combination of state-sponsored charity and government activity within the grain markets.[22]

The second widespread drought occurred in 1803 during the extended period of warfare that culminated in the Second Anglo-Maratha War. The famine resulting from the combination of military activities and a deficient monsoon apparently caused a crisis in the Deccan. An EIC Lieutenant named Jasper Nicolls, travelling through the Deccan in April 1804, reported, 'All around Poonah looks dreary & desolate, grain bears a very high price ... The river is very low and the water bad'.[23] Newspapers from Bombay reported a high price of grain due to 'the great failure of the crops in Salsette'.[24] Although mortality figures are absent, it is apparent that the famine caused significant depopulation, either through death or migration. In December of 1804 Nicolls commented on 'tracts of country ... from 10 to 20 miles in extent ... in which there are not 10 families, frequently not one soul remaining'.[25]

After 1804, the then *Peshwa* Bajirao II sought several measures to return the land to cultivation. The first was to relocate 'many people' from the Konkan into the land to the south of Poona. Whether this migration was enforced by coercion, enticement or forced resettlement is not specified. However, the measure was a success and the area became prosperous.[26] The second was to 'farm' the revenue of all regions under his control to the highest bidder,[27] a measure that had been adopted previously within the *Gaekwad* territories in Gujarat.[28] This process incentivised the collection of extortionately high land rents; however, it also served to increase the amount of land under cultivation,[29] with the country becoming 'very populous and prosperous'.[30] The only region that suffered long-term depopulation was Khandesh[31] in present-day north-western Maharashtra, where inhabitants migrated to areas under the control of neighbouring chiefs.[32]

There are reports that King Shivaji of Sholapur instigated charitable support in 1803 in the form of donations of grain.[33] A similar 'Subscription' was set up by the rich classes of Bombay, in order to collect funds to be 'appropriated in such manner as may be most conductive to the alleviation of misery'.[34] As had occurred in 1792, the Government of Bombay formed a Grain Committee,[35] which bought up all rice exported into Bombay and sold it at a fixed and advertised price.[36] Exportation of grain from Bombay was only allowed to British territories,[37] and in EIC-controlled territories in Gujarat import duties were cancelled and exportation of grain was prohibited.[38] The EIC also set up grain stores in the *Peshwa's* territories by the Bombay Army, although the purpose of these was apparently not benevolence but to attract 'those wandering, undecided Marathas' into the EIC army.[39] A 'humane hospital' was provided for immigrants into Bombay 'who from age, infancy or infirmary were unable to earn a livelihood'.[40] These were provided with 'sufficient rice for maintenance'.[41] The reported death rate was, however, around 50 %.[42]

Famine and the Market: 1812 and 1824

The famine of 1812 is the first for which detailed reports are available, predominantly through a report by the EIC Resident in Baroda (Vadodara) J.R. Carnac, which he presented to the Bombay Literary Society. Carnac's report asserts that the famine was initiated by a deficient monsoon in 1811 in Marwar (present-day southern Rajasthan), which was accompanied by a locust swarm. This led to substantial migration into Gujarat during late 1811 and early 1812. Locusts also destroyed crops in Kathiawar and southern Gujarat during early 1812,[43] resulting in significant grain shortages in these areas by the early part of the monsoon of 1812.

Crop failures were reported throughout western India at the end of 1812 following a widespread deficient monsoon. The effects of harvest shortfalls were apparently felt most strongly in Gujarat.[44] Refugees from Marwar placed an additional pressure on the resources, with the most severe consequences of the drought reported within the immigrant population.[45] Carnac wrote, 'In the vicinity of every large town, you perceived suburbs surrounded by these creatures ... some furnished with a scanty covering, others almost reduced to a state of nudity ... a life-less corpse at intervals brought to notice by the bewailings of a near relative'.[46] Reports also describe high disease incidence,[47] including an epidemic of smallpox and venereal disease amongst women, 'who to obtain food on their entry into the country had prostituted themselves'.[48] The total mortality at Ahmedabad was estimated to be 'a hundred thousand souls', whilst at Baroda, 'the numbers of Marwarees who died in a single day ... could scarcely be counted, and the return of burials in twenty-four hours often exceeded five hundred bodies'.[49] Reportedly 'not more than one in a hundred [immigrants from Marwar] ever returned to their native country'.[50]

The year 1812 witnessed a shift in the response to drought within the Bombay Government, brought about by the spread of the *laissez-faire* policies espoused by Adam Smith. This advocated the action of free markets in drought response, established through total non-intervention by government in the grain trade and policies to encourage the free movement of grain—a reversal of previous responses. The effect of this policy was to encourage huge levels of grain exportation into Gujarat, causing price rises in areas otherwise unaffected by the famine such as Bombay. The apparent distress this caused was the subject of a petition submitted to the Government of Bombay on 24 August 1812.[51] Reports from this period also frequently complain that the lack of government intervention in the grain trade was encouraging the *sahukars* to speculate on their own stock and deliberately hold back from selling in order to artificially raise grain prices.[52] This complaint was repeated in subsequent droughts, as will be discussed further later.

Bombay witnessed substantial immigration from famine-hit districts in 1812–13, amounting by January 1813 to 'between 110 and 120 Souls per month'.[53] The response of the colonial government was to build 'humane hospitals' in Kaira (Kheda)[54] and Bombay.[55] Relief was also provided through a charitable fund established by the 'Native inhabitants' of Bombay, which was extended to the European population on 19 December.[56] In 1813, the *Bombay Gazette* published an article praising the donations of Sunderji Sewji, a merchant based at the town of Porbandar, who had contributed two lakhs of rupees (2,00,000)[57] to various charitable funds at Bombay and on the Kathiawar peninsular.[58] The drought of 1812–13 was ended by heavy rainfall during the monsoon of 1813, although this also caused damage due to widespread flooding.[59]

The next widespread drought in the region occurred in 1824. As had been the case in 1812, this drought again followed a localised drought in 1823, predominantly affecting the *rabi* (winter) harvest at Poona.[60] The country around Poona was therefore already partially desiccated before the deficient monsoon on 1824. As early as 3 July 1824, the *Bombay Courier* reported 'cattle are dying in great numbers' in the Deccan,[61] which represented a threat to agricultural activities.[62] Crop failures were reported at Ahmedabad and Kathiawar in September 1824.[63] The effect of the drought was, however, apparently curtailed by heavy rainfall during October 1824.[64] This averted the threat of famine, but did not prevent a 'scarcity'.[65]

The first drought during which the whole of western India was under EIC governance was in 1824, following the Third Anglo-Maratha War in 1817–18. *Laissez-faire* drought response was therefore adopted across the region. Grain price rises due to exportation to Gujarat were again the subject of a petition to Government from the inhabitants of Bombay in September 1824,[66] although this was again ignored. Migration from Kutch into Bombay was reported; on this occasion the migrants were put to work drilling wells and widening ditches, at the pay of 'the smallest sum which may be sufficient to enable them to provide for their subsistence in grain [and] considerably less than a laborer [sic] would consume in time of plenty'.[67] This action was put forward partially in response to a major water shortage in Bombay in 1824, which resulted in a further three petitions to Government.[68] This eventually led to the creation of a committee to review the options available for clearing tanks and a proclamation limiting water for uses other than drinking.[69] Prices of grain remained higher than normal in Gujarat until July 1825, with heavy deaths amongst cattle,[70] and severe water shortages in Bombay and the Deccan.[71]

Riot and *Ryots:* 1832–33

The drought of 1833 occurred following three to four years of deficient monsoon, with Revenue reports recording crop failures throughout western India, particularly at Poona.[72] In November 1832, the Subcollector of Sholapur reported that grain was selling for 'three times the average price for the period since we got the country'.[73] An article in the *Gazette* on 17 November stated that in Ahmednagar 'the grain which was sold from 14 to 15 Paylees at that station for one Rupee, is now hardly procurable at from 3 to 4 Paylees'.[74] A 40 % grain price rise was noted in Bombay in August 1833 in the *Bombay Courier*. This was attributed chiefly to exportation of grain to Kutch, where severe drought conditions were being experienced.[75] In Sholapur, 'forty cases of deaths from want of food in the last four days' were reported on 15 July, with 'cases ... of young Children being left destitute in the Streets by the death as abandonment of their parents'.[76] A major water shortage was also reported in the Deccan in 1834, the response of the Bombay Government to drill new wells being apparently largely ineffectual.[77]

Reports of price speculation amongst the grain dealers and the deliberate holding back of stock increased significantly in 1832. In Ahmednagar, for example, it was reported that 'the Cultivators have suffered much less than in other parts of the Deccan [however] the high prices are attributable in a great measure, to an extensive combination of the Grain Merchants'.[78] On 14 November 1832, the *Bombay Gazette* reported 'in consequence of the small quantity of rain which has fallen this year, the cruel grain Merchants, and persons who deal in this article have raised the prices, although every eatable has ripened well this year on the Island and there are moreover large imports from different places on the Coast. By this, the destitute will suffer much'.[79] The American missionary Caroline Farrar wrote in Ahmednagar in November 1832 'the poor people are greatly troubled. Tho' there is plenty of grain, but the Baneyans ask such enormous price that the poor are not able to purchase'.[80]

The culmination of these reports was a series of grain riots in November 1832 at Poona, Ahmednagar and Malegaon. 'Ryots of all descriptions' attacked the shops of the grain dealers and looted merchandise valued at between 10,000 and one lakh of rupees.[81] The riots were attributed to grain price rises disproportionate to the actual availability of stock, and 'the monopoly of the necessaries of life by a few rich Hindoos'.[82] The degree to which the grain dealers were to blame for price rises is debatable. The *sahukars* were of a separate caste and religion (usually Jainism), and were reported to indulge in religious rituals to prevent rain and justify price rises.[83] This, combined with the fact previously mentioned that the *sahukars* were moneylenders, may have led to an inherent sense of distrust. Furthermore, it would have suited the EIC Government to place the weight of blame upon the *sahukars* rather than their own policies. However, the presence of a report by an American missionary suggests that such accounts were not merely British exaggerations to alleviate their own culpability in providing a lack of support.

The Bombay Government did not desist from the policy of *laissez faire* after the 1832 grain riots. The response of the government was instead to put through measures to increase the openness of the market. This took the form of fortnightly reports of *The Weather and the Crops*: summaries of local weather conditions and agricultural reports sent to Bombay by the Collectors in the various regions. These were published in Bombay newspapers during every monsoon from July 1833, allowing the literate population to make informed decisions regarding the likely availability of grain. This apparently had some success, at least in the sense that price rises in future droughts were generally attributed by the European population to shortage of supply rather than speculation by the grain dealers.[84]

Charity and Water Shortage: 1838 and Beyond

The drought of 1838–39 was apparently fiercer than that of 1833—particularly in Gujarat—with the most severe reports pertaining to the Kathiawar Peninsular. On 3 October, the *Bombay Gazette* reported, 'that there is every probability of an almost total failure of the crops of every description, in consequence of the long continued drought'.[85] Crop failures and 'a great loss of Cattle' were reported in 1839 in Kathiawar and Broach (Bharuch).[86] Three thousand refugees arrived at Bombay by early October 1838. However, the Government of Bombay offered no assistance on this occasion although the European population offered support in the form of a charitable fund.[87] Water shortages led to a petition from 'the heads of the native community' to the government requesting wells on the Esplanade to be unblocked,[88] with the government responding by passing laws limiting the amount of water to be used,[89] as well as drilling six new wells on the Esplanade.[90]

Drought was reported somewhere in western India in every year between 1838 and 1845. Famine was reported in Kathiawar throughout 1839–42,[91] and in June 1843, the *Bombay Courier* wrote of the 'despair' of the 'cultivators of the soil', who 'dread a late and untimely fall of rain'.[92] In March 1844, 'a great deficiency in the harvest' was reported in Ahmednagar due to a failure of the November rains.[93] In early 1845, newspapers were already reporting apprehensions for a drought in Bombay 'from the defective fall of rain during the past two seasons'.[94] The predominant complaint during this period was a shortage of water throughout the Bombay Presidency.[95] The response of the Bombay Government was apparently slow, and plans to create reservoirs in Salsette in December 1845 were cancelled following the heavy monsoon of 1846.[96] However, in 1844 a Parsee merchant Sir Jamsetjee Jeejeebhoy donated 1,17,000 rupees to build a dam and reservoir on the river Mula Mutha, in order to supply water for Poona. It is unclear when the works were completed, but reports state that they were operational by 1854.[97]

A localised famine was reported in Ahmednagar in 1846 due to three years of deficient rainfall, and the price of grain in Poona reported as 'three times the price it was two years ago'.[98] In 1846, a letter to the *Bombay Times* complaining of 'enormously high' grain prices, despite a good monsoon. On this occasion, discussions occurred within the pages of the *Bombay Times* regarding the creation of a European-funded joint-stock company for warehousing and distribution of grain, in the absence of a government response. A bank for the *ryots* (peasant farmers) was also suggested, although this was rejected, as 'its advances would be most likely to fall ultimately into the hands of the very soucars whose profits were desired to be curtailed'.[99] The company began collecting shares, but as there is no reference to the company after 31 October 1846, it is likely that it did not start trading. Its failure was attributed by another correspondent to the *Times* to the nature of the economy existing between the *ryots* and the grain dealers, that is, to acquire grain before the *Baniyas*, the company would have to pay in advance, '*which is not contemplated*' [emphasis in original].

Social Responses to Climatic Variability in Western India

Social Vulnerability to Climatic Variability

This chapter has outlined the periodic droughts of early nineteenth-century western India. Attention will now turn to factors affecting social vulnerability and adaptive measures. The first point of mention is that climatic stress did not seem to adversely affect the European population, with the exception of occasional shipwrecks during the early part of the monsoon. There is only one report of drought conditions directly impacting upon the livelihoods of British colonists, and then only in the military outstation of Porbandar in 1812.[100] The response of the European population of Bombay to the drought of 1824 was to open a theatre, 'to divert the attention, now and then, from the horrors of expected drought and famine'.[101] On Michaelmas Day 1824 (29 September), the wife of the Chief Justice of Bombay, Lady Lucretia West, 'insisted upon having a roast goose for dinner',[102] six days after reporting immigrants into the city from Kutch offering their children for sale in exchange for food.[103] The Governor of Bombay, Mountstuart Elphinstone, wrote in his diary of a threatened famine, 'the most horrible of evils'. However, he noted his own 'general exemption from the calamity', causing him to consider himself 'a fool of reproach'.[104]

The archival information consulted for this study makes it difficult to determine which groups suffered the highest vulnerability. In many cases, all rural communities and castes were referred to under the term *ryots*. The grain riots in 1832 were described as containing 'ryots of all descriptions'.[105] This does enlighten us much on the degree to which castes and classes of *ryots* were affected, as the inflated grain prices were likely to have been a nuisance to all, regardless of the degree to which livelihoods were affected. However, it is likely that cultivators under debt obligations to *sahukars* would have been particularly vulnerable due to the inability to store their own crops. This would have necessitated further loans of grain during scarcity, creating a cycle of debt from which it would have been difficult to escape.

Reports indicate that communities within the Kathiawar peninsular may have been most vulnerable, with several reports of migration from this region—often to rivers, with individuals accompanied by cattle. This suggests a planned adaptation option and the prevalence of pastoralist or semi-pastoralist communities.[106] However, it is clear that in many cases migration represented a 'last resort' option, particularly during periods of severe drought.[107] On these occasions, migration—generally into cities—was often accompanied by the 'selling' of individuals or their children.

On a practical level, damage to tanks and wells through neglect and poor repair reportedly increased vulnerability.[108] Damages to tanks in urban areas such as Bombay were felt especially hard, particularly in years of deficient rainfall.[109] In some cases, damage to crops caused by other factors exacerbated climatic stress. Swarms of locusts were reported in Gujarat in 1811,[110] in the land surrounding Ahmedabad in 1825,[111] in the Deccan in November 1832[112] and in Deesa (Disa) in September 1833.[113] In all of these cases, the locusts contributed to harvest failures that had occurred in that year due to a deficient rainfall, or exacerbated deficiencies in monsoon rainfall in the following or preceding year.

Adaptation Measures

The importance of local systems of patronage and clientage has been highlighted as the most important instruments in coping with subsistence crises elsewhere in early colonial India.[114] It is likely that such networks also existed in western India, although opportunities for grain storage by individuals or families would have been limited due to the peasant–*sahukar* relationship. It is unclear as to what extent financial capital was available at the village level. Previous studies[115] suggest that cultivators 'sold' their crop in kind to the local *sahukars* without any need for exchange of payment. Furthermore, before the introduction of the railway in the 1850s, markets available to peasant farmers were small, so there was little opportunity to exchange crops for capital. These small markets may have actually led to *improved* adaptive capacity, however, in discouraging the cultivation of non-food crops and high-yield, high-risk export crops.[116]

Agricultural methods were designed to combat scarcity, at least before the introduction of the railway and colonial exportation. Methods adopted included the mixing of crops in a field, the planting of drought-resistant crops and the banking of rice fields to store monsoon rains.[117] The consumption of 'famine crops' was occasionally recorded during heavy droughts, including 'wild Root',[118] berries and 'jungle trees'.[119] Irrigation was used in some areas, particularly around Ahmedabad, Surat and Ahmednagar, although apparently not extensively. Methods were occasionally rudimentary: in Rajkot, a method was described whereby wells were positioned

Fig. 4.2 Irrigation well near Deesa, Gujarat. From diary of M. Emma Walter, 19 March 1840 (Image courtesy of the British Library)[129]

between every three to four fields and water was drawn throughout the day using buckets (Fig. 4.2).[120] Tanks and wells were generally communal, although rich and high-caste individuals occasionally owned their own, creating significantly higher resilience to drought than members of the lower castes. Indeed, religious traditions dictated in some areas that high-caste individuals 'wash their crops several times a day', necessitating a considerable supply of water.[121] It may be that this religious tradition developed in order to provide justification for the monopolisation of water sources by higher caste Hindus. A detailed exploration of the origins of this tradition would be a valuable field of study.

As may be expected within pre-modern cultures,[122] supernatural forces were frequently held responsible for changes in weather patterns. Prayers for rain were often utilised either by individuals or by members of the *Brahman*[123] caste. Offerings to the gods of food, including animal sacrifices, were reported,[124] as well as rituals involving weather gods.[125] The European population also adopted religious adaptation measures, often resorting to prayers in times of drought, and fasting was reported amongst the 'Arab' and 'Moormen' populations during the drought of 1824.[126] British and American missionaries were often blamed for changing the season supernaturally,[127] a source of regular tension between missionaries and members of the *Brahman* caste regarding which gods were 'responsible' for the return of the rain after drought periods.[128]

Conclusions

This chapter has outlined responses to the droughts and famines during the early nineteenth century in western India. Adaptive capacity evidently broke down with regularity, and at least six famines (or at least proto-famines) occurred during the period, although none apparently as widespread as those of the later nineteenth century.[130] The primary measures to buffer against climatic shocks were apparently crop rotation, rogation and recourse to loans from local moneylenders, with migration and consumption of 'famine crops' resorted to during periods of exceptionally strong stress. During such periods, where adaptive capacity broke down, it is apparent that wealthy individuals—either within government or wealthy merchants—provided a role in maintaining support. This support system partially broke down when the EIC adopted *laissez-faire* drought response policies, culminating in the grain riots of 1832.

In particular, this study has outlined the highly differentiated impacts of drought during the period, particularly between the European and indigenous populations. However, this is an introduction to social responses to climate only, and more analysis is needed. Why, for example, did reports of famines decrease after 1812 when there is no indication that rainfall increased after this date?[131] This suggests either that colonial *laissez-faire* drought policies were more effective adaptive measures than have previously been suggested,[132] or that some other factor was buffering adaptive capacity during the middle years of the nineteenth century. A detailed analysis of institutional responses to drought during this period, including land tax policies and degree of cultivation, will be necessary to resolve this. This is particularly pertinent given the occurrence of major famines in Deccan in the 1870s. These issues are partially discussed in Adamson (2014).

The degree of differential vulnerability at the village level also requires further analysis. This has only been suggested through this study, and is likely to require the analysis of vernacular sources and/or narrative histories. Responses to drought by the Maratha governments also require further investigation. This chapter has relied entirely on English-language sources, and the limitations of such an approach are recognised. The author is aware, however, of substantive Gujarati- and Marathi-language records pertaining to drought in the period. The utilisation of these sources is essential to fully unravel social vulnerability to drought in the early colonial Bombay Presidency.

Notes

1. Webster, P.J., V.O. Magana, T.N. Palmer, J. Shukla, R.A. Thomas, M. Yanai, and T. Yasunari, 'Monsoons: Processes, predictability and the prospects for prediction', *Journal of Geophysical Research*, CIII (1998): 14451–14510; Ding, Y., 'The Variability of the Asian Summer Monsoon', *Journal of the Meteorological Society of Japan*, LXXXV (2007): 21–54.
2. Adamson, G.C.D. and D.J. Nash, 'Long-term variability in the date of monsoon onset over western India', *Climate Dynamics*, XL (2013): 2598–2603.
3. Endfield, G.H.,'Archival explorations of climate variability and social vulnerability in colonial Mexico', *Climatic Change*, LXXXIII (2007): 9–38.
4. cf. Meyer, W.B., K.W. Butzer, T.E. Downing, B.L. Turner II, G.W. Wenzel and J.L Wescoat, 'Reasoning by analogy' in S. Rayner and E. L. Malone (eds.) *Human Choice & Climate Change – Volume 3, Tools for Policy Analysis* (Columbus: Battelle Press, 1998): 253.
5. See for example: Srivastava, H.S., *The History of India Famines, 1858–1918* (Agra: Sri Ram Mehra and Co., 1968); Klein, I., 'When the rains failed: famine, relief, and mortality in British India', *Indian Economic Social History Review*, XXI (1984): 185–214; Brennan, L., 'The development of the Indian famine code' in Currey, B. and G. Hugo (eds.) *Famine as a Geographical Phenomenon* (Dordrecht: Reidel, 1984): 91–111; Bhatia, B.M., *Famines in India: A Study in Some Aspects of the Economic History of India with Special Reference to Food Problem, 1860–1900* (Delhi: Konark Publishers PVT Ltd, 1991); Catanach, I.J., 'India as metaphor: Famine and disease before and after 1947', *South Asia*, XX

(1998): 243–261; Davis, M., *Late Victorian Holocausts: El Niño Famines and the Making of the Third World* (London and New York: Verso, 2001); Hall-Matthews, D.N.J., *Peasants, Famine and the State in Colonial Western India* (London: Palgrave Macmillan, 2005); Roy, T. 'State, society and market in the aftermath of natural disasters in colonial India: A preliminary exploration', *The Indian Economic and Social History Review*, XLV (2008): 261–294.

6. Meyer et al., 'Reasoning by analogy'.
7. Borgaonkar, H. P., A.B. Sikder, S. Ram and G.B. Pant, 'El Niño and related monsoon drought signals in 523-year-long ring width records of teak (Tectona grandis L.F.) trees from south India', Palaeogeography, Palaeoclimatology, Palaeoecology, CCLXXXV (2010): 74–84; Adamson, G.C.D and D.J. Nash, 'Documentary reconstruction of rainfall variability over western India, 1781–1860', *Climate Dynamics*, XLII (2014): 749–769; Adamson and Nash, 'Long-term variability in the date of monsoon onset over western India'.
8. DAGM REV 7/1091 Bullard to Vibart 31 December 1839.
9. Kaiwar, V., 'The colonial state, capital and the peasantry in Bombay presidency', *Modern Asian Studies*, XXVIII (1994): 793–832; Hardiman, D., 'Usury, dearth and famine in western India', *Past and Present*, CLII (1996): 113–156.
10. Hardiman, 'Usury, dearth and famine in western India', 121–23.
11. BL SM 73 Letter to the Editor from 'The Natives' Friend' 14 October 1846.
12. Hardiman, 'Usury, dearth and famine in western India', 127–28.
13. Hardiman, 'Usury, dearth and famine in western India', 120.
14. Adamson and Nash, 'Documentary reconstruction of rainfall variability over western India, 1781–1860', 749–769.
15. Grove, R.H. 'The great El Niño of 1789–93 and its global consequences: Reconstructing an extreme climate event in world environmental history', *The Medieval History Journal*, X (2007), 75–98.
16. BL IOR/F/4/428/10490 24 December 1790.
17. NLS MS.13695 8 April 1804.
18. BLNC MC 1112 19 October 1793.
19. Etheridge, A.T., *Report on Past Famines in the Bombay Presidency* (Bombay: Government of the Bombay Presidency, 1868): 87, 95.
20. Etheridge, *Report on Past Famines in the Bombay Presidency*, 87.
21. BL IOR/F/428/10490 24 December 1790.
22. DAGM REV 34/501 G. Grant to J. Bax 13 November 1833; Sen, S.N., *Administrative System of the Marathas* (Calcutta: Calcutta University Press, 1976): 383; Mahajan, T.T., *Aspects of Agrarian and Urban History of the Marathas* (New Delhi: Commonwealth Publishers, 1991), 63–70.

23. BL MSS Eur F175/7 1 October 1804.
24. BLNC MC 1112 1 October 1803.
25. BL MSS Eur F175/7 6 April 1804.
26. '*The Peshwa recruits half his revenue to all these settlers if not to all the inhabitants of the country that have suffered by the famine*' BL MSS Eur F88/370; BL MSS Eur F88/370 15 July 1812.
27. Wink, A., 'Maratha revenue farming', *Modern Asian Studies*, XVII (1983), 591–628.
28. NLS MS.136950 11 January 1804; NLS MS.136950 8 April 1804.
29. Wink, 'Maratha revenue farming', 593, 625.
30. Elphinstone, M. *Report on the Territories Conquered from the* Paishwar (Cambridge: Cambridge University Press, 2011), 299.
31. MSS Eur F88/363 16 July, 23 July 1818.
32. Wink, 'Maratha revenue farming', 623.
33. Etheridge, *Report on Past Famines in the Bombay Presidency*, 101–102.
34. BLNC MC 1112 28 July 1804.
35. Rabitoy, N., 'The control of fate and fortune: The origins of the market mentality in British administrative thought in South Asia', *Modern Asian Studies*, XXV (1991): 753.
36. DAGM Grain Scarcity 320/466 Minute of Duncan 10 June 1805.
37. DAGM Grain Scarcity 320/466 Minute of Duncan 10 June 1805.
38. Rabitoy, 'The control of fate and fortune': 754.
39. BL MSS Eur E293/14 2 September 1812.
40. BL IOR/F/428/10490 10 August 1804.
41. BL IOR/F/428/10490 10 August 1804.
42. BL MSS Eur F175/8 26 September 1804.
43. Carnac, J.R., 'Some account of the famine in Guzerat in the years 1812 and 1813, in a letter to William Erskine, Esq', *Transactions of the Literary Society of Bombay*, I (1819): 297.
44. BL MSS Eur D666 30 September 1812.
45. Reports by the colonial Government of Bombay suggest that the entire province of Marwar may have been depopulated during 1812, as well as '*the greatest part of* [the] *population*' of the Kathiawar peninsular and Kutch. During this period, the death rate in Bombay rose to between '*110 and 120 Souls per month, principally from the very emaciated state in which the greatest part of* [the migrants] *reached the Island.*' BL IOR/F/428/10490 10 October 1812.
46. Carnac, 'Some account of the Famine in Guzerat in the years 1812 and 1813', 298.
47. BLNC MC 1112 4 December 1813.
48. 'Some account of the famine in Guzerat in the years 1812 and 1813', 301.
49. 'Some account of the famine in Guzerat in the years 1812 and 1813', 300, 301.

50. 'Some account of the famine in Guzerat in the years 1812 and 1813', 302.
51. DAGM PUB 339 Petition by 'Sundry native inhabitants of Bombay' 21 August 1812.
52. BL MSS Eur E293/14 2 September 1812.
53. BL IOR/F/428/10490 Extract Public Letter from Bombay 14 October 1812.
54. DAGM PUB 343 Minute by Government 9 December 1812.
55. BL IOR/F/428/10490 14 October 1812.
56. BLNC MC 1112 19 December 1812.
57. South Asian numbering system. 1,00,000 = one lakh (one hundred thousand). All prices are stated using this numbering system if stated as such within the archive material.
58. BLNC MC 1122 13 August 1813.
59. Adamson and Nash, 'Documentary reconstruction of rainfall variability over western India, 1781–1860', 749–769.
60. BL IOR/F/4/919/25867 H. D. Robertson to W. Chaplin 25 November 1823.
61. BLNC MC 1112 3 July 1824.
62. BLNC MC 1112 17 July 1824.
63. DAGM GEN 11/67 D. Barnecuall to Farish 29 September 1824.
64. BLNC MC 1112 9 October 1824; BLNC MC 1112 16 October 1824.
65. DAGM GEN Minute by Mountstuart Elphinstone 27 October 1824.
66. DAGM GEN Petition by 'The committee of inhabitants of Bombay' 28 September 1824.
67. DAGM GEN Farish to H. Bellasis 18 October 1824. The response of the Bombay Government was not entirely malignant, however, and the Governor of Bombay agreed to pay the expenses of the emigrants on their return to Kutch in 1825: DAGM GEN 11/95 Letter from Government to the Collector in the Northern Concan 3 June 1825.
68. DAGM GEN 11/67 Petition by 'The Bheestees of Bombay' N.D. October 1824; Petition by 'The House Holders and Inhabitants of the Village of Mazagan'; Petition by the Parsee Inhabitants of Bombay 20 March 1825.
69. DAGM GEN 11/67 Proclamation by the Governor 15 October 1824.
70. BLNC MC 1112 30 July 1825.
71. BLNC MC 1112 3 July 1824; BL MSS Eur D888/1 18 August 1824; BL MSS Eur F88/426 6 April 1825.
72. DAGM REV 2/406 10 August 1831; 3/407 G. Gilberne to Dunlop 4 September 1830; 3/407 Gilberne to Dunlop 14 August 1831; DAGM REV 3/407 10 August 1831; DAGM REV 3/407 M. Jackson to J. Vibart 30 September 1831; DAGM REV 3/407 J.M. Read to J. Dunlope 7 September 1831, DAGM REV 8/476 M. Arbuthent to

Gilberne 30 July 1831; S. M. Townsend to H. D. Robertson 15 March 1833.

73. BL IOR/F/4/1439/56798 R. K. Pringle to T. Williamson 7 November 1832.
74. BLNC MC 1112 17 November 1832.
75. BLNC MC1112 24 August 1833.
76. DAGM REV 34/501 Pringle to H. D. Robertson 15 July 1833.
77. BL MC 1112 12 April 1834.
78. BL IOR/F/4/1439/56798 9 January 1833.
79. BLNC MC1122 14 November 1832.
80. HLH ABC 16.1.1 2 November 1832.
81. BLNC MC 1122 10 November 1832; 21 November 1832, letters to the Editor from 'A Correspondent' and 'the Jami-Jenshed'.
82. BLNC MC 1122 Translated extract of a Marrhatta [sic] letter from Poona 10 November 1832.
83. In Ahmednagar in 1835, the American missionary Rev. D.O. Allan described an incident involving a '*Jaina priest*' who was interrupted by villagers in the middle of a spell to prevent rain. '*He had several sticks into the ground and in a state of perfect nudity was walking around them, muttering to the clouds & to the winds and throwing hands of grain & water in different directions, sometimes in the air and sometimes to the earth*'. The priest was accosted by the local population and brought before the local magistrate, who eventually released him. HLH ABC 16.1.1 15 August 1835; Hardiman, 'Usury, dearth and famine in western India'.
84. '*others again are eating grain which the Marwaree merchants have had on hand many years, and are now forced to sell, and that at exorbitant prices, though at any other time buyers would reject such black musty food*'. BLNC C 1114 2 February 1846, report from Ahmednagar, Bombay Bi-Monthly Times.
85. BL MC 1122 3 October 1838.
86. BLNC MC 1112 Letter to the Editor from 'A Resident in Kattywar' 19 January 1839; DAGM REV 17/1101 G. Cerles to W. C. Andrews 28 September 1839.
87. BL MC 1112 31 March 1838; BL MC 1112 17 November 1838.
88. BL MC 1112 13 October 1838.
89. BL MC 1122 23 November 1838.
90. BL MC 1112 30 March 1839.
91. BLNC MC 1122 5 June 1839; BL MSS Eur B625 21 December 1840; BL MSS Eur B625 27 July 1841; BL MSS Eur C570 14 August 1842.
92. BLNC MC 1112 16 June 1843.
93. HLH ABC 16.1.1 9 March 1844.
94. BLNC C 1114 1 February 1845.

95. BLNC C 1114 2 February 1846.
96. BL SM 73 10 December 1845.
97. MSS Eur F87/195 28 September 1854.
98. BL SM 73 Letter to the Editor from 'Simon Sly' 2 February 1846.
99. BL SM 73 9 September 1846.
100. BL MSS Eur D666 30 September 1812.
101. BLNC MC 1112 31 July 1824.
102. BL MSS Eur D888/1 29 September 1824.
103. BL MSS Eur D888/1 23 September 1824.
104. BL MSS Eur F88 / 426 15 August 1824.
105. BLNC MC 1122 10 November 1832.
106. BLNC MC 1122 19 November 1838. Pastoralism is still prevalent in Kutch amongst the *Rebari* group, although it is no longer present in modern-day Saurashtra (Kathiawar) Sharma V.P., Köhler-Rollefson, I., Morton, J., *Pastoralism in India: A Scoping Study* (Ahmedabad, India: Centre for Management in Agriculture, Indian Institute of Management; League for Pastoral Peoples, Ober-Ramstadt, Germany; and Natural Resources Institute, University of Greenwich, UK; 2003); Dyer, C., 'Literacies and discourses of development among the Rabaris of Kutch, India'. *Journal of Development Studies*, XLIV (2008): 863–879.
107. BL MSS Eur F175/7 6 April 1804; BL IOR/F/428/10490 14 October 1813; BLNC MC 1112 1 November 1823; DAGM REV 34/501 Williamson to Reid 16 July 1833; BLNC MC 1112 4 October 1838; BLNC MC 1122 30 November 1838; BLNC C 1114 2 February 1846.
108. BL IOR/V/23/212 No. 16 20 April 1788.
109. BLNC MC 1112 20 December 1828.
110. BL IOR/F/428/10490 14 October 1813.
111. BLNC MC 1112 3 December 1825.
112. DAGM REV 16/424 Townsend to Norris 20 November 1832.
113. BLNC MC 1112 21 September 1833.
114. Scott, J. C., *The Moral Economy of the Peasant* (London: Yale University Press, 1976); Ahuja, R, 'State formation and 'famine policy' in early colonial south India', *The Indian Economic and Social History Review*, IXL (2002): 351–380.
115. Kaiwar, 'The colonial state, capital and the peasantry in Bombay presidency'; Hardiman, 'Usury, dearth and famine in western India'.
116. The following report was given to the Indian Famine Commission in 1898 by the Collector of Bijapur, immediately to the south of the study area: *'In olden times when there were no facilities of communication and export, cultivators, it is said, used to sow jowari in large quantities and very little land was given to growing the other food-grains, such as wheat, gram, etc. The area under cotton was also small. But with the opening of the rail-*

way and the greater facilities afforded for communication and export the ryots, with a desire to make money, devote a larger area than before to such crops the yield of which is exported. The consequence is that they have very small stocks of jowari in reserve to fall back upon in times of scarcity like the present when the year's harvest happens to fail.' (Indian Famine Commission, 1898, p. 6); Scott, *The Moral Economy of the Peasan*, 15–26.

117. BL IOR/V/23/212 No. 16 16 August 1787; BL MSS Eur F175/4 19 June 1803; McAlpin M. B., 'Dearth, famine and risk: The changing impact of crop failures in western India, 1870–1920', *Journal of Economic History*, XXXIV (1979): 147.
118. DAGM REV 7/1091 Bullard to Vibart 31 December 1839.
119. BLNC C 1114 2 February 1846.
129. BL MSS Eur B265 19 March 1840.
120. BL MSS Eur C570 13 November 1840.
121. IOR/V/23/212 No. 16. 27 November 1787.
122. Hardiman, 'Usury, dearth and famine in western India', 134.
123. A priestly caste in India, the highest-ranking caste.
124. UBL C I3/O79/16 29 August 1837; C I3/O79/19 25 July 1838.
125. HLH ABC 16.1.1 25 July 1824; UBL C I3/O30/26 15 September 1834.
126. BL MSS Eur D888/1 27 July 1824.
127. UBL C I3/O30/5 11 September 1833.
128. This behaviour has also been documented in some regions of nineteenth-century Africa: Grove, R.H., *Ecology, Climate and Empire: Colonialism and Global Environmental History, 1400–1940* (Cambridge: White Horse Press, 1997); Endfield, G.H. and Nash, David, 'Missionaries and morals: climatic discourse in nineteenth-century central Southern Africa', *Annals of the Association of American Geographers*, XCII (2002): 727–742.
130. cf. Davis, *Late Victorian Holocausts.*
131. Adamson and Nash, 'Documentary reconstruction of rainfall variability over western India, 1781–1860'.
132. cf. Bhatia, *Famines in India*; Davis, *Late Victorian Holocausts.*

Bibliography

Adamson, G.C.D. 2014. Institutional and community adaptation from the archives: A study of drought in western India, 1790–1860. *Geoforum* LV: 110–119.

Adamson, G.C.D., and D.J. Nash. 2013. Long-term variability in the date of monsoon onset over western India. *Climate Dynamics* XL: 2598–2603.

———. 2014. Documentary reconstruction of rainfall variability over western India, 1781–1860. *Climate Dynamics* XLII: 749–769.

Ahuja, R. 2002. State formation and 'famine policy' in early colonial south India. *The Indian Economic and Social History Review* IXL: 351–380.

Bhatia, B.M. 1991. *Famines in India: A study in some aspects of the economic history of India with special reference to food problem, 1860–1900*. Delhi: Konark Publishers PVT Ltd.

Borgaonkar, H.P., A.B. Sikder, S. Ram, and G.B. Pant. 2010. El Niño and related monsoon drought signals in 523-year-long ring width records of teak (*Tectona grandis* L.F.) trees from south India. *Palaeogeography, Palaeoclimatology, Palaeoecology* CCLXXXV: 74–84.

Brennan, L. 1984. The development of the Indian famine code. In *Famine as a geographical phenomenon*, eds. B. Currey, and G. Hugo, 91–111. Dordrecht: Reidel.

Burgess, R., and D. Donaldson. 2010. Can openness mitigate the effects of weather shocks? Evidence from India's famine era. *American Economic Review: Papers and Proceedings* C: 449–453.

Carnac, J.R. 1819. Some account of the famine in Guzerat in the years 1812 and 1813, in a letter to William Erskine, Esq. *Transactions of the Literary Society of Bombay* I: 296–303.

Catanach, I.J. 1998. India as metaphor: Famine and disease before and after 1947. *South Asia* XX: 243–261.

Davis, M. 2001. *Late Victorian Holocausts: El Niño famines and the making of the third world*. Verso: London and New York.

Ding, Y. 2007. The variability of the Asian summer monsoon. *Journal of the Meteorological Society of Japan* LXXXV: 21–54.

Dodwell, J. 1924. Transport and the Second Mysore War. *Journal of the Society for Army Historical Research* III: 266–272.

Dyer, C. 2008. Literacies and discourses of development among the Rabaris of Kutch, India. *Journal of Development Studies* XLIV: 863–879.

Edwardes, S.M., and J. Campbell. 1909. *The gazetter of Bombay city and island*. Bombay: Government of the Bombay Presidency.

Elphinstone, M. 2011. *Report on the territories conquered from the Paishwar*. Cambridge: Cambridge University Press.

Endfield, G.H. 2007. Archival explorations of climate variability and social vulnerability in colonial Mexico. *Climatic Change* LXXXIII: 9–38.

Endfield, G.H., and David Nash. 2002. Missionaries and morals: Climatic discourse in nineteenth-century central Southern Africa. *Annals of the Association of American Geographers* XCII: 727–742.

Etheridge, A.T. 1868. *Report on past famines in the Bombay presidency*. Bombay: Government of the Bombay Presidency.

Grove, R.H. 1997. *Ecology, climate and empire: Colonialism and global environmental history, 1400–1940*. Cambridge: White Horse Press.

——— 2007. The great El Niño of 1789–93 and its global consequences: Reconstructing an extreme climate event in world environmental history. *The Medieval History Journal* X: 75–98.

Hall-Matthews, D.N.J. 2005. *Peasants, famine and the state in colonial western India*. London: Palgrave Macmillan.

Hardiman, D. 1996. Usury, dearth and famine in western India. *Past and Present* CLII: 113–156.

Kaiwar, V. 1994. The colonial state, capital and the peasantry in Bombay presidency. *Modern Asian Studies* XXVIII: 793–832.

Klein, I. 1984. When the rains failed: Famine, relief, and mortality in British India. *Indian Economic Social History Review* XXI: 185–214.

Mahajan, T.T. 1991. *Aspects of agrarian and urban history of the Marathas*. New Delhi: Commonwealth Publishers.

McAlpin, M.B. 1979. Dearth, famine and risk: The changing impact of crop failures in western India, 1870–1920. *Journal of Economic History* XXXIV: 143–157.

——— 1983. *Subject to famine: Food crises and economic change in western India, 1860–1920*. Princeton University Press: Princeton.

Messerli, B., M. Grosjean, T. Hofer, L. Núñez, and C. Pfister. 2000. From nature-dominated to human-dominated environmental changes. *Quaternary Science Reviews* XIX: 459–479.

Meyer, W.B., K.W. Butzer, T.E. Downing, B..L. II Turner, G.W. Wenzel, and J.L. Wescoat. 1998. Reasoning by analogy. In *Human choice & climate change – Volume 3, tools for policy analysis*, eds. S. Rayner, and E.L. Malone, 217–289. Columbus: Battelle Press.

Orlove, B. 2005. Human adaptation to climate change: A review of three historical cases and some general perspectives. *Environmental Science and Policy* VIII: 589–600.

Rabitoy, N. 1991. The control of fate and fortune: The origins of the market mentality in British administrative thought in South Asia. *Modern Asian Studies* XXV: 737–764.

Roy, T. 2008. State, society and market in the aftermath of natural disasters in colonial India: A preliminary exploration. *The Indian Economic and Social History Review* XLV: 261–294.

Scott, J.C. 1976. *The moral economy of the peasant*. London: Yale University Press.

Sen, S.N. 1976. *Administrative system of the Marathas*. Calcutta: Calcutta University Press.

Sharma, V.P., I. Köhler-Rollefson, and J. Morton. 2003. *Pastoralism in India: A scoping study*. Ahmedabad: Centre for Management in Agriculture, Indian Institute of Management; League for Pastoral Peoples, Ober-Ramstadt, Germany; and Natural Resources Institute, University of Greenwich.

Srivastava, H.S. 1968. *The history of India famines, 1858–1918*. Agra: Sri Ram Mehra and Co..

Webster, P.J., V.O. Magana, T.N. Palmer, J. Shukla, R.A. Thomas, M. Yanai, and T. Yasunari. 1998. Monsoons: Processes, predictability and the prospects for prediction. *Journal of Geophysical Research* CIII: 14451–14510.
Wink, A. 1983. Maratha revenue farming. *Modern Asian Studies* XVII: 591–628.

CHAPTER 5

Philippine Typhoons Since the Seventeenth Century

James Francis Warren

Typhoons: Introduction

The Philippine Archipelago is situated between Taiwan and Borneo, with the waters of the Pacific Ocean and China Sea washing its coastline. Due to its geographical location in the 'typhoon belt,' cyclonic storms and floods have been the principal natural hazard in the Philippines—based on frequency of occurrence and intensity.[1] For the Philippines, major weather catastrophes like extreme storms and huge floods are a painfully real and common occurrence of life. The variability of climate and weather, especially cyclonic storms, has impacted negatively on the livelihoods of peasants, fishers, merchants and traders, as well as on the aspirations and careers of colonial and post-colonial administrators, entrepreneurs and meteorologists since 1565.

In this chapter, I elaborate briefly on the two types of cyclonic storms that affect the Philippine Archipelago annually and then give a description of the typhoon phenomenon: how it is formed, where it originates from and historical difficulties with forecasting the approach or trajectory of any given typhoon. I discuss problems associated with the seasonal patterns, increasing wind speeds and the successive nature of typhoons that are associated with the Philippines, before turning my attention to

J.F. Warren
School of Arts, Murdoch University, Perth, Australia

G. Bankoff, J. Christensen (eds.), *Natural Hazards and Peoples in the Indian Ocean World*, DOI 10.1057/978-1-349-94857-4_5

important analysis associated with the El Niño and La Niña phenomena. In conclusion, I discuss the impact of typhoons on agriculture and explore the suitability of various responses to the inevitable humanitarian crises and the dire associated costs as more such disasters will increasingly occur in this century.

Throughout this discussion, I interweave the associated death, destruction, heartache and hunger wrought on the Visayan Islands by Super Typhoon Haiyan (named Yolanda in the Philippine Area of Responsibility, or PAR) on 8 November 2013. Haiyan is estimated to be the most ferocious typhoon ever to make landfall anywhere in the world. The storm packed winds of over 350 kilometres per hour (km/h) (190–195 mph), and with an accompanying storm surge, it centred on and flattened Tacloban city (see Map 5.1).[2] Haiyan also left a 600-km-wide swathe of destruction across the islands of Panay, Bohol and Leyte and substantially elevated the annual percentage of GDP spent on typhoon damage for 2013. The Philippines suffers an estimated $US1.6 billion in losses each year from the calamities,[3]but Haiyan's 'economic losses alone were valued at $15 billion.'[4]

Definition of a Typhoon

The Jesuit meteorologists of the early twentieth century made clear the exact meaning they gave to the term typhoon. Louis Froc S.J., in his *Observatory Atlas of the Tracks of 620 Typhoons 1893–1918*, clarified that the term 'typhoon' explained a dynamic–physical understanding of:

> all the depressions, or cyclones of oceanic origin, born more generally under the tropics, but not always so far, which as a rule advance from the open sea, towards the Asiatic coasts, or towards the long line of islands developed in a large arc, roughly circular, from Borneo and the Philippines to Japan, the Kurile and Aleutian Islands and Kamchatka. It is a class of phenomena entirely distinct from the group of the continental squalls or depressions, born on the land and progressing, in an opposite way, towards the Ocean.[5]

In modern meteorological terminology, a distinction is made between three kinds of tropical cyclones. The Philippines has adopted the World Meteorological Organisation classification based on maximum wind speeds within the centre of the cyclone: a tropical depression has maximum winds up to 63 km/h (34 knots or 39 mph); a tropical storm has maximum

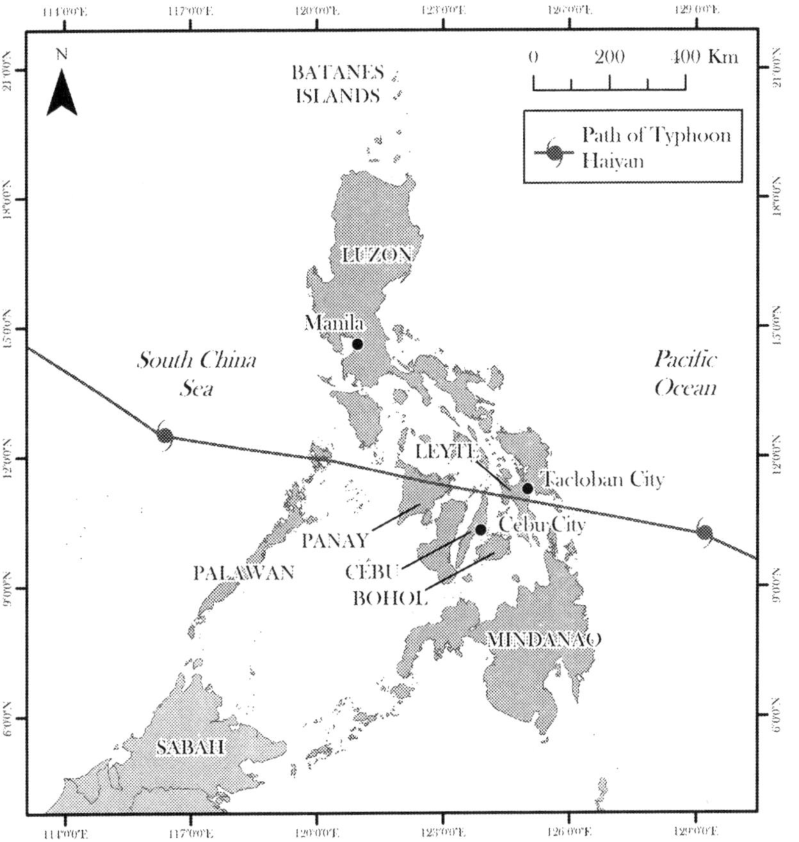

Map 5.1 Path of Super Typhoon Haiyan, November 2013

winds ranging from 64 to 118 km/h (35–64 knots or 40–70 mph); while a typhoon has maximum winds of over 118 km/h (64 knots or 74 mph).[6] Every year an average of 20 typhoons hits the Philippines. *Baguio* is the Filipino term for such a storm.

In terms of weather models and forecasting techniques, the early Jesuit meteorologists and the Philippine Weather Bureau have recognised two types of typhoons: *bean* (or rain) typhoons and wind typhoons. Rain typhoons make landfall during the spring months of April to August and generally cross northern Luzon and the Batan Islands. These rain-bearing typhoons, moving in a north-north-westerly direction, deliver large quantities of precipitation to the northernmost regions of the Philippines.

The second type of typhoon, the more intense wind typhoon, generally strikes in summer and autumn—the months of October, November and December are when these form and hit the coasts of southern Luzon, the Visayas and Mindanao. According to José Coronas S.J., these remarkable typhoons are 'one of the greatest natural calamities that may occur in any place.'[7] The wind typhoons tend to move on a west-north-westerly course. They bring hurricane-force winds and extreme disaster in their deepening or mature stages to the Bikol region, the Visayas—particularly Leyte and Samar—and the regions of northern Mindanao. Wind speeds of these cyclonic storms often readily reached up to 322 km/h (200 mph) and were sometimes accompanied by 4.5–6 metre/15–20 feet storm surges.[8] In November 2013, Super Typhoon Haiyan, moving on a west-north-westerly trajectory, surpassed these statistics by bringing wind speeds of an estimated 380 km/h and a storm surge of up to six metres.[9]

Where They Come From

The vast waters of the Pacific Ocean are ideal for the incubation of typhoons. They form in low-pressure areas east of the Philippines where warm sea temperatures of at least 26 degrees centigrade, a thick layer of moist air extending to a height of at least three kilometres and sufficient latitude favour the development of tropical storms. Typhoons cannot form at the equator where the deflective force of the earth's rotation prevents the development of the spiral winds of the storm. Indeed, typhoons rarely form within five degrees of the equator.[10] The number of typhoons formed in the China Sea is relatively insignificant when compared to the many spawned in the Pacific Ocean.[11]

In the life trajectory of a typhoon, the beginning and end are hard to determine—a characteristic that has made it difficult to accurately calculate the duration of a typhoon from its inception till its decline, especially before the recent advent of vastly improved forecasting techniques and technological systems. While a typhoon can last for days, it is usual for a single date to be given to identify its passage—the day the chronicler, meteorologist or official experienced the storm.[12] After its birth in the western Pacific, the typhoon tracks from its site of origin to other regions along its trajectory[13]—a curving almost parabolic track that in general whips across the Philippines into the coastal regions of China, islands of Southeast Asia or Japan.

In the early years of the scientific study of typhoons, a careful examination of the daily course of atmospheric pressure in the Philippines was crucial in order that meteorologists could reliably forecast the arrival of a typhoon—at least several days in advance. The daily barometric oscillations in the archipelago are normally quite regular and therefore the least irregularity can signal the proximity of some atmospheric disturbance.[14] Clouds also provided additional information to Filipinos and early twentieth-century meteorologists as to the possible existence of a typhoon before it reached an observable spot. Hence, cloud formations were considered true harbingers of approaching storm fronts, while other clouds tended to reveal, by the direction of their motion and form, the exact place where the centre, or eye, was located.[15] Before modern technological advances, cirrus and cirro-stratus clouds exhibited a predictive value, especially at sunrise and sunset when a deep red tint resembling a great fire within the clouds heralded an impending typhoon.[16]

The storm swell caused by a typhoon is also a genuine early sign, as its velocity is far greater than the rate of advance of the typhoon. As Fr José Algué explained, 'Its great import for the mariner lies in the fact that it is a truly precursory sign, warning him, whether on the high sea or near the coast, sufficiently in advance of the vicinity of a typhoon.'[17] The crucial point about the swell as a storm sign is the great distance away from the storm from which it could be observed. The swell of the sea caused by a storm is often felt before any other indication of a typhoon.

The diameter of a typhoon may range from 30 to 40 miles (48–64 km) to a thousand more miles (1600 km or more), with a typical width of about 300 miles (482 km). In the early years of meteorological forecasting, the small storms frequently escaped the notice of the meteorological observers, unless some ship blundered into them, whereas the larger typhoons could affect the weather simultaneously all over North and Southeast Asia.[18] In terms of calculating the area of the body of a typhoon, the Jesuit meteorologists estimated size only when the storm actually entered the archipelago.

Generally, typhoon strength is unrelated to overall size, and very strong typhoons usually have relatively small eyes. The most violent activity of the cyclonic storm takes place in the area around the comparative calm of the eye, known as the eye wall. Because they could rarely predict their passage, it was the smaller typhoons which the Jesuit meteorologists felt posed extreme danger. A big typhoon took a day or two to alter its course, whereas a small one could change direction fairly quickly. Hence, the small typhoon arrived seemingly out of nowhere, whipped across unsuspecting island and coastal communities and caused sudden loss of life, property and livestock. Unsuspecting people caught in their boats or in flood-prone areas were often lost (see Map 5.3). If a large typhoon crossed the coast, the loss of life was usually not as great because people were usually warned and they could take appropriate measures to protect themselves and their property—but this was not the case with Super Typhoon Haiyan. Not only was it big (600 km in diameter), but it had the lowest recorded central barometric pressure of a typhoon that had made landfall,[19] and its wind gusts packed 380 km/h. Although people were warned well in advance, the heavy bricks some put on the roofs of their corrugated iron dwellings to hold them down were no match for the destructive forces of the wind and accompanying storm surge that demolished all in their path.[20]

Irrespective of size, many typhoons 'conk out' or gradually lose energy when they pass over land. On occasion, some of the most powerful typhoons recorded in the Pacific Ocean in the last century all but fizzled out as they crossed the coasts of the Philippines. As they moved inland, wind gusts tended to drop and the storms weakened rapidly. However, over the water, where they rely upon the warm ocean surface for their fuel, typhoons strengthen in intensity. Philippine typhoons draw their strength from the heat of the Pacific Ocean and China Seas, converting that heat into mechanical energy in the form of strong winds. After an average of eight to ten days, the typical typhoon dies when it travels too far from the moist tropical latitudes of its birth or makes landfall, which hinders the free flow of the winds, and does not offer the supply of moisture needed to intensify its energy to keep the storm going.

When typhoons move very slowly across the Pacific Ocean to the north-east or north-west of Manila, or when they follow one another without interruption in the Pacific itself, or whenever a typhoon recurves somewhere north of the Philippines there can be catastrophic consequences.[21] Across the centuries, slow-moving storms have often proved the most destructive because of the threat of flooding and landslides. By the 1930s, Jesuit meteorologists agreed that the typhoons of the Visayas and Mindanao, which have a strong westerly component, tended to progress at a slower rate with the result that the rains and cloudy weather induced by the low velocity of these storms persisted for several days.[22]

These stationary typhoons occurred chiefly at the end of May and during June and July.[23]

Seasonal Patterns

Cyclonic storms are generated in the tract of ocean known as the 'typhoon alley' or 'typhoon belt' in which the Philippines is located.[24] This belt stretches between 10 and 40 degrees latitude and is notorious for its high incidence of storms, particularly within the areas of southern Luzon and the eastern Visayas, targeting a rough square ranging between Daet in Camarines Norte and Borongan in eastern Samar. The northern Philippines is the buckle in the Pacific typhoon belt. Sweeping up from the south-east, about half the powerful storms slam into northern Luzon and the Batanes Islands which sit 217 km (135 miles) into the area where the Philippine Sea meets the South China Sea.[25] But, the highest annual frequency of typhoon passage is in the Ilocos region as some storms also form west of the Philippines and tend to move towards the north-east. At the opposite end of Luzon, over the Bikol region, there were 111 typhoons between 1948 and 1990, ranking fourth among 15 regions in terms of typhoon passage, with an average of about three per year.[26] One remarkable typhoon alone every year is more than enough to make the lives of the inhabitants of several provinces miserable for many months, if not years.

In terms of zones of formation, between 1880 and 1901, 82 % of the typhoons were formed in the Pacific Ocean, 15 % in the China Sea and 4 % in the Sulu Sea, or inter-insular seas south of Luzon.[27] A century later, based on the increasing accuracy of reportage, the number of typhoons originating over the Pacific had risen to almost 90 %, while those coming from the South China Sea had declined to only 6.5 %.[28]

By 1918, Jesuit scientists of the Manila Bureau had identified at least five main typhoon tracks: one that crosses to the north of Manila; one that passes south of the capital; one that traverses east or north-east of the archipelago, either disappearing or recurving in the Pacific; one that forms in the China Sea to the west of the Philippines; and another that recurves in the China Sea between the parallel 10 degrees and 20 degrees (see Map 5.2). This important work was further enhanced after Fr Charles Deppermann became assistant Director of the Weather Bureau in December 1932. He applied the theoretical knowledge he had acquired from his study in Norway to the Philippines situation, focussing his theory

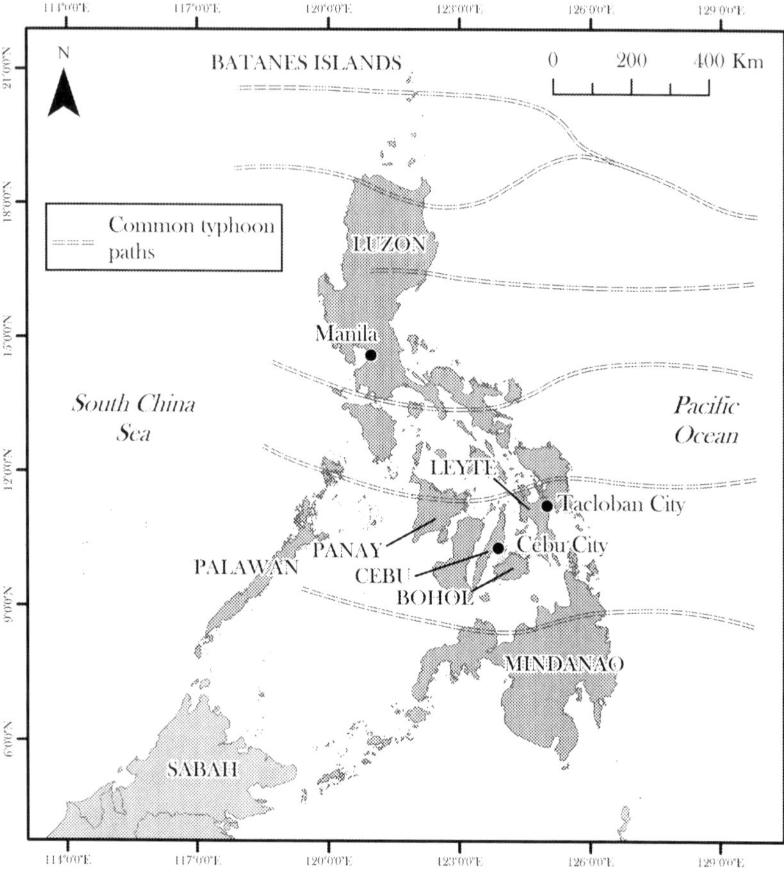

Map 5.2 Common typhoon paths in the Philippines

and practice of meteorology upon the 'genesis and paths of typhoons.' His promising work advanced the forecasting techniques of typhoons in the Philippines and 'tropical orient' on behalf of agriculture, shipping and aviation on the eve of the Pacific War.[29] Nevertheless, 50 years after Deppermann's path-breaking studies there were still no ready answers when it came to predictable typhoon tracking.

One of the most remarkable typhoons of the twentieth century, Yoling, started with a slight early morning drizzle on Tuesday, 19 November

1970. There was no warning of an impending catastrophe, except for media reports of an approaching cyclonic storm. Tragically, the typhoon tracking of the Weather Bureau in the hours just before Yoling struck was incorrect, and people in Greater Manila did not get any warning that a typhoon bearing 200 km/h winds was on its way. The meteorological projections indicated a trajectory way below Manila. Yet, by 10:15 a.m., Yoling was battering the city. In retrospect, it seems rather odd that with more than half a century of typhoon tracks compiled and recorded, no one at the Weather Bureau had ever done a statistical study of the mean gradient in the typical typhoon curve. Following the catastrophe, the newspapers recommended the Weather Bureau's usual advisory be rewritten as it had become too hackneyed.[30]

From study of the distribution of ordinary typhoons according to provinces, Weather Bureau scientists eventually concluded that no province in the archipelago is absolutely free from ordinary typhoons or depressions; and the typhoons that often cross the northern part of Luzon are remarkable typhoons, while Samar and Leyte are affected by an almost equal number of ordinary and remarkable typhoons. The meteorologists stressed that to give a correct interpretation of the statistical data and its graphic representation of the percentage of typhoons by areas, the size and geographical orientations of the islands needed to be taken into account. Samar and Leyte, for example, from the very fact that they are fairly large islands and oriented almost north–south, were likely to be frequently struck broadside by typhoons. Marinduque, on the other hand, on account of its small size presents a relatively small surface target to the path of typhoons.[31]

Monthly Distribution

Generally, the Philippines has two distinct seasons, the wet and dry, but the seasonal variation can actually be differentiated into the hot wet, cold dry and hot dry regimes. During the hot wet season from June to October, the south-west monsoon prevails, bringing rain to the western portions of the archipelago. This season has also facetiously been termed 'wet and wild' since it accounts for the higher frequency of typhoons. The coming of the monsoon rains signals the start of rice planting in the archipelago. The cold dry season follows, starting from November and lasting up to February. The hot sunny dry season with its slow trade winds lasts

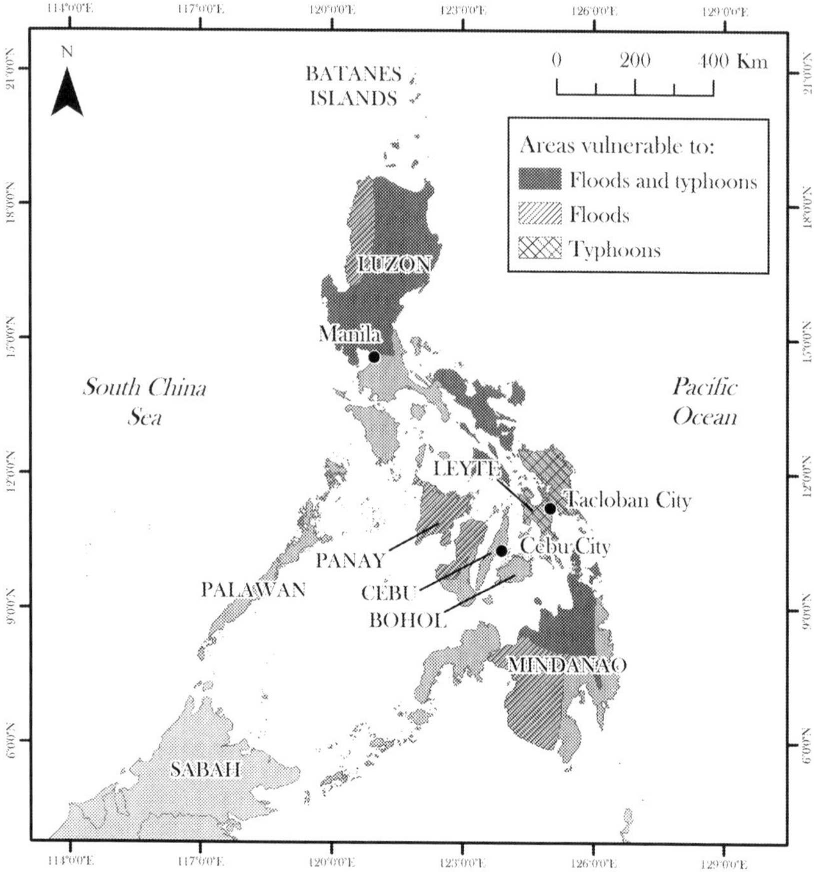

Map 5.3 Areas vulnerable to floods and typhoons, Philippines

from March to May.[32] However, some parts of the islands, namely on their eastern side, have virtually no dry season whatsoever (Map 5.3).

Set against this climatological background, the general paths and seasonal variation of Philippine typhoons differ across regions of the archipelago. In January, only a few typhoons reach the islands' shores as they usually curve northwards before reaching the country. February and March are almost entirely free from typhoons. A few small cyclonic storms sometimes start in April and affect mainly the eastern coasts stretching from central Luzon to Mindanao. From May until August, the number of typhoons increases and they can hit northern Luzon. Due to the passage of the south-west monsoon across the islands at this time, the entire archipelago with the exception of parts of Mindanao can also experience copious rains and sudden or violent winds. From September to December, typhoon paths are most threatening for the islands from central Luzon down to northern Mindanao. Northern Mindanao is rarely directly hit by a typhoon, central and southern Mindanao hardly ever; but the eastern coast at the beginning and end of the typhoon season may be struck by small, fast typhoons which quickly dissipate.[33] However on 3 December 2012, Category 5 Super Typhoon Bopha, a typhoon that originated unusually close to the equator, caused widespread death and damage to the Mindanao provinces of Davao Oriental and Compostela Valley.[34]

At the end of the nineteenth century, it was of particular scientific and economic importance for the Meteorological Observatory in Manila to try to establish the months that had the greatest frequency of storms. For this purpose, the Jesuit meteorologists included in their tables and graphs those typhoons which either crossed the archipelago or passed through it for some distance and whose trajectory they were able to trace.

Observations recorded by Jesuit meteorologists between 1880 and 1898 included 397 typhoons—an average of almost 21 per year.[35] Not one of these 397 was observed in the month of February, and only three in March. Their frequency then increases from April, when nine occurred, to July, when 66 occurred. There was then a slight decrease in August, and the maximum of 79 was reached in September. The number gradually diminished between October and January, when only six were recorded.[36]

In a Jesuit survey of the monthly and annual distribution of remarkable typhoons in the Philippines in the 15 years between 1903 and 1918, similar findings occurred: between January and April, only two remarkable typhoons were registered, with none recorded for either February or March. Then the frequency gradually increased with three in May, four in June and six in July. The peak season occurred between August and October with nine in August and the maximum of 13 each in September and October. In November, the number of remarkable typhoons diminished by more than half to only six and four were recorded in December (Table 5.1).

The annual frequency of remarkable typhoons (storms that caused the barometer, in or near the cyclonic centre, to fall below 742 mm) ranged from as few as one (1916) or two (1904, 1907, 1914, 1917, 1918) to as many as five (1909, 1912), six (1913) and a maximum of seven in 1908 and 1911 (Table 5.2).[37]

Table 5.1 Number of remarkable typhoons per month, 1903–1918. Census of the Philippine Islands: Taken under the Direction of the Philippine Legislature in the Year 1918, (Manila: Philippine Islands Census Office, Bureau of Printing, 1920–1921), vol. 1, pp. 448–51

January	1	July	6
February	0	August	9
March	0	September	13
April	1	October	13
May	3	November	6
June	4	December	4

Table 5.2 Annual remarkable typhoon frequency, 1903–1918. Census of the Philippine Islands: Taken under the Direction of the Philippine Legislature in the Year 1918, (Manila: Philippine Islands Census Office, Bureau of Printing, 1920–1921), vol. 1, pp. 448–51

1903	3	1911	7
1904	2	1912	5
1905	4	1913	6
1906	4	1914	2
1907	2	1915	4
1908	7	1916	1
1909	5	1917	2
1910	4	1918	2

According to Philippine Atmospheric, Geophysical and Astronomical Service Administration (PAGASA), during the post-war period, 1948–1994, the Philippines was hit by the largest number of storms of varying intensities in the month of October, with an accumulation of 37 typhoons, 21 tropical storms and 15 tropical depressions. Early twenty-first-century analysis of monthly variations of typhoons between the end of the nineteenth and the beginning of the twenty-first centuries suggests that the 'typhoon season' is lengthening into the latter part of the year. Greg Bankoff notes that a historical comparison of events between 1880 and 1994 shows that the percentage of typhoons in August and September had decreased while the proportion in December had noticeably increased.[38]

Wind Systems and Wind Speeds

During the early Spanish period, letters were written giving accounts of the way sailing vessels prevailed against or were destroyed by the force of a typhoon's winds. Letters also contained information about the scale of the destruction done by remarkable wind typhoons. One such letter, written by Fr Juan J. Delgado, a Spanish Jesuit missionary, referred to a typhoon that occurred on 27 June 1639:

> It overturned houses, and did great damage in all the others and in the churches. It blew the tiles through the air as if they were bits of paper. The galleons along the shore were a great cause for anxiety.... Two hundred houses were overthrown in the village of the Indians. But what caused most fear to those natives (and the old men say that they have never seen such a thing, or heard it told by their ancestors), is that the hurricane carried into the air the small boats that they use, which are called bancas and resemble canoes. It is said that they were blown about like paper, and that when they fell again they were broken into pieces.[39]

Without any doubt, the epithet 'scourge' should be used to describe not only the intensity of the typhoon winds that 'lashed' the archipelago, but also the terrifying sound—the constant high-pitched roaring of the wind—that accompanied the widespread havoc.

The Jesuits measured the monthly maximum velocities of the wind in Manila at single hourly intervals during the period from 1885 to 1898. The maximum velocity registered over the entire period occurred on the morning of 30 September 1890, at which time the wind reached 100 km/h when a violent typhoon crossed the centre of Luzon, north of Manila. The next greatest velocity recorded at the end of the nineteenth century was 96 km/h, on the afternoon of 7 November 1885.[40] However, in the nineteenth and twentieth centuries, some of the highest wind velocities in Asia occurred in Japan and Taiwan. In the 1930s, a velocity of 370.8 km/h (230 mph) was reported from Tsukuba, an elevated station, and the Taiwanese records show velocities of 252 km/h (157 mph). No such velocity was ever recorded in the Philippines prior to the Second World War. But since then some of the highest wind speeds in the Asian region have occurred in the Philippines. Velocities of over 200 km/h (124 mph) became increasingly commonplace, particularly in the 1980s and 1990s. In the twenty-first century, Super Typhoon Bopha (December 2012) recorded ten-minute sustained wind speeds of 185 km/h (115 mph) with gusts of 280 km/h (175 mph),[41] a death toll of 1146 and a damage bill of $US1.04 billion. Haiyan, (November 2013) formally described as the strongest typhoon to make landfall anywhere in the world, recorded maximum sustained winds of about 315 km/h (196 mph) with gusts exceeding 379 km/h (236 mph).[42] By 9 December 2013, 5924 deaths had been recorded with the number expected to rise.[43]

The increased frequency of typhoons with extraordinary wind velocities is demonstrated in Table 5.3, listing the 25 most intense typhoons in the Philippines between 1951 and 1998, in the order of wind speed.

Between 1970 and 1998, there were 16 typhoons with wind speeds in excess of 200 km/h (124 mph), eight typhoons with velocities ranging between 220 and 240 km/h (137–149 mph) and four with maximum wind speeds of between 250 and 275 km/h (155–171 mph). What is clear is that the majority of Philippine typhoons with the highest wind velocities have occurred in the Bikol region with Virac on Catanduanes having been battered four times by these remarkable typhoons in the decade 1988–1998. Table 5.3 also indicates that since 1976 the wind speed of the typhoons has increased by 10 km/h (6.2 mph) or more per decade and nearly doubled in velocity between 1974 and 1998.

In the final third of the twentieth century, increasing numbers of cyclonic storms rated as 'super storms,' with one-minute-sustained winds of more than 241 km/h (150 mph), have struck the country on an annual basis. Ruth, which skirted northern Luzon in November of 1991, had sustained winds of 251 km/h (156 mph) and gusts of up to 286 km/h (178 mph) as it moved north-west at 19 km/h (12 mph). It was dubbed a 'super' typhoon because it had more than twice the minimum wind speed required for ordinary typhoon status. The previous decade had proved to be a particularly excruciating period for the Philippines with four super typhoons with wind speeds between 240 and 260 km/h (149 and 162 mph) hitting the archipelago between 1981 and 1987, while five other extremely strong storms struck the Philippines during the 1980s with maximum wind speeds of between 210 and 230 km/h (130.5 and 143 mph). However, both Bopha (2012) and Haiyan (2013) exceeded maximum wind speeds experienced in the 1990s with Bopha packing winds of 280 km/h (175 mph) and Haiyan exceeding 350 km/h (190 mph).

The vulnerability of buildings to high winds has historically depended upon the strength of materials and design, the age of the facility and its site exposure. High occupancy buildings such as town halls, churches, convents and schools were often used as storm shelters and evacuation centres. However, their damage and destruction due to the growing intensity of typhoon winds have caused considerable loss of life and adverse effects on the social life of communities with the rapid rise in population since the end of the eighteenth century.[44]

In the twentieth century, the force of the winds of remarkable typhoons has destroyed entire communities as population densities increased, and

Table 5.3 The 25 most intense tropical cyclones in the Philippines (1951–2001), in order of wind speed

Type	*Name*	*Date*	*Year*	*Max wind speed (km/h)*	*Max wind speed (mph)*	*Observed at*	*Casualties*	*Damage in billion pesos*
STY	Sening/Joan	11–15 Oct	1970	275	170.9	Virac	583	1.89
STY	Rosing/Angela	30 Oct–4 Nov	1995	260	161.6	Virac	881	3.50+
STY	Anding/Irma	21–27 Nov	1981	260	161.6	Daet	176	0.62
STY	Loleng/Babs	15–24 Oct	1998	250	155.4	Virac	156	UD
TY	Amy	6–19 Dec	1951	240	149.2	Cebu	991	0.56
STY	Sisang	23–27 Oct	1987	240	149.2	Legaspi	650	UD
STY	Herming/Betty	7–14 Aug	1987	240	149.2	Catarman	200+	UD
STY	Saling/Dot	15–20 Oct	1985	240	149.2	Daet	118	0.28
TY	Undang/Agnes	3–6 Nov	1984	230	142.9	Tacloban	862	UD
STY	Nitang/Ike	31 Aug–4 Sep	1984	220	136.7	Surigao	1363	2.46
STY	Gading/Peggy	6–10 July	1986	220	136.7	Vigan	89	0.65
STY	Ruping/Mike	10–14 Nov	1990	220	136.7	Cebu	200+	UD
TY	Trix	16–23 Oct	1952	215	133.6	Legaspi	440	0.72
TY	Unsang/Ruby	21–26 Oct	1988	215	133.6	Virac	UD	UD
STY	Aring/Betty	2–7 Nov	1980	210	130.5	Casiguran	UD	1.61
STY	Yoling/Patsy	17–20 Nov	1970	200	124.3	Manila	611	0.11

(*continued*)

Table 5.3 (continued)

Type	*Name*	*Date*	*Year*	*Max wind speed (km/h)*	*Max wind speed (mph)*	*Observed at*	*Casualties*	*Damage in billion pesos*
STY	Kading/Rita	25–27 Oct	1978	185	115	Virac	400	1.9
TY	Dinang/Lee	23–28 Dec	1981	175	108.8	Catarman	2764	0.64
TY	Monang/Lee	2–7 Dec	1993	170	105.7	Virac	200+	UD
TY	Didang/Olga	12–27 May	1976	150	93.2	Iba	215	1.16
TY	Weling/Nancy	11–15 Oct	1982	130	80.8	Tuguegaro	309	0.63
TY	Bebeng	12–16 Jul	1983	120	74.6	Recon	115	0.46
TY	Bising/Nelson	22–29 Mar	1982	100	62.2	Maasin	288	0.59
STY	Titang/Kate	16–23 Oct	1970	95	59	Cuyo Island	526	1.75
TS	Openg/Vera	18–24 Nov	1973	90	55.9	Guian	162	0.59

Source: This table is based on the table '25 most intense tropical cyclones in the Philippines,' found on Typhoon2000.com: http://www.typhoon2000.ph/25most.htm, accessed 5 March 2012

high-risk structures built of light materials were constructed in old built-up residential neighbourhoods, slum areas and zones of mixed development. In the 1990s, the impact of cyclonic winds on major cities and commercial centres in the Philippines was grim. On 13 November 1990, when Super Typhoon Mike slammed into Cebu, with peak winds of 254 km/h (150 mph), the nation's second largest city and a thriving economic area was largely cut off from the outside world. The furious winds knocked out power, phones, factories and post facilities. The devastating typhoon left 370,000 homeless and hundreds of millions of dollars worth of damage. 'It's like an atomic bomb has hit us,' Sen. John Osmena stated after visiting the battered city. 'There is no electric post standing, no tree stands, no light, no water, no telephone, no gasoline.'[45] But in the twenty-first century, when Super Typhoon Haiyan hit Tacloban city on 8 November 2013, one newspaper noted that 'one of the most intense typhoons on record whipped The Philippines yesterday, terrifying millions as monster winds tore roofs off buildings and giant waves washed away flimsy homes.'[46] As at 16 December 2013, the United Nations announced that, 'more than 14.9 million people have been affected by Typhoon Haiyan, up to five million of them children.... More than 4.3 million remain displaced [and] the livelihoods of 2 million have been destroyed.'[47]

El Niño and La Niña

Over the past century, particular years have been active in terms of typhoon frequency while others have been relatively calm (see Fig. 5.1).

This pattern could have something to do with the workings of the El Niño Southern Oscillation (ENSO), and global warming. Climate not only varies from one year to the next but also can change to a new multi-year average and then stay that way for longer periods of time.[48] The link between El Niño events and typhoons is still somewhat speculative, but climatologists have pointed out that El Niño years usually have fewer typhoons, while anti-El Niño years (La Niña) have more of them. Like global warming, El Niño—or more precisely the climate cycle that produces El Niño—does not generate weather per se; rather, it promotes the growth of the context in which weather takes place.[49] Meteorologists are now convinced that high atmospheric pressure in the eastern Pacific sends trade winds blowing to the west. These winds push water before them like an invisible plough, causing the sea's surface to rise about half a metre higher around Indonesia and Australia than it does off the coast of Peru. When the pressure drops and the trade winds slacken, the water reverses and sloshes back downhill across the Pacific to the east.[50] However, while El Niño climate cycles seem to involve both the atmosphere and ocean, increasingly scientists believe that the air currents are too fickle to orchestrate the complex cycles. Alternatively, ocean surface currents that follow a sweeping circular route, taking anywhere from 10 to 20 years to complete a single journey, make them the ideal vehicle for transmitting El Niño's high-frequency messages.[51]

Historically, climatic indicators of an El Niño event in the Philippines include weak typhoon activity, fewer numbers of typhoons entering the PAR and less intense typhoons. Table 5.4 indicates that annual typhoon numbers between 1880 and 1898 ranged from as few as 11 tropical cyclones in 1880, 1882 and 1885, all El Niño years, to as many as 28 in 1887, a strong La Niña year. Interestingly, in the last two decades of the nineteenth century, there were some exceptions, concerning El Niño

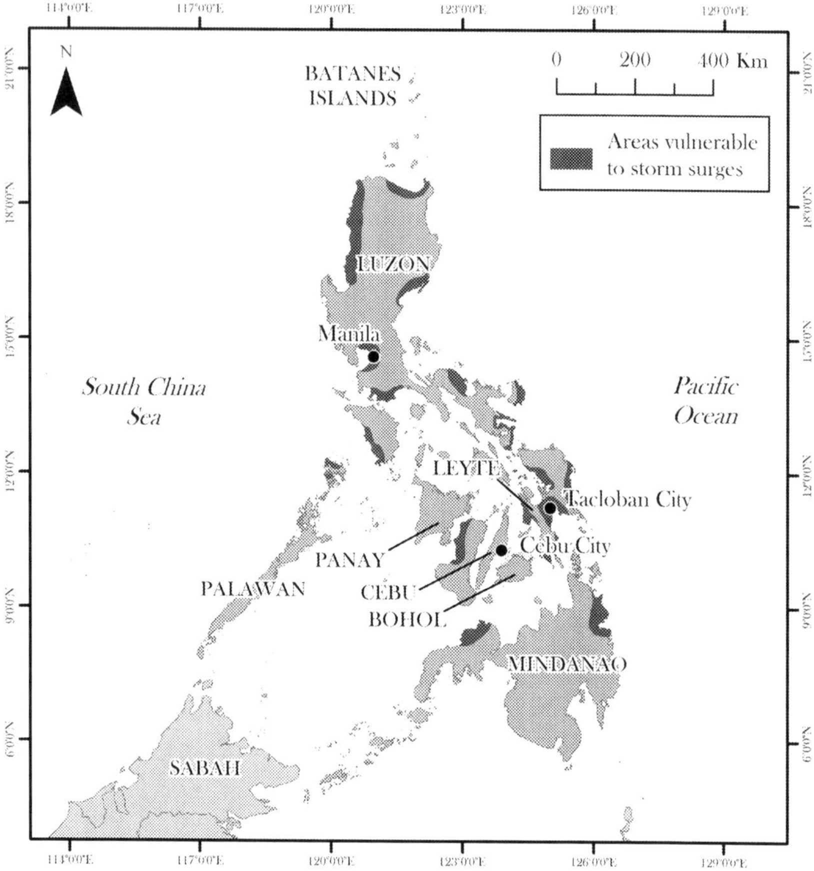

Fig. 5.1 Annual frequency of tropical cyclones within the PAR (1948–2000) (Source: This table was constructed from data provided by PAGASA on www.pagasa.dost.gov.ph)

events and typhoon frequency. For example, there was no decrease in typhoons following a particular La Niña when an El Niño event appeared in the following year, as occurred in 1890–1891 and 1893–1894.

The annual tropical cyclone frequency between 1880 and 1898 in Fig. 5.2 indicates that there was less typhoon activity during five El Niño events (1880, 1882, 1885, 1888, 1896) with the exception of the years 1891 and 1894. The El Niño of 1891 had followed closely on the heels

Table 5.4 Annual tropical cyclone numbers

Date	*Cyclone nos*	*Date*	*Cyclone nos*
1880	11	1890	27
1881	21	1891	28
1882	11	1892	24
1883	22	1893	24
1884	21	1894	34
1885	11	1895	24
1886	16	1896	20
1887	28	1897	20
1888	16	1898	25
1889	14	Total	397

Source: This table is compiled from data provided in the *Report of the Philippine Commission Philippine Commission to the President, 1900*, Vol. IV (Washington: Government Printing Office, 1901), p. 292

of the impact of a strong La Niña event in 1890 marked by 27 typhoons. La Niña years favoured tropical cyclone formation over the western Pacific which tended to increase the annual frequency of typhoons in the Philippines from 16 storms in 1886 to 28 storms the following year. There were 27 typhoons in 1890, and 24 in 1893, which was then followed by a record 34 typhoons in 1894.

Interestingly, meteorologist Keqin Dong's investigation into the relationship between El Niño and typhoon frequency shows that:

- The typhoon activity was suppressed by El Niño events in the basin west of 160°S, especially in the longitude interval 120°E–160°E, but increased in the extreme eastern part of the Pacific Basin (east of 160°E).
- A clear reduction of typhoon frequency has been found during the El Niño year period (about 0–2 months lagged) in comparison with those before and after the period. It is clear that typhoon formation in the north-western Pacific (west of 160°E) is consistently suppressed during each of the nine El Niño year periods between August 1950 and September 1983, without exception.
- The reduction in numbers of typhoons during stronger El Niño events is somewhat greater than during weak ones.[52]

In Dong's study, a concept of El Niño periods and a division of their intensity are introduced and determined by the period of maximum warm

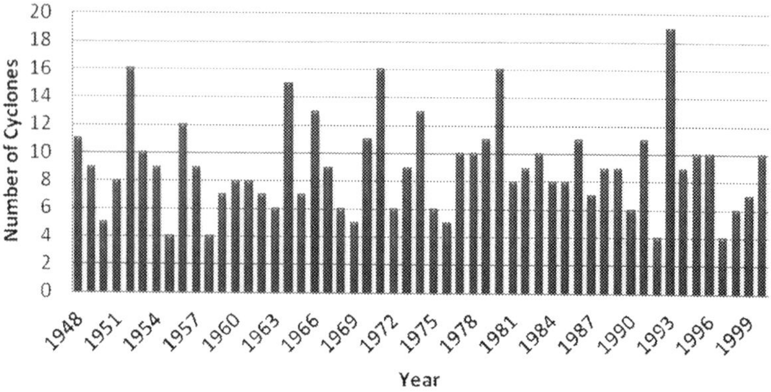

Fig. 5.2 Annual tropical cyclone El Niño/La Niña frequency, 1880–1898 (Source: Tropical cyclone data is based on the *Report of the Philippine Commission 1900*, p. 292)

12 months and the degree of the surface temperature anomaly in the eastern equatorial Pacific.

The typhoon data gathered from initial investigations undertaken by the Jesuit meteorologists for the American colonial government, and more recent recordings from PAGASA, delineate a set of approximate patterns about the frequency and intensity of large typhoons, with particular reference to peak years and cycles over five centuries. A significant finding, from the standpoint of cyclical movement and time spans, is that major typhoon activity involving super storms occurred at 50-year intervals—cycles marked by the opening years of either of the two decades prior to the mid-century which were then followed by major tropical cyclone activity, and a period of alternating flood and drought events in the last two decades of the century.

Figure 5.3 sets out the extreme typhoons that have hit Manila and neighbouring provinces, causing thousands of deaths and massive damage over the centuries. These remarkable typhoons occurred in 50-year cycles, with clusters of powerful storms and highly variable climatological activity also occurring towards the end of each century. While it is not yet altogether clear what caused these cycles, the integration of the earlier meteorological record with historical accounts and the latest information and statistics on weather and climate will help us to better understand how El Niño periods and other climatological conditions effect such major weather phenomena and activities.

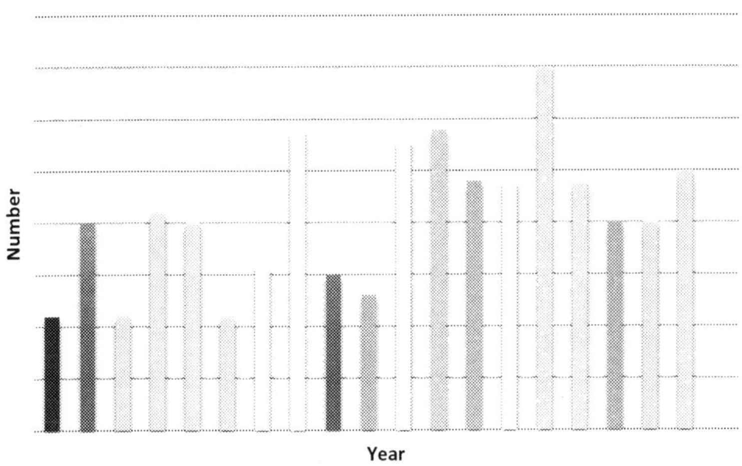

Legend. White=La Niña; Light Grey=neutral; Dark Grey= El Niño/La Niña; Black=El Niño.

Fig. 5.3 Frequency and intensity of Super Typhoons, 1600–2000. Based on an index compiled from PAGASA's records, Miguel Selga, *Charts of Remarkable Typhoons in the Philippines 1902–1934*, Miguel Selga, *Catalogue of Typhoons 1348–1934* (Manila: Bureau of Printing, 1935) and archival records found in Spain and the Philippines

One of the main factors in terms of understanding the significance of these cycles is that the human and material losses from typhoon disasters in the Philippines have steadily increased over the course of the twentieth and twenty-first centuries, but there has been no increase in the frequency of typhoon strikes since the 1990s. Table 5.5 shows that the number of typhoons affecting the Philippines seems to have peaked in the 1910s and 1920s with declining levels of crossings occurring at the middle and end of each century.

But this seems somewhat paradoxical, because it was precisely in those decades at the end of each century that remarkable typhoons emerged, along with some of the worst floods, droughts and other extreme weather. One point that Mike Davis makes is that El Niño events should be dated seasonally or monthly. Davis states, that in the last quarter of the nineteenth century, El Niño activity increased and La Niña activity decreased and the same pattern seems to have taken a frightful hold upon the last quarter of the twentieth century in the Philippines.[53]

Table 5.5 Number of typhoons affecting the Philippines from 1880 to 1898. *Report of the Philippine Commission to the President*, 1900, Vol. IV (Washington: Government Printing Office, 1901), p. 292

Decades	*No. of TCs*	*Source*
1880–1889	171	RPC 1900
1890–1898	226	RPC 1900
1910–1919	227	Selga
1920–1929	241	Selga
1950–1959	182	PAGASA
1960–1969	205	PAGASA
1970–1979	203	PAGASA
1980–1989	203	PAGASA
1990–1999	185	PAGASA

It is important to know how this fluctuating pattern of El Niño events has affected typhoon activity. Clearly, as Fig.5.3 shows, the intensity of destructive tropical cyclones increased in the 1980s as the wind velocity and the size of the storms grew quite dramatically, spawning unprecedented destructive winds and unrelenting rains. In terms of frequency, 1970 was the worst year (three destructive typhoons in one year), the 1980s was the worst decade (13 typhoons), followed by the 1970s (six typhoons). The 1990s have proved equally devastating with millions of people affected by super typhoons in the final three years of the twentieth century. To make matters even worse, these very powerful typhoons appeared at a time when the sea levels were rising around the Philippines, due to global warming. Higher sea levels and more intense typhoons can only further exacerbate the increased level of destruction in those places historically hit by storm surges, especially coastal areas on the eastern side of the Philippines (see Map 5.4).[54] In the second decade of the twenty-first century, typhoons Bopha and Haiyan demonstrate that as the workings of weather and climate continue to change, such extreme disasters will loom ever larger on the horizon.[55]

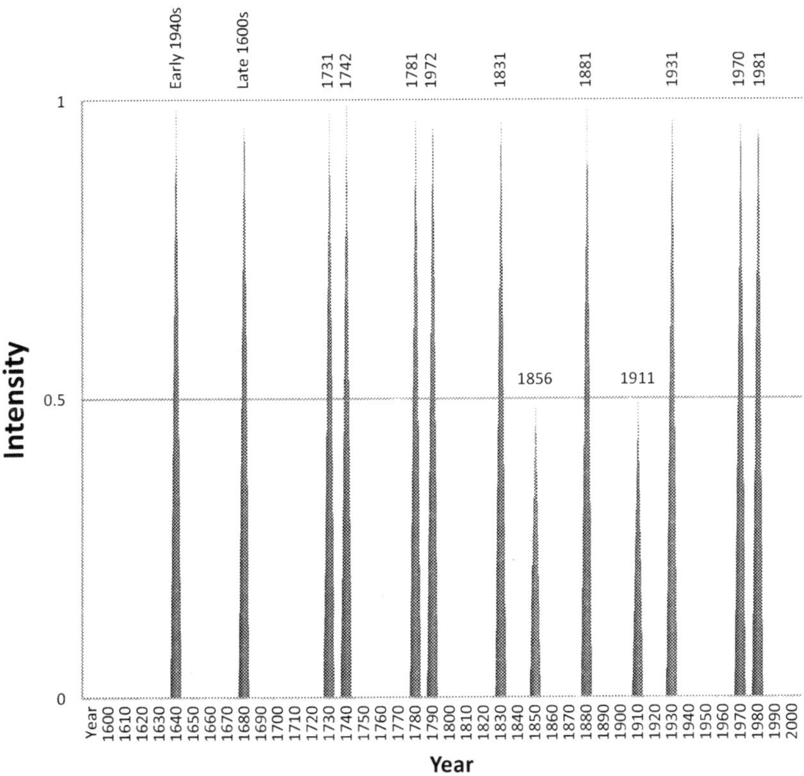

Map 5.4 Areas vulnerable to storms surges, Philippines

Impact on Agriculture

In the colonial period, Spanish and American provincial reports repeatedly highlighted the devastation wrought by the passage of successive typhoons on agriculture—the major activity of the vast majority of people and the economic mainstay of the archipelago. Over the centuries, the Philippines has suffered from typhoons coming ashore in rapid sequence from June onwards, during both the planting and harvesting cycles for particular crops, but especially in the period between October and December. Not only are these late-season storms deadlier than their earlier counterparts, but they also generally make landfall within weeks of each other. The pattern of this rapid sequence of tropical storms has caused some of the worst flooding in certain parts of the Philippines since 1850. Visayans, in regard to accurately forecasting the weather, popularly believed that unless a typhoon ended with winds from the south, it was sure to be followed by another one.[56]

Throughout the first three decades of the twentieth century, typhoons making landfall in the latter part of the year were usually not only more destructive than those crossing the Philippines earlier in the year but also frequently came ashore in rapid succession causing widespread destruction of buildings and agricultural land.

The Philippines continued to suffer the recurrent effects of tightly spaced extreme typhoons throughout the 1990s while the cost of damage caused by their regular recurrence escalated. In early September 1998 when Typhoon Babs, which came on the heels of typhoons Vicki and Zeb, swept through the Philippines, the three storms destroyed more than 734,700 tons of rice worth 1.74 billion pesos ($US42.55 million).[57] Typhoon Babs was the third storm in just a few weeks, and the ninth of the year. It caused the worst devastation on the island of Catanduanes where old growth trees were uprooted and agricultural lands were inundated.[58] Many areas remained flooded for an extended period, aggravated by seasonal monsoon rains during which hungry survivors were forced to subsist on banana shoots.

Traditional agriculture in the Philippines could be characterised as an advanced polyculture, with annual crops mixed and/or diversified with a large number of shrub and tree species. In certain key areas in the Philippines, polyculture was replaced by cash crop agriculture as farmers were compelled by their colonial overlords to take risks by developing or investing in monocrop economies. But the specific climatic and weather conditions under which the farmers had to operate, especially the frequent typhoons and floods covering large areas which adversely affected agricultural practices, led to widespread cash crop damage and outbreaks of hunger.[59] In the coconut- and hemp-growing areas of the southern typhoon belt, people did not produce enough food for their own consumption, but had to buy it with the money earned from cash crop production. Typhoons generally deprived them of their source of income from coconut and hemp production for at least two years. The recovery period of cash crop plantations from disasters of this sort inevitably took a long time because of the damage caused to the slow-growing trees and plants. Without sustained relief, under both Spanish and American rule, the farmers faced a grave risk of destitution. They could no longer afford to purchase rice or buy feed for their animals. But hunger and famine due to the lingering impacts of a typhoon would often occur months after the storm itself had passed.[60]

The same evolving situation, under both colonial economies, of hope, prosperity and want had also existed historically in the Cagayan Valley where, for several centuries, mainly tobacco was grown for export. In addition, in many of these export-driven monocrop-producing areas, particularly on Samar, southern Luzon and the Cagayan Valley, the respective populations have made the least progress of any parts of the Philippines. The frequency and severity of the typhoons did not just destroy current output; they also destroyed potential future output because the storms also destroyed the capital goods and the local–regional infrastructure. Education and public works lagged behind other regions of the archipelago, and a sizeable proportion of the people were undernourished, if not chronically malnourished, and impoverished in certain areas where tobacco, pepper, sugar, hemp and coconuts were cultivated.[61]

I want to briefly examine here the damage wrought by two closely spaced extreme typhoons nearly a century apart. General conditions on Samar towards the end of 1821 were extremely distressful. The cholera epidemic that had caused so much havoc in Manila in September and October of 1820 first appeared in towns throughout Samar on 1 September 1821. In spite of the measures taken by civil and ecclesiastical authorities, the epidemic decimated the province until January 1822. But the suffering of the Samareños was exacerbated by the appearance of three consecutive typhoons that 'whipped' across the island in November 1821 causing death and destruction—especially the last two, which on 14 and 26 November knocked down churches, parish houses and levelled villages. Almost impossible to tally was the destruction caused in the fields and orchards to the coconut, sugarcane, coffee, cocoa, abaca, cotton and fruit-bearing trees.[62] Since there was an obvious economic advantage for the inhabitants of Samar to cultivate the more profitable export crops of hemp, copra and sugar, neither rice nor corn were produced in quantities sufficient to meet the needs of the Samareños—hence rice was imported. This inherent food dependence that developed with the nineteenth-century commercialisation of agriculture often resulted in widespread social and economic dislocation when a rapid sequence of typhoons occurred, like the three that caused such destruction in November 1821.[63]

In a similar case, almost a century later, in a 'rice bowl' region, four typhoons crossed in quick succession in October and November 1934, which caused extensive flooding that, 'inundated and submerged thousands of hectares of lands, particularly rice lands...[and washed] great quantities of sand and eroded rocks and big stones from the slopes of Mayon volcano, rendering those lands unproductive now and for many years to come and perhaps permanently.'[64] The human disaster and crop damage associated with such widespread flooding and devastation also caused hunger and death. Rice plants, destroyed or damaged by high winds and torrential rains, led to drastically reduced crop yields and left tens of thousands of people dependent on government-sponsored food aid and famine relief.

Culture of Response/Culture of Mitigation

When the Spanish arrived in the Philippines in the sixteenth century, they soon recognised that changes had occurred to the landscape in particular areas due to the impacts of cyclonic storms and floods. Over the centuries, this was apparent to the Spanish as a consequence of experiencing the processual or recurrent nature of the storms year after year, particularly with respect to the impacts of typhoon-related flooding, and/or the changing course of rivers due to the storms and storm surge activity.[65] As more intense typhoons progressively altered the environment over the next three centuries, so too did the Spanish and their colonised subjects. They increasingly perpetrated environmental destruction with the advent and/or imposition of commercial export agriculture and the sustained and rapid growth of population from the end of the eighteenth century.

The evidence compiled from the historical records and the recorded meteorological data points to the emergence of cyclical patterns of half-century peaks in intense typhoon activity and periods of severe flood and drought events, which have occurred within the context of the long pendulum swing of El Niño/La Niña periods spaced up to a century apart, namely, between 1791 and 1798, 1891 and 1898 1991 and 1998. The devastating impact of these events associated with the El Niño cycle was also linked to colonial rule, the spread of population and commodity export-driven economic development and mineral and resource exploitation.

The Philippine government has regularly called on the Red Cross and other members of the International family to provide relief for typhoon-affected areas.[66] In the 12 chaotic years, from 1994 to 2006, the United Nations repeatedly concentrated on attempting to provide relief to tens of millions of Filipinos most in need of food and shelter in the worst-affected areas of the archipelago.[67] Since 1994, the United Nations and its partners have frequently launched consolidated appeals for assistance for the massive number of people tragically affected by Philippine typhoons.

Since 2001, the Philippine government has constantly stressed the importance of being prepared to deal with the increasing uncertainty linked with recent weather and global climate change, which has now become an inevitable aspect of unbounded, complex typhoon disasters like the extreme storms that descended on the Philippines in December 2004 and 2006, December 2012 and November 2013. In the wake of the Southeast Asian tsunami of 2004, the President of the Philippines directed the National Disaster Coordinating Council (NDCC) to further strengthen the existing disaster preparedness and mitigation programme by upgrading the archipelago's early-warning system. The growing awareness of the population and various sectors of the government of the need to invest more on mitigation and preparedness than on response has been gaining ground.[68] Indeed, in the twenty-first century, capacity building, particularly in the area of mitigation and preparedness, has been placed right at the top of the government's emergency preparedness agenda.[69] But in these times of global warming and climatic change, the government needs to do more than create a culture of anticipation, which can cope reasonably well with routine and even non-routine emergencies.

In the past, there has existed among governments, donors and humanitarian organisations in the Philippines, a well-developed culture of response to disasters, rather than a deep-seated, or embedded, culture of mitigation. That kind of short-term thinking no longer prevails in government disaster-management circles in the Philippines to the same extent as it once did.[70] The Philippines has managed to develop over the past several decades, out of extreme adversity and sheer necessity, an established institutional and legal framework for disaster management. This includes built-in mechanisms of participation of people nationwide, at every level of government and NGOs in decision making and emergency programme implementation.[71] In 2002, a comprehensive disaster-management framework was developed by the NDCC to serve as a guide for all disaster coordinating councils in pursuing typhoon and flood disaster-management activities in the four phases of disaster management: mitigation and preparedness in the pre-disaster stage, and response, recovery and rehabilitation in the post-disaster stage.[72] But further government support and cross-sectoral cooperation are needed to expand these efforts to ensure environmental sustainability and to manage the risk of the emerging enormous storms, such as Super Typhoons Bopha and Haiyan, of the twenty-first century.

The growing awareness of the population and various sectors of the government, of the need to invest more on mitigation and preparedness than on response, has been gaining ground. However, localising the process to make communities at the grassroots' level more resilient to typhoon and flood disasters has not been without contention. NGOs have questioned the capacity and commitment of some local élites, who often dominate local government units, to actually implement capacity-building approaches based on community-initiated managed and sustained efforts.[73] The NGOs have had reason for scepticism. In the past, especially during the Marcos years, the record shows in certain areas systematic obstruction of relief aid, wilful acts of theft and the sale of relief supplies. The cycle of poverty and typhoon disasters has not only created conditions where the poor become poorer, but also, as Bankoff notes, whereby some of the élite have become wealthier and more powerful, profiting politically and financially from the provision of emergency-related services or through graft and corruption.[74] But the Philippine government is now depending more than ever before on the successful application of lessons learned from past experiences, and the basic realisation that sound implementation of the nation's disaster mitigation programmes will be reflected in the preparedness and resiliency of its people in increasingly calamitous and trying times.

The recent repeated decade-long hammering by sets of consecutive strong typhoons indicates that global warming will make—and undoubtedly already is making—those cyclonic storms that form in the southwestern Pacific far more destructive than they would have been in the past. Hence, the populations of certain areas of the Philippines will become even more vulnerable over the course of the next several decades, while attempting to cope with the challenges of changes in climate, demography and livelihoods. The case of collaborative, multi-sectoral mitigative action to limit the hazardous impacts of the current workings of the weather and climate change, especially in confronting the increasing intensity and scale of these extreme typhoons, is thus highly compelling, and urgent.[75]

This has been powerfully demonstrated by Super Typhoon Haiyan that wrought so much devastation upon Tacloban and surrounding areas. As Tarique Niazi expressed so thought-provokingly, 'Super-storm Haiyan... buffeted the most vulnerable of Filipinos, 40% of whom live below the poverty line (i.e., \$1.25 a day).' Many were fisher people whose livelihood compelled them to live 'dangerously close to the shoreline of the western Pacific… [in some cases] just one meter above sea level.'[76] He went on to say, 'The cumulative losses in lives and livelihoods, homes and hearths, businesses and infrastructure have no parallel in Philippines history, just as Haiyan has no precedent in the annals of meteorology.'[77]

Notes

1. Alan L. Pineda, 'The Philippines: common haven of natural disasters,' in *International Centre for Disaster Mitigation Engineering*, Institute for Industrial Science, University of Tokyo, vol. 2, no. 3 (October–December 1993:pp. 1–7, p. 1.
2. Mark Fischetti, 'Was Typhoon Haiyan a record storm,' *Scientific American*, Blog (12 November 2013), online: http://blogs.scientificamerican.com/observations/2013/11/12/was-typhoon-haiyan-a-record-storm/, accessed 5 December 2013.
3. 'The small get it all: $15bn Philippines typhoon more painful than Sandy for US,' in *RT Question More* (12 November 2013).
4. Tarique Niazi, 'The Asia-Pacific in the eye of super-storms,' *The Asia-Pacific Journal*, vol. 11, issue 48, no. 5 (2 December 2013), online: http://www.japanfocus.org/-Tarique-Niazi/4043, accessed 12 December 2013.
5. Louis Froc S.J., *Zi-ka-wei Observatory Atlas of the Tracks of 620 Typhoons 1893–1918* (Shanghei: Zi-ka-wei, 1920), p. 1.
6. V.F. Balagot, *Review of Tropical Cyclones in the Philippine Area of Responsibility 1948–1972* (Manila: PAGASA, 1973), p. 2.
7. José Coronas S.J., 'The climate and weather of the Philippines,' in *Census of the Philippine Islands 1918*, vol. 1 (Manila: Bureau of Printing, 1920), p. 446.
8. David Longshore, *Encyclopedia of Hurricanes, Typhoons and Cyclones* (Chicago: Fitzroy Dearborn Publishers, 1998), pp. 260–63.
9. Jim Gomez, 'Philippine typhoon death toll could top 10,000,' in *AP The Big Story* (9 November 2013), online: http://bigstory.ap.org/article/over-100-dead-typhoon-onslaught-philippines, accessed 5 December 2013.
10. Alberto Rous, 'Typhoon season – what is there to know about the ill winds that bring floods, death and destruction,' in the *Sunday Times Magazine* (17 August 1962).
11. José Algué S.J., *The Cyclones of the Far East* (Manila Bureau of Printing, 1904), p. 24.
12. Miguel Selga S.J., 'Catalogue of Philippine Typhoons: 414–1703,' in *Philippines Studies*, vol. 20 (1972): 12–39, p. 13.

13. Algué, *The Cyclones of the Far East*, p. 11.
14. Ibid., pp. 38–39.
15. Ibid., p. 100.
16. Algué, *The Cyclones of the Far East.*
17. Ibid., p. 161.
18. Frank G. Haughwout, 'The destructive forces on typhoons,' in *Philippine Magazine*, vol. 35, no. 6 (June 1938), pp. 282–83 and 296–303, p. 303.
19. Fischetti, 'Was Typhoon Haiyan a record storm?'
20. *SBS News* (Australia), 7 November 2013.
21. Algué, *The Cyclones of the Far East*, p. 15.
22. Miguel Selga S.J., 'The baguios of Zamboanga,' Meteorological Note no. 19, *Meteorological Bulletin* (July 1932), p. 240.
23. Algué, *The Cyclones of the Far East*, p. 231.
24. 'Lessons from Typhoons,' in the *Manila Bulletin* (11 June 1998).
25. Ibid., pp. 306–11.
26. Bernardo M. Soriano Jr., 'Tropical cyclone statistics in the Bicol region,' in *AngTagamasid* (July–August, 1992): 12–15.
27. Algué, *The Cyclones of the Far East*, p. 86.
28. Soriano Jr., 'Tropical cyclone statistics on the Bicol region,' p. 13.
29. W.C. Repetti, *The Manila Observatory: Manila Philippines* (Washington DC: Edwards and Brothers, 1948), pp. 36–37.
30. Edward R. Kiunisala, 'The typhoon of the century,' in *Philippine Free Press* (28 November 1970); Jose V. Barrameda Jr., 'Typhoon Yoling: a cold grey November day,' in *Graphic* (9 December 1970).
31. Miguel Selga S.J., 'Percentage and distribution of typhoons by provinces and sub-provinces of the Philippines,' Meteorological Note 7, *Meteorological Bulletin for July 1931*, p. 222.
32. Frederick L. Wernstedt, and J. E. Spencer, *The Philippine Island World* (Berkley, Los Angeles, London: University of California Press, 1967), pp. 39–62.
33. Charles Deppermann, S.J., 'The climate of the Philippines,' Archives of the Manila Observatory (AMO), Box 36, item 17, no. 4, p. 8.
34. Jeff Masters, 'Typhoon Bopha hits the Philippines at Cat 5 strength; at least 40 killed,' *Dr. Jeff Masters' WunderBlog*, 4 December 2012, online: http://www.wunderground.com/blog/JeffMasters/comment.html?entrynum=2304, accessed 6 December 2013.
35. *First Report of the Philippine Commission to the President* (Washington, DC: Government Printing Office, 1901), vol. 4, p. 154.
36. Ibid., p. 156.
37. *Census of the Philippine Islands: Taken Under the Direction of the Philippine Legislature in the Year 1918*, (Manila: Philippine Islands Census Office, Bureau of Printing, 1920–1921), vol. 1, pp. 448–51.

38. Greg Bankoff, 'Storms of history: water hazard and society in the Philippines 1565–1930,' in *A World of Water: Rain, Rivers and Seas in Southeast Asian Histories*, ed. Peter Boomgaard (Leiden: KITLV Press, 2007), pp. 153–86, p. 162. See also Vicente Manalo, Vic Leaño and Ernie Verceles, 'Frequency of tropical cyclones by intensity crossing the Philippines,' in *AngTagamasid*, vol. 23, no. 2 (1995): 7–8.
39. Juan J. Delgado, 'Religious conditions of the islands,' Historia general sacro-profanapolitica y natural de las Islas ponientellamadas Filipinas,' 1751–54, in Blair and Robertson (eds), *The Philippine Islands*, vol. 28, pp. 166–67.
40. 'Climatology and meteorological features,' in *Report of the Philippine Commission to the President*, vol. 4 paper no. 21 (Washington: Government Printing Office, 1901), p. 253.
41. 'Typhoon Bopha,' NASA Earth Observatory, http://www.earthobservatory.nasa.gov/NaturalHazards/view.php?id=79959, accessed 6 Novemberr 2015.
42. 'Day of hell as super-typhoon tears through Philippines at 379 km/h,' the *Australian* (9 November 2013).
43. Chris Larano, 'Death toll in the Philippines nears 6,000 one month after typhoon,' *Wall Street Journal Southeast Asia* (9 December 2013), online: http://blogs.wsj.com/searealtime/2013/12/09/death-toll-in-the-philippines-nears-6000-one-month-after-typhoon/, accessed 12 December 2013.
44. See Charlotte Benson, *The Economic Impact of Natural Disasters in the Philippines*, Overseas Development Institute Working Paper no. 99 (London, 1997), pp. 85–89; Piers Blaikie, Terry Cannon, Ian Davis and Ben Wisner, *At Risk: Natural Hazards, People's Vulnerability and Disasters* (London: Routledge, 1994), pp. 32–37.
45. Bob Drogin, 'Typhoon dims a Philippine bright spot economy,' in *Los Angeles Times*, 17 November 1990; Lawrence McDonald, 'Cebu still digging out from typhoon – bright spot in Philippine economy takes hard hit,' in the *Wall Street Journal* (17 December 1990).
46. 'Day of hell as super-typhoon tears through Philippines at 379 km/h.'
47. 'Typhoon Haiyan (Yolanda),' *World Vision* (2013), online: http://worldvision.com.au/Issues/Emergencies/Current_Emergencies/Typhoon-Haiyan-Philippines.aspx?source=ASIA_APPEALFY13_SEM_SEM, accessed 16 December 2013.
48. Reid A. Bryson and Thomas J. Murray, *Climates of Hunger: Mankind and the World's Changing Weather* (Canberra: ANU Press, 1977), p. xii. See also, Michael H. Glantz, *Currents of Change El Nino's Impact on Climate and Society* (Cambridge: Cambridge University Press, 1996), pp. 1–9, 90–108.

49. 'Fire and rain,' in *Times International*, vol. 150, no. 34 (20 April 1998), p. 34; Glantz, *Currents of ChangeEl Nino's Impact on Climate and Society*, pp. 90–109.
50. 'Fire and rain,' p. 34; Glantz, Currents of Change El Niño Impact on Climate and Society, pp. 90–109.
51. 'Fire and rain,' p. 36.
52. Keqin Dong, 'El Niño and tropical cyclone frequency in the Australian region and Northwest Pacific,' in *Australian Meteorological Magazine*, vol. 36, no. 4 (1988): 219–25.
53. Mike Davis, *Late Victorian Holocausts: El Niño, Famines and the Making of the Third World* (London: Verso, 2001), pp. 253–54.
54. Charles Seabrook, 'Hugo's fury may become commonplace as earth warms up,' in *Atlanta Journal and Constitution* (8 October 1989).
55. Niazi, 'The Asia-Pacific in the eye of super-storms.'
56. William Henry Scott, *Barangay: Sixteenth Century Philippine Culture and Society* (Quezon City: Ateneo de Manila University Press, 1994), p. 126.
57. 'Philippine Red Cross seeks foreign assistance for typhoon's damage,' *Associated Press Newswires* (28 November 1998).
58. 'Red Cross rescue operations in the Philippines, July 12, 1920,' *International Federation of the Red Cross and Red Crescent Societies – News Archives* (12 July 2002), p. 3.
59. Johan Colding, Thomas Elmqvist and Per Olsson, 'Living with disturbance: building resilience in social-ecological systems,' in *Navigating Social-Ecological Systems: Building Resilience for Complexity and Change*, ed. FikretBerke, Johan Colding and Carl Folke (New York: Cambridge University Press, 2003), pp. 163–86, pp. 165–66.
60. *Annual Report of the Governor General of the Philippine Islands 1928* (Washington: Government Printing Office, 1930), p. 16.
61. 'Hurricanes in the Philippines,' Bureau of Insular Affairs, National Archives Records Administration, Record group 350, File 2711, Record Card 79–83.
62. Archivo General Indias, Filipinas 819, officio del Alcalde Mayor de Samar sobre la aparicion de la coleraensuprovincia y sobrelosdestroyozenedificios y sembradosen la mismacausadospor las dos bagious del 14 y 26 de Noviembre, 30 March 1822; Miguel Selga, S.J., 'Los baguios de Noviembre de 1821 en Samar,' Meteorological Note 45, *Meteorological Bulletin for December 1939*, p. 301; Philippine National Archives, Ereccion del pueblos–Samar 1793–1835, Exp. 18, Alcalde Mayor de Samar to Senor Gefe Politico y Superior de estos Islas, 5 February 1822.
63. Ibid.
64. The dates of the typhoons were 14 October, 28 October, 15 November and 29 November 1934. See 'Accounts of eruptions of various volcanoes

1839–1911 in P.I.,' Archive of the Manila Observatory (AMO), Box 13, item 8, Selga collection.

65. Miguel Selga, S.J., 'The baguios and floods of the Cagayan Valley,' Meteorological Note 39, *Meteorological Bulletin for June 1939*, p. 140; 'Floods,' Archive of the Manila Observatory, Selga Collection, Box 10, Item 37, pp. 5, 20; 'Typhoons 16 October, 15 November and 29 November 1934, Visayas, Luzon, Cagayan Flood, December 8, 1934,' *News Service of the American Red Cross*, Washington, DC.
66. International Federation of Red Cross and Red Crescent Societies, 'Typhoons and floods,' appeal no. 32/98, situation report no. 1, 29 October 1998, p. 6; International Federation of Red Cross and Red Crescent Societies, 'Philippine typhoons: focus on Albay, Catanduanes and Camarines Sur,' appeal no. MDRPH002, 11 December 2006, p. 1.
67. United Nations assisting the Philippines after typhoon,' *Official Website of the United Nations in the Philippines*, 5 December 2006, online: http://unphilippines.wordpress.com/2006/12/, accessed 17 January 2014; Agnes Aliman, 'UN seeks 46 million for Philippines typhoon relief,' *UN Information Center for the Philippines*, 21 December 2006.
68. Emmanuel M. Luna, 'Disaster mitigation and preparedness: the case of NGOs in the Philippines,' *Disasters*, vol. 25, no. 3 (2012): 216–26.
69. Agnes Palacio, 'Country perspectives: Philippines,' Small Group Workshop on Preparing for Large Scale Emergencies (Manila: Asian Development Bank, 5–6 July 2007), p. 2.
70. Janet N. Abramovitiz, 'Averting unnatural disasters,' in *State of the World 2001: A Worldwatch Institute Report on Progress Toward a Sustainable Society*, ed. Lester R. Brown, Christopher Flavin, Hilary French (New York, W.W. Norton, 2001), pp. 123–42, pp. 134–37.
71. See, Luna, 'Disaster mitigation and preparedness: the case of NGOs in the Philippines.'
72. Palacio, 'Country perspectives: Philippines,' p. 1.
73. Mary Racelis, 'Contributions and contradictions: Philippine Civil Society and the Millennium Development Goals,' unpublished paper (Quezon City: Institute of Philippine Culture, Ateneo de Manila University, 2005), p. 7.
74. Greg Bankoff, 'A history of poverty: the politics of natural disasters in the Philippines, 1985–1995,' *The Pacific Review*, vol. 12, no. 3 (1999): 381–420.
75. Tom Pedersen, 'Climate change fore and aft: where on earth are we going?' *Global Change Newsletter*, no. 44 (December 2000): 3–4.
76. Niazi, 'The Asia-Pacific in the eye of super-storms.'
77. Ibid.

Bibliography

Abramovitiz, Janet N. 2001. Averting unnatural disasters. In *State of the world 2001: A worldwatch institute report on progress toward a sustainable society*, eds. Lester R. Brown, Christopher Flavin, and Hilary French, 123–142. New York: W.W. Norton.

Aliman, Agnes. 2006. UN seeks 46 million for Philippines typhoon relief. *UN Information Center for the Philippines*, 21 December 2006.

AMO. 1982. Accounts of eruptions of various volcanoes 1839–1911 in P.I. Archive of the Manila Observatory (AMO), Box 13, item 8, Selga collection.

AMO. 1986. Floods. Archive of the Manila Observatory, Selga Collection, Box 10, Item 37.

Annual Report of the Governor General of the Philippine Islands 1928. Washington: Government Printing Office, 1930.

Archivo General Indias, Filipinas 819, officio del Alcalde Mayor de Samar sobre la aparicion de la colera en su provincia y sobre los destroyoz en edificios y sembrados en la misma causados por las dos bagious del 14 y 26 de Noviembre, 30 March 1822.

Balagot, V.F. 1973. *Review of tropical cyclones in the Philippine area of responsibility 1948–1972*. PAGASA: Manila.

Bankoff, Greg. 1999. A history of poverty: The politics of natural disasters in the Philippines, 1985–1995. *The Pacific Review* 12(3): 381–420.

———. 2007. Storms of history: Water hazard and society in the Philippines 1565–1930. In *A world of water: Rain, rivers and seas in Southeast Asian histories*, ed. Peter Boomgaard, 153–186. Leiden: KITLV Press.

Barrameda Jr., Jose V. 1970. Typhoon Yoling: A cold grey November day. *Graphic* 9 December.

Benson, Charlotte. 1997. The economic impact of natural disasters in the Philippines. Overseas Development Institute Working Paper No. 99, London.

Blaikie, Piers, Terry Cannon, Ian Davis, and Ben Wisner. 1994. *At risk: Natural hazards, people's vulnerability and disasters*. London: Routledge.

Bryson, Reid A., and Thomas J. Murray. 1977. *Climates of hunger: Mankind and the world's changing weather*. Canberra: ANU Press.

Ignacio Villamor. *Census of the Philippine Islands: Taken under the Direction of the Philippine Legislature in the Year 1918*. Manila: Philippine Islands Census Office, Bureau of Printing, vol. 1, 448–451. 1920–1921.

Climatology and meteorological features. Report of the Philippine Commission to the President, vol. 4, paper no. 21. Washington: Government Printing Office, 1901.

Colding, John, Thomas Elmqvist, and Per Olsson. 2003. Living with disturbance: Building resilience in social-ecological systems. In *Navigating social-ecological systems: Building resilience for complexity and change*, eds. Fikret Berke, Johan Colding, and Carl Folke, 163–186. New York: Cambridge University Press.

Davis, Mike. 2001. *Late Victorian Holocausts: El Niño, famines and the making of the third world*. London: Verso.

Day of hell as super-typhoon tears through Philippines at 379 km/h. *The Australian* 9 November 2013.

Dong, Keqin. 1988. El Niño and tropical cyclone frequency in the Australian region and Northwest Pacific. *Australian Meteorological Magazine* 36(4): 219–225.

Delgado, Juan J. 1903. 'Religious conditions of the islands,' Historia general sacro-profana politica y natural de las Islas poniente llamadas Filipinas,' 1751–54, eds. Emma Blair and James Alexander Robertson, *The Philippine Islands*, vol. 28, pp. 166–167.

Deppermann, Charles, E. 1939. Climate of the Philippines. Manila. Bureau of Printing. 31 pages. Archives of the Manila Observatory (AMO), Box 36, item 17, no. 4.

Drogin, Bob. 1990. Typhoon Dims a Philippine Bright Spot. *Los Angeles Times* 17 November 1990. http://articles.latimes.com/1990-11-17/news/mn-4291_1_national-economy.

Fire and rain. *Times International* 150(34),20 April 1998.

Fischetti, Mark. 2013. Was Typhoon Haiyan a record storm. *Scientific American*, Blog 12 November 2013. http://blogs.scientificamerican.com/observations/2013/11/12/was-typhoon-haiyan-a-record-storm/. Accessed 5 Dec 2013.

Glantz, Michael H. 1996. *Currents of change El Nino's impact on climate and society*. Cambridge: Cambridge University Press.

Gomez, Jim. 2013. Philippine typhoon death toll could top 10,000. *AP The Big Story* 9 November. http://bigstory.ap.org/article/over-100-dead-typhoon-onslaught-philippines. Accessed 5 Dec 2013.

Haughwout, Frank G. 1938. The destructive forces on typhoons. *Philippine Magazine* 35(6): 282–303.

International Federation of Red Cross and Red Crescent Societies. 1998. Typhoons and floods. Appeal No. 32/98, situation report no. 1, 29 October.

———. 2006. Philippine typhoons: Focus on Albay, Catanduanes and Camarines Sur. Appeal No. MDRPH002, 11 December.

José Algué, S.J. 1904. *The cyclones of the far east*. Manila: Bureau of Printing.

José Coronas, S.J. 1920. The climate and weather of the Philippines. In *Census of the Philippine Islands 1918*, vol 1. Bureau of Printing: Manila.

Kiunisala, Edward R. 1970. The typhoon of the century. *Philippine Free Press* 28 November.

Larano, Chris. 2013. Death toll in the Philippines nears 6,000 one month after typhoon. *Wall Street Journal Southeast Asia* 9 December. http://blogs.wsj.

com/searealtime/2013/12/09/death-toll-in-the-philippines-nears-6000-one-month-after-typhoon/. Accessed 12 Dec 2013.

Lessons from Typhoons. *The Manila Bulletin* 11 June 1998.

Longshore, David. 1998. *Encyclopedia of hurricanes, typhoons and cyclones.* Chicago: Fitzroy Dearborn Publishers.

Louis Froc, S.J. 1920. *Zi-ka-wei observatory Atlas of the tracks of 620 typhoons 1893–1918.* Zi-ka-wei: Shanghei.

Luna, Emmanuel M. 2012. Disaster mitigation and preparedness: the case of NGOs in the Philippines. *Disasters* 25(3): 216–226.

Manalo, Vicente, Vic Leaño, and Ernie Verceles. 1995. Frequency of tropical cyclones by intensity crossing the Philippines. *Ang Tagamasid* 23(2): 7–8.

Masters, Jeff. 2012. Typhoon Bopha hits the Philippines at Cat 5 strength; at least 40 killed. *Dr. Jeff Masters' Wunder Blog*, 4 December. http://www.wunderground.com/blog/JeffMasters/comment.html?entrynum=2304. Accessed 6 Dec 2013.

McDonald, Lawrence. 1990. Cebu still digging out from typhoon – Bright spot in Philippine economy takes hard hit. *The Wall Street Journal* 17 December.

Miguel Selga, S.J. 1931. Percentage and distribution of typhoons by provinces and sub-provinces of the Philippines. Meteorological Note 7. *Meteorological Bulletin* July: 222.

——— 1932. The baguios of Zamboanga. Meteorological Note No. 19. *Meteorological Bulletin*July: 240.

——— 1939a. The baguios and floods of the Cagayan Valley. Meteorological Note 39. *Meteorological Bulletin*June: 140.

———. 1939b. Los baguios de Noviembre de 1821 en Samar. Meteorological Note 45. *Meteorological Bulletin*December): 301.

——— 1972. Catalogue of Philippine Typhoons: 414–1703. *Philippines Studies* 20: 12–39.

National Archives Records Administration. Hurricanes in the Philippines. Bureau of Insular Affairs, National Archives Records Administration, Record group 350, File 2711, RG 350 T 1905 - 1914, Record Card 79–83.

Niazi, Tarique. 2013. The Asia-Pacific in the eye of super-storms. *The Asia-Pacific Journal* 11(48), no. 5,2 December. http://www.japanfocus.org/-Tarique-Niazi/4043. Accessed 12 Dec 2013.

Palacio, Agnes. 2007. Country perspectives: Philippines. Small Group Workshop on Preparing for Large Scale Emergencies. Manila: Asian Development Bank, 5–6 July 2007: 2.

Pedersen, Tom. 2000. Climate change fore and aft: where on earth are we going?*Global Change Newsletter*, (44): 3–4.

Philippine National Archives, Ereccion del pueblos–Samar 1793–1835, Exp. 18, Alcalde Mayor de Samar to Senor Gefe Politico y Superior de estos Islas, 5 February 1822.

Philippine Red Cross seeks foreign assistance for typhoon's damage. *Associated Press Newswires* 28 November 1998.

Pineda, Alan L. 1993. The Philippines: Common haven of natural disasters. *International Centre for Disaster Mitigation Engineering* 2(3): 1–7, Institute for Industrial Science, University of Tokyo.

Racelis, Mary. 2005. Contributions and contradictions: Philippine Civil Society and the Millennium Development Goals. Unpublished paper. Quezon City: Institute of Philippine Culture, Ateneo de Manila University. http://www.die-gdi.de/fileadmin/_migrated/content_uploads/Racelis_26_04_2005_Volltext_Contributions_and_Contradictions_Philippine_Civil_Society_and_the_Millennium_Development_Goals.pdf. Accessed 19 May 2016.

Red Cross rescue operations in the Philippines, July 12, 1920. International Federation of the Red Cross and Red Crescent Societies – News Archives (12 July 2002): 3.

Repetti, W.C. 1948. *The Manila observatory: Manila Philippines.* Washington, DC: Edwards and Brothers.

Rous, Alberto. 1962. Typhoon season – What is there to know about the ill winds that bring floods, death and destruction. *The Sunday Times Magazine* 17 August.

Scott, William Henry. 1994. *Barangay: Sixteenth century Philippine culture and society.* Quezon City: Ateneo de Manila University Press.

Seabrook, Charles. 1989. Hugo's fury may become commonplace as earth warms up. *Atlanta Journal and Constitution* 8 October, p.E/1.

Soriano Jr., Bernardo M. 1992. Tropical cyclone statistics in the Bicol region. *Ang Tagamasid*July–August: 12–15.

The small get it all: $15bn Philippines typhoon more painful than Sandy for US. *RT Question More* 12 November 2013.

Typhoon Bopha. NASA Earth Observatory. http://www.earthobservatory.nasa.gov/NaturalHazards/view.php?id=79959. Accessed 6 Nov 2015.

Typhoon Haiyan (Yolanda). *World Vision* 2013. http://worldvision.com.au/Issues/Emergencies/Current_Emergencies/Typhoon-Haiyan-Philippines.aspx?source=ASIA_APPEALFY13_SEM_SEM. Accessed 16 Dec 2013.

Typhoons 16 October, 15 November and 29 November 1934, Visayas, Luzon, Cagayan Flood, December 8, 1934. News Service of the American Red Cross, Washington, DC.

United Nations assisting the Philippines after typhoon. Official Website of the United Nations in the Philippines, 5 December 2006. http://unphilippines.wordpress.com/2006/12/. Accessed 17 Jan 2014.

Wernstedt, Frederick L., and J.E. Spencer. 1967. *The Philippine Island World.* Berkley, London: University of California Press.

CHAPTER 6

Bushfire in Madagascar: Natural Hazard, Useful Tool, and Change Agent

Christian A. Kull

Introduction

The island of Madagascar can be a dangerous place. It lies directly in the pathway of many Indian Ocean tropical cyclones, which regularly devastate one part or the other of the island, flooding settlements and crop fields, tearing roofs off houses, and interrupting road, rail, electricity, and communications networks. Weather patterns provide further hazards, including floods and droughts, both particularly stressful to a population largely dependent on food and income from the land. The island is also no stranger to epidemics—cholera recurred recently; plague is still present—and locusts swarm out of the south-west every decade or so, devastating crops and pastures. And of course, the hazards of poverty and political strife make life challenging for all.[1]

Bushfire does not usually figure on lists of natural hazards and disasters in Madagascar, despite being an 'Isle of Fire'.[2] This is quite a contrast to other parts of the world—particularly Australia, California, and the Mediterranean rim—where bushfire is one of the most high-profile hazards. Fire on Madagascar is usually written about as a 'hazard for

C.A. Kull
Institut de géographie et durabilité, l'Université de Lausanne, Lausanne, Switzerland

G. Bankoff, J. Christensen (eds.), *Natural Hazards and Peoples in the Indian Ocean World*, DOI 10.1057/978-1-349-94857-4_6

nature', not as a 'natural hazard' for humans. This is because it is the proximate cause of much habitat loss and deforestation on an island with a rare, endemic, and threatened flora and fauna.

Madagascar is prone to bushfire because most of the island is neither too wet, nor too dry to burn. A generous wet season drops an average 1500 mm of rain on much of the island between December and March, promoting rapid vegetation growth. This is typically followed by a long dry season of six to eight months. If, by October, people have not burnt the dry grasses, lightning from convective thunderstorms ahead of the wet season may provide the spark. In my previous work, I showed that in this fire-prone landscape, farmers and herders harness fire to achieve a variety of land and vegetation management goals—from pasture renewal and fuel management to field clearance and pest control. This parallels peoples' reliance on fire for diverse uses around the Indian Ocean, whether in Indonesia, Western Australia, or Africa.[3] I also showed how these local practices have long clashed with the anti-fire views and policies of government officials, foresters, and conservationists, leading to tensions and illicit burning. Again, such tensions are paralleled around the Indian Ocean.[4]

However, largely unexplored in these accounts is the fact that fire can be hazardous to humans and their property. On the Great Red Island, people's homes—with their thatched roofs, and in coastal areas with thatched walls—are particularly susceptible and frequently go up in flames. Norwegian missionary Johan Einrem tells of often hearing of houses or villages that burned, even of drunken Sakalava villagers partying as their hamlet burned.[5] In 1969, eight houses were ravaged in villages in the commune of Analalava.[6] Just recently, in June 2013, 51 houses in the village of Tsaratanàna in the Vohémar region burned catastrophically.[7] Many of the more dramatic fires that destroy houses and buildings are domestic, or urban, fires, not bushfires. A kitchen fire, for instance, allegedly caused the 2013 village inferno mentioned above. Yet bushfires do affect houses, often by 'spotting', that is, by burning embers lifted by the hot air only to land on flammable rooftops. According to official statistics from the Forest Service, in 1968 bushfires in Madagascar destroyed 16.6 km^2 of forest plantations, 38 km^2 of natural forests, 236 houses, 52 granaries, two churches, two government farms, 6000 tons of cotton, one hectare of vineyards, 300 orange trees, and 3000 pineapple plants; the fires also burned 283 people and caused one death.[8]

These hazards, while impressive, of course do not rival the disasters linked to specific cyclonic or drought events, but nonetheless fire can be a damaging hazard on the island. So why then is there little attention to fire as a hazard on the island? This chapter argues that the lack of attention to fire as a hazard comes from the fact that fire is multivalent, ambiguous, and flexible. It is both human and natural, and it is both useful and problematic. Fire has multiple interpretations in rural landscapes. It is simultaneously an occasional 'hazard' for thatch-roof huts or crop fields; a 'useful tool' that farmers and pastoralists use to manage vegetation; and a major transformational force, or 'change agent', that damages land and forest.

The rest of this chapter builds a story of fire in Madagascar. In order to structure the narrative and analysis, it builds on the concept of *pyric phases*.[9] This concept suggests that the human relationship with fire changes in tandem with transformations to our cultural, social, and economic systems. In Madagascar, we might identify six important pyric phases. First, natural biospheric fires characterized landscapes before humans began to set their own fires (and continued in remote areas until recently). A second pyric phase corresponds to the fires lit by hunter-gatherers that were presumably the first residents on the island (perhaps 4000 years ago, continuing until recently in some areas). After intensive periods of trans-Indian Ocean contact and migration, two new pyric phases appeared on the island linked to agriculturalists (ca. 1500 years ago to present) and pastoralists (ca. 1100 years ago to present). The fifth and sixth pyric phases correspond with attempts to control fire in the context of colonial and post-colonial governance. These phases are associated with the goal of a modern, industrial economy (ca. 1900 to present) and efforts towards nature conservation (1920s to present). These phases are sequential yet also concurrent, particularly for the latter four categories which are all present in different parts of the landscape today. In each pyric phase, fire serves, or is perceived to serve, different roles as hazard, tool, or change agent. What follows thus traces fire in Madagascar through these pyric phases, emphasizing the different role of fire in each phase, and telling some of the historical story of how governments have sought to manage the complex phenomenon of fire (Fig. 6.1).[10]

Biospheric Fire Before the First Humans

Before humans arrived on the island, there were, for lack of a better term, natural biospheric fires. These fires were not a *hazard* nor were they *tools*, at least not in the anthropogenic sense in which these terms are usually used, for there were no humans. They were, however, *change agents*, pro-

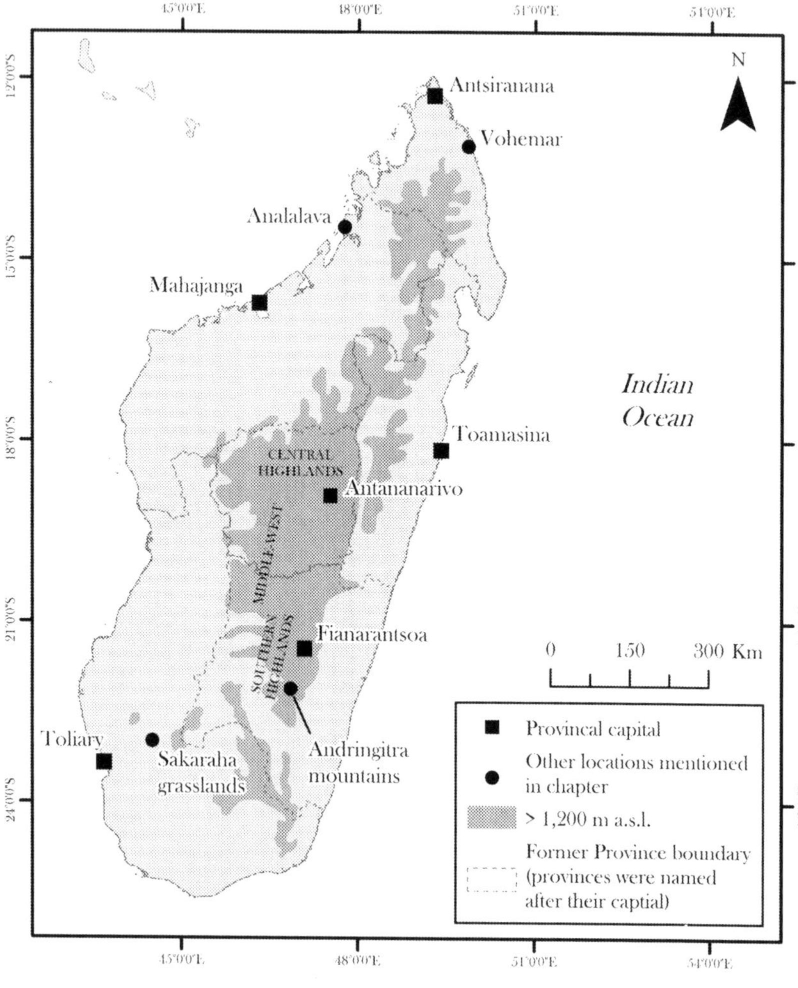

Fig. 6.1 Madagascar, showing locations mentioned in the chapter

cesses crucial to the maintenance and transformation of the vegetation cover of the island. Palaeoecological research suggests that pre-human Madagascar was characterized by a mosaic of forests, woodlands, savannas, and heath, shifting in response to climatic swings. Fire is evident throughout the sediment core record, presumably lit by lightning as well as—in certain regions and time periods—by volcanic activity. The specific composition and distribution of vegetation types, shaped in the shorter term by fire, grazing megafauna (like large lemurs, elephant birds, and pigmy hippopotamuses), and periodic cyclones, changed over the longer term in response to fluctuating climate conditions.[11] Fires possibly burned quite big when they did burn, due to fuel build-up between lightning strikes. One could argue that such 'natural' fires characterizing this pyric phase persisted for many centuries after human settlement, in areas such as mountaintops remote from human influence. Lightning strikes continue to this day, though they touch down mostly in vegetation already shaped through anthropogenic influences.

Foraging Fires (ca.4000 Years Ago to Recent)

The early settlement of Madagascar is a contentious topic, but several recent archaeological finds suggest the possibility of people hunting and gathering on the island as far as 4000 years ago.[12] The details of their arrival and subsequent movements aside, it is unquestioned that people with a foraging lifestyle have—at different times and places—occupied the island. These could be the hypothesized early arrivals in the research cited above, the semi-mythical 'vazimba' who preceded the rice-farming Indonesians in the highlands, or the Mikea who persist to this day as a cultural group in south-western forests. People living with a subsistence foraging lifestyle would have been associated with a particular pyric phase, what David Bowman and colleagues called 'wildland anthropogenic fire'.[13] Like hunter-gatherer cultures elsewhere around the world—Australian aborigines being a primary example[14]—they would have used, shepherded, and lit fires to manage habitats to favour certain habitats, certain plants, and certain game animals. Fire would have been a *tool*. But it would also have been a *hazard*, one suspects, in terms of damage to shelters, food stocks, life and limb. Finally, from an environmental perspective, foraging fires would have been a *change agent*, with more frequent fires, probably in different seasons, leading to sometimes subtle, sometimes major alterations in previous 'natural' vegetation cover.

Agricultural Fires (ca.1500 Years Ago to Present)

The earliest archaeological evidence of fixed settlements and associated farming activities dates back to 500–700 CE, with both African and Indonesian heritage.[15] The spread of agriculture on the island heralded a new pyric phase, 'agricultural anthropogenic fire' in the terms of Bowman and colleagues. This phase is visible in the palaeoecological and archaeological records, which, in the central highlands for instance, suggest a tenfold increase in fires from previous levels between 600 and 800 CE, and show evidence of cultivated crops and permanent settlements from ca. 1000 CE, with grasses replacing woody vegetation across the landscape. A second wave of settlement took place early in the Second Millennium, arriving, it is thought, in successive waves of colonization from present-day Indonesia. By the thirteenth and fourteenth centuries, evidence of their well-established villages is widespread. These settlers found a largely grass-dominated landscape, which they maintained through fire and grazing; they also introduced irrigated rice-based agriculture to the floodplains and valleys. Over time, centralized political structures emerged in a number of highland river valleys.[16]

In the agricultural pyric phase, fire is very clearly used as a *tool*, primarily to clear fallow fields or forest patches for cultivation, but also as a means of wildfire prevention, pest control, and fertilization. Fire's use in agriculture in Madagascar parallels similar uses in traditional farming systems around the world.[17] Crop fires tend to be smaller in extent than pasture fires, but relatively numerous. In grassland zones, people burn the standing vegetation in the plot they intend to cultivate—it may be uncultivated grassland, long-fallow fields covered with grass, ferns, or bushes, or short fallows with crop stubble. After burning, a team of two or three farmers labours with spades to prepare the soil for cultivation. Sometimes, farmers collect additional fuel to burn in their fields—cut grass, rice straw, and scrub. Other fires include those used to 'clean' field edges and canals, facilitating travel, water flow, and pest control. In more humid forest zones, 'slash-and-burn' or shifting cultivation is more common, usually called *tavy* in Madagascar. Farmers choose a plot—usually an overgrown long-fallow field called *savoka*, secondary forest, or more rarely today, in primary forest. They cut the standing vegetation, let it dry for a few weeks, prepare a fire break, then set fire just before the rainy season, typically in October or November. The plot is then cultivated with rainfed rice and other crops for one or more years before being allowed to fallow. Farmers perceive fire-as-a-tool to be an integral part of the agricultural system.[18]

As useful as they are, fires in this pyric phase can also be *hazards*. A poorly mastered fire can, given the right conditions, damage life and property. In farming areas, it is possible to lose not just food stores but also standing crops to fire.[19] Most dramatically, perhaps, certain agricultural fires are important *change agents.* Fire is used as the proximate tool, after all, to render an area of forest, bush, or grassland cultivable. Thus, fire is associated with the advancing agricultural frontier, particularly in the forest zones that encircle the island. Repeated cycles of agricultural fire for subsistence and cash crops are the proximate cause of the replacement of native forests with farm-, bush-, and grassland.[20] There are even legends in Madagascar of a 'great fire', or *afotroa*, that is associated with forest clearance during the original settlement.[21]

Pastoral Fires (ca.1100 Years Ago to Present)

The earliest evidence of a proliferation of zebu cattle on the island dates to around the tenth century, in the north-west, with expansion to the central highlands within a century or two.[22] While most scholars assert that cattle were introduced from Africa, some hypothesize direct links to India.[23] The expansion of a pastoral lifestyle heralded a 'pastoral pyric phase'. Pastoral fires continue to affect the largest surface area of any fires in Madagascar. These fires are used to manage grasslands used for extensive—that is, low density—grazing. Madagascar's cattle herds, limited by dry season forage resources, cannot keep up with the growth in rainy season grasses. People light fires to keep bushes from encroaching, to remove lignified dry grass, to encourage a 'green bite' of new sprouts, and to guard against fuel accumulation.[24] When asked to justify pasture fires, herders I interviewed retorted 'what would the cattle eat without fire?' and stated that the fresh green shoots are needed '*mba voky ny omby*—so that the cattle have eaten enough'. While walking through the hills with my informants, it was apparent that they appreciate a landscape kept free of bush encroachment and accumulated dry grass fuel, calling such a landscapes '*madio*' or 'clean'.[25]

These useful fires are key *tools* in the pastoral economy. In the past, fires served to maintain and renew grassland resources. This was recognized by colonial district officers, as I report below. The colonial cattle industry relied on fires as well. Ranches in the Middle West burned their lands for pasture renewal,[26] and cattle industry experts promoted the experience of South African ranchers with fire.[27] Today, pasture fires continue to play an important role in the open spaces where herders maintain herds for subsistence and prestige[28] as well as in more densely settled areas where mixed agropastoral economies dominate.

While these fires serve to manage fuel loads in landscapes where a highly productive rainy season is followed by a hot, dry season, they can also become *hazards* when they burn grass roofs, enclosures, or crop stores. And these fires are certainly *change agents*, expanding grassy land covers at the expense of woody ones.[29] But the emphasis in this pyric phase is on fire as a useful tool.

Fire Suppression for a 'Modern' Economy (ca.1900 to Present)

While the previous three pyric phases co-existed in the landscape for many centuries, the colonial period brought a new phase to Madagascar. During the 1800s, a period in which the island became unified politically, a number of proclamations and rules addressed forest cutting and forest burning, yet they did not generally regulate crop field or grassland fires.[30] The highlands and savannah zones of the west were characterized by a free-burning, pasture-oriented fire regime. The landscape was dominated by endless grass-covered hills; trees only grew near hilltop villages. In eastern forest zones, *tavy* fires provided the rice upon which people depended for food.

From 1896, France colonized Madagascar.[31] The French intended to establish a profitable economy linked to agricultural production and resource extraction, and their policies reflect that goal. They abolished slavery, developed transportation infrastructure, requisitioned rice from farmers, and established commercial crop (and later forest) plantations. While many officers and settlers recognized the utility of fire in certain circumstances, the emphasis was increasingly on fire exclusion. The advent of industrialization and modernity brought by colonialism led to increased use of internal combustion to replace what used to be done by fire (field clearance, fertilization) and gave the mechanical tools and ideological frameworks necessary to circumscribe fire. In this pyric phase, fire is only grudgingly seen as a *tool*, and instead is largely seen as a *hazard*, a threat to productive assets and resources (like timber, soils, and infrastructure). The *change agent* aspect of fire is less emphasized; this comes with a focus on conservation as I profile later.

During the first decade of French control, authorities focussed on 'pacifying' the island and establishing administrative structures. Their approach to fire policy, marked by hesitation, centred on economic issues. The Forest Service's mission was to protect forests for 'rational' exploitation and to protect soils from degradation.[32] Colonial district officers, faced with the difficult task of actually administering the island for economic profit and political stability, did not share the foresters' anti-fire enthusiasm.

In 1897, Governor Joseph Gallieni ordered his officers to do their best to stop the damage done by fires and peasant cultivators to the forest.[33] The Forestry Decree of 1900 banned dry season fires within 200 metres of forests, except by authorizations granted to private landowners (which thus excluded indigenous *tavy* cultivators).[34] This proved unpopular. District officers wrote to the Governor that fires are critical to cattle husbandry, useful in controlling insects, rats, and locusts, and are used from China to South Africa.[35] As a result, Gallieni, recognizing that 'modern' techniques of mowing and haying were not yet realistic, re-authorized much grassland burning, but reiterated a strict ban on fires in forest.[36]

The tone of the following two decades was set by the primordial economic concerns of the colony. The central government's position against fire, especially *tavy*, hardened, as it sought to protect the logging industry and commercial agricultural plantations. Governors Victor Augagneur, Albert Picquié, and Hubert Garbit each sought to tighten the reins, issuing circulars to this effect in 1907, 1909, 1913, 1915, and 1920. However, district officers continued to defend peasant burning, and in a compromise, the government 'temporarily' began to regulate and authorize burning until 'modern' agricultural and pastoral techniques could be introduced.[37]

In the meantime, the government increased its efforts to enforce the new regulations. Some fires burned into forests, tree plantations, or colonists' lands; others destroyed houses or entire villages. As a culprit could rarely be found, officers relied on collective fines. Enforcement was uneven. Burning continued, with or without all the detailed prescriptions of local fire regulations. Fire legislation remained largely non-functional, due to the number of exceptions, the weakness of the Forest Service, and the lack of will on the part of some authorities.[38] The logic behind this modernist pyric phase continues to this day. Fire is repeatedly seen as antithetical and hazardous to a modern economy. However, from the 1920s, these ideas were complemented by those centred on nature conservation.

Fire Suppression for Nature Conservation (1920s to Present)

A corollary of the modern, industrial pyric phase and its preference for fire exclusion is the 'protect nature' pyric phase. The environmental ravages of the industrial revolution were one of the reasons, after all, for the emergence of the modern conservation movement and its fundamentally modernist project of separating 'wild nature' from productive landscapes. Given that dominant ecological ideas in the twentieth century saw fire as antithetical to the conservation of wild nature, the conservation movement promoted fire suppression. In this pyric phase, fire is hardly seen as a *tool*, instead it is emphasized as a *hazard*, not for humans, but for wildlife and habitats. Most crucially, fire is seen as a *change agent* that lies behind the deforestation and degradation that threatens the island's natural resources.

This pyric phase appeared in Madagascar in the 1920s as a complement to the economic arguments outlined above. The tone of state-led fire repression stiffened from the mid-1920s into the 1940s. Naturalists concerned with the loss of forests gained an increasingly prominent role in policymaking. The number of district officials defending local fire practices shrunk significantly (though not completely). A key moment was the new Forest Decree, which appeared in January 1930 after a decade of consultation and debate.[39]

The 1930 Decree arose out of foresters' concerns with the pace of extractive logging, and naturalists' alarm over the loss of forest to fire.[40] Prominent naturalists Henri Perrier de la Bâthie and Henri Humbert actively promoted the decree.[41] Perrier believed that the Malagasy burn 'out of simple habit and without any reason'. Humbert's influential monograph is tellingly subtitled 'the destruction of an island flora by fire'. Louis Lavauden, who in 1928 became Chief of the Forest Service, blamed the 'natives' and their incendiary practices for the disappearance of the forest and the resulting climatic desiccation and degradation. He promoted extremely repressive policies and ridiculed the compromising colonial administration. The administration's views reflect this harder anti-fire line (even if field officers were still more pragmatic). Governor Hugues Berthier stated in 1929 that the goal of colonial agrarian policy was to move people out of the hills and mountains, stop shifting cultivation, and to institute a modern agricultural regime.[42] *Tavy* fires, he declared, were a nefarious practice that would destroy the forest and ruin the country.

As a result, the 1930 Forest Decree significantly tightened regulations for forest exploitation.[43] *Tavy* in forestland was simply illegal, as were all fires within 500 metres of any forest. The decree prohibited bushfires and prairie fires in non-forest lands, except when authorized by the governor's delegated authorities. It also legislated a more severe repression of infractions, increased the powers of forest agents, and reinforced punitive measures.

The 1930 decree set the tone for the following three decades. It was strengthened through amendments in 1937 and 1941, tightening the rules and broadening enforcement measures.[44] However, the repressive legislative record masks continued debate in which fire use was defended.[45] These debates hint at a recognition of the different logics of the concurrent pyric phases—the use of fire for agricultural or pastoral activities versus the need to modernize agriculture and conserve nature through fire suppression. For instance, Governor Léon Cayla (1930–1939), echoing his predecessors, suggested that Madagascar was (still) in a transition to modern pasture maintenance (mowing, seeding), and thus that pasture fires had to be tolerated, while Governor Armand Annet (1940–1942) likewise asked districts to designate pasture areas where regulated fire could be permitted.[46]

Towards a Compromise: 1940s to 1960s

The Second World War and the 1947 anti-colonial uprising led to a change in the emphasis. It was increasingly recognized, in some circles at least, that a number of 'pyric phases' were clashing. Farmers and herders continued burning, while the modernizing colonial state sought to promote economic growth and rational (non-fire) exploitation on the one hand, and nature conservation (without fire) on the other. Unsurprisingly, during the 1947 rebellion, foresters—who represented the government's anti-fire activities—were a prominent target of violence. Many forestry posts were destroyed, forest plantations were burned, and a massive increase in fires and *tavy* was noted.[47]

In the 1950s, the administration harshly suppressed political activity and concentrated on economic development. Enforcement of fire-related infractions was relatively effective. Foresters received bonuses for successful citations, military patrols made tours of the countryside, fines were collected, and permits duly issued. New rules prohibited grazing in pastures burned without authorization, and prohibited cultivation in cut or burned state forests, giving the Forest Service a better means to enforce illegal burns.[48]

Fire policies were actively discussed. On the one side, hardliners including prominent botanists, Forest Service chiefs, and rangeland ecologists pushed for fire exclusion for economic and environmental reasons.[49] These opinions were strengthened by botanist André Aubréville, influential head of the French colonial forest service, who strongly argued for forest protection by fire exclusion.[50] In opposition, several fire-tolerance advocates emerged. They included district administrators, some range specialists, and anthropological and geographical researchers.[51]

The fact that French rule after 1947 was more tenuous, and the appearance of new voices for fire tolerance, led to a slight softening of the repressive regime against fires. Fire came to be seen as a 'necessary evil' and several ideas emerged.[52] The first, not particularly new idea was to delimit areas where fires could be authorized. The second was that of legalizing counter-season fires or early fires, which would remove accumulated fuel but run a lesser danger of total combustion, escape from control, and subsequent soil and forest damage.[53] A third new idea related to *tavy*. While the government had always pushed farmers to engage in irrigated rice cultivation, it had never helped them do so, so the Forest Service began investing in building small irrigation dams and canals and in instructing farmers on different techniques for soil conservation and intensive cultivation. A final idea was the use of pre-emptive controlled fires to reduce wildfire danger (this had some uptake across Africa, but never took hold in Madagascar).

Madagascar regained its independence on 26 June 1960, and within less than four months the new country had written its own fire laws.[54] *Ordonnance* 60–127 differed from colonial laws in two significant ways: it contained no general ban on fires, and it was the first to clearly distinguish between different types of fire and set different policies for each. Fires for clearing forest for cultivation, that is, *tavy*, were declared illegal in all state forests, and required authorization elsewhere, restricted to flat lands or the lower thirds of hills. Fires in permanent crop fields or along field edges were allowed at all times outside of forests. Pasture fires were allowed in officially delimited zones during the rainy season, and could 'exceptionally' be authorized at other times. Finally, wildfires burning without control or limits, in any type of vegetation, without any economic utility, were always illegal.

This legislation was considerably more realistic than colonial legislation, as it recognized the utility of fire as a *tool*. However, it was still framed within an official position that highlighted fire as a *hazard* and a negative *change agent* causing environmental degradation. Its moderate view is reflected in publications of the time by sociologists and rangeland specialists.[55] The effect of the legislation is hard to determine. Pasture fire and *tavy* authorizations were given in certain areas throughout this period, but the actual number of fires likely far outstripped the authorizations. From the farmer's point of view, the necessity to get authorizations for burning often involved prohibitive bureaucracy. Enforcement of fire infractions continued. [56]

Resurgent Fire Exclusion for Nature Conservation, 1969 to Present

The exclusion of fire for nature conservation—a normative type of pyric phase that was aspired to but not often implemented—regained dominance in Madagascar in the late 1960s and remains dominant until today. This phase predominantly sees fire as a problematic *change agent*, sometimes as a *hazard*, and downplays any aspect of fire's useful role as a *tool*.

From the late 1960s, fires were at the top of the national agenda.[57] Several drought years and incendiary fire seasons caught the attention of President Philibert Tsiranana (1960–1972). His *Conseil de Cabinet* immediately banned *all* fires, and a high-level Interministerial Commission for the Fight Against Bushfires was convened. Concurrently, Madagascar became increasingly concerned with nature conservation. In 1970, the government hosted an international conference on conservation. Scientist after scientist condemned the degradation caused by *tavy* and pasture fires.[58] All this attention on fires resulted in three new themes. First, the primary approach that was proposed—after legislative changes were abandoned—was public awareness through an expensive propaganda campaign. Second, there were repeated attempts to make rural communities responsible for fire management, so that by 1971 there were 5000 'anti-fire committees' created—at least on paper.[59] Third, for the first time, the use of heavy machinery for firefighting was proposed.[60]

Despite the regime changes during the revolutionary period of 1972–1975, and the supposedly populist dictatorship that followed, the anti-fire agenda intensified.[61] New decrees strengthened the rules for enforcement of illegal fires, by 1977 even making them capital offences. These 'brutal' rules hardly helped, as cattle raisers and forest farmers felt attacked by the unfair legislation[62], and as the administration largely failed to enforce them.[63] Anti-fire programmes and reports proliferated. From 1976 to 1979, the government spent 40–90 million Malagasy francs per year on an anti-fire programme of propaganda, field enforcement tours, and firefighting machinery.[64] In his 1980 New Year speech, President Didier Ratsiraka spoke of the need to fight without fail for the complete eradication of bushfires. This led to another Interministerial Commission, which criticized the apathy, laziness, and pyromania of the rural populations, grudgingly recognized the reasons behind the uses of fire, and paved the way for a seven-year *Opération Danga* focussed on awareness and enforcement.[65]

From the early 1980s, however, Madagascar's economy sank into turmoil, and the attention of most state agents turned away from fire. Despite the repressive laws and anti-fire programmes, enforcement became more and more inconsistent. From an average of over 1000 infractions per year between 1969 and 1975, the mid-1980s only saw 200–300 infractions per year. At the same time, district Forest Service offices continued to authorize pasture fires as demanded by cattle raisers. Authorizations were given for 150,000–850,000 hectares of fire each year, in the 1970s, and this system persisted to some extent into the 1980s. Politicians bent to the wishes of their constituencies, and repression was in name only.[66]

In the mid- to late 1980s, the momentum for conservation lost since the 1970 conference resumed. Environmental concerns and activities boomed, yet the government was financially and politically paralysed. Fires were denounced with renewed vigour. In 1984, the country adopted a National Strategy for Conservation and Development. A year later, it convened a second international conference on conservation to discuss the implementation of the Strategy and to showcase the Malagasy environment. Diverse programmes were initiated as a result, centred on conservation, forest management, and environmental awareness, funded by the World Bank, bilateral donors, WWF, and UNESCO. These were succeeded by an ambitious 15- to 20-year Environmental Action Plan, launched in 1990 under the leadership of the World Bank. With initial funding commitments of over $100 million, the Plan included a wide variety of programmes designed to protect biodiversity, stop environmental degradation, and establish the institutional structures necessary for environmental monitoring and management.[67]

The situation on the ground in this period was a strong contrast to the grand ideas, intentions, and programmes launched at the national and international level. The economic crisis and IMF austerity measures of the mid-1980s had paralysed the nation, and political unrest in 1991 and 1992, including a six-month general strike, led to the first of a number of tumultuous regime changes in the two decades that followed. As far as fires, the rules had not changed, yet fires burned more or less uncontrolled and enforcement lagged. As unpaid or striking government agents became less motivated to do their work, the rural population largely ignored the requirements and stopped seeking authorizations.[68]

From the mid-1990s, the government and foreign donors opened a new chapter in the history of fire politics, seeking to reconcile environmental goals and poorly functioning regulations through community-based resource management. Legislation passed in 1996 allowed for the transfer of resource management responsibilities and rights to community-based associations, and could, in theory, allow for community-based fire management.[69] However, the contracts for local management had to conform to existing legislation, to Forest Service enforcement of that legislation (it had, for instance, ceased giving fire permits in 1990), and to the dominant discourse that saw fire as a hazardous change agent. As a result, local management contracts were more successful in managing forests than in addressing fire.[70]

From the mid-1990s, the first glimmers of a new 'pyric phase' also appeared. This could be characterized by the expression 'fire for ecosystem management', or the recognition—belated and incomplete on the Great Red Island—that fire is an integral component of many ecosystems. This approach, used for instance in park or forest management in South Africa, Australia, and the USA, sees fire as a useful *tool*, but for certain ecosystem management goals, not specifically for human utilitarian use.[71] It also tends to recognize, and circumscribe if appropriate, fire's potential to be a *hazard* and a *change agent*.[72] In a way, this pyric phase is a new, human-piloted version of the 'natural biospheric fire' pyric phase that existed before human settlement. Or it is a post-modern nature conservation pyric phase. In Madagascar, it has been recognized by scientists and managers working in specific conservation landscapes that fire plays a particular role in maintaining certain types of biodiversity, including orchids in the montane prairie of Andringitra, or the grasslands of Sakaraha.[73] Yet such recognition remains rare.

The above initiatives, which gently poked holes at different aspects of the government's strong fire exclusion approach, were further marginalized with the regime change in 2002. The goal-driven government of Marc Ravalomanana revitalized a top-down attitude of stopping all fires. Incentives were added to the rules and fines: municipal budgets depended on their fire-stopping performance. Satellite-based fire monitoring was provided by foreign conservationists,[74] and some high-profile cases of illegal slash-and-burn fires resulted in strict gaol sentences. This strictness may, among a number of other factors, have contributed to the general concern over Ravalomanana's rule that led to his overthrow in 2009. In the years since, however, the fragile 'transitional' government of Andry Rajoelina has not fundamentally changed the emphasis on fire first and foremost as an undesirable *change agent* to the environment, second as a *hazard* to economic assets and production, both ahead of any recognition of fire as a useful *tool*.

Conclusion

In Madagascar's turbulent landscape of cyclones, droughts, floods, and political unrest, bushfires rank low as a priority for officials and agencies engaged in disaster risk management. It ranks very high, however, for environmental managers and conservationists seeking to preserve the highly valued native biodiversity of the island. That is, bushfire, while clearly a *hazard* at times for certain physical assets, rarely takes on disastrous proportions. However, attention focusses on it as a *change agent* leading to deforestation and habitat modification. Meanwhile, its use as a *tool* for pastoralists, farmers, and other land managers is largely relegated to an afterthought necessitated through political compromise.

In the end, it is important to recognize that fire is multivalent, ambiguous, and flexible. Malagasy villagers know this. One day in 1998, I walked with four boys in the hills above their highland village. In response to a bothersome swarm of little black flies, one of the boys lit the grass adjacent to the path, and we all stood in the fly-repelling smoke. As the grass was still green, it burned poorly, and soon all three boys were busy lighting little fires, laughing and chatting. One schoolboy chastised his brother 'hey, don't burn too much, the mayor might see us'. He continued, with a hint of irony, 'look, it's burning, we're destroying the environment!' Fire is both human and natural; it is both useful and problematic. Traditional farmers in Madagascar, as in the rest of the world, have dealt with fire's multiple meanings and manifestations for centuries—using fire yet fearing it, keeping it at bay yet taking advantage of it. Successive governments in Madagascar have misunderstood peasant burning practices, criminalizing a practice important to peasant livelihoods and landscape management due to their concerns about fire as a dangerous hazard and environmental change agent. Yet the limited resources of the state, the vast terrain, the inevitability of fire in wet–dry landscapes, and the inherent trump card held by those who want to burn allowed peasants to continue to light fires despite decades of bans, restrictions, fines, and anti-fire speeches. On the ground, the pastoral and agricultural pyric phases continue to trump, in most places, the fire exclusion pyric phases associated with modern economies and nature conservation. And the perception of fire as *hazard* always plays second fiddle to its utility as a *tool* for rural farmers and its threat as a *change agent* to conservationists. The hazard of fire to human life and property in Madagascar is punctual, localized, and a relatively small price to bear—when viewed at an aggregate level—compared to its benefits in agropastoral land management. For conservationists, fire matters less as a natural hazard to humans, but more as a hazard to nature, an instrument of habitat conversion. As the population continues to grow and the island modernizes, it remains to be seen what sorts of ambiguous, multivalent, and flexible fire regimes emerge in the future.

Notes

1. Donque, Gérald. 'Les Cyclones Tropicaux des Mers Malgaches.' *Madagascar Revue de Géographie* 27, (1975): 9–63; Olson, Sherry. 'Environments as Shock Absorbers, Examples from Madagascar.' *Environmental Review* 12, no. 4 (1988): 61–80; Ballet, Jérôme, and Mahefasoa Randrianalijaona, eds. *Vulnérabilité, Insécurité Alimentaire et Environnement à Madagascar*. Paris: L'Harmattan, 2011. It is striking how little published research exists on natural hazards and disasters in Madagascar.
2. Kull, Christian A. *Isle of Fire: The Political Ecology of Landscape Burning in Madagascar*. Chicago: University of Chicago Press, 2004. The historical narrative in the present chapter is substantially drawn from Chapter 7 of *Isle of Fire*.
3. For Indonesia, see Fowler (2013), for Western Australia, see Bird, Douglas W., Rebecca Bliege Bird, and Christopher H. Parker. 'Aboriginal Burning Regimes and Hunting Strategies in Australia's Western Desert.' *Human Ecology* 33, no. 4 (2005), 443–6; for Africa see Eriksen, Christine. 'Why do they Burn the 'Bush'? Fire, Rural Livelihoods, and Conservation in Zambia.' *Geographical Journal* 173, no. 3 (2007), 242–56; Laris, Paul. 'Burning the Seasonal Mosaic: Preventative Burning Strategies in the Wooded Savanna of Southern Mali.' *Human Ecology* 30, no. 2 (2002): 155–86.
4. Sivaramakrishnan, K. 'The Politics of Fire and Forest Regeneration in Colonial Bengal.' *Environment and History* 2, no. 2 (1996), 145–94; Pyne, Stephen J. *Vestal Fire*. Seattle: University of Washington Press, 1997; Kull, Christian A., and Paul Laris. 'Fire Ecology and Fire Politics in Mali and Madagascar.' In *Tropical Fire Ecology: Climate Change, Land Use, and Ecosystem Dynamics*, edited by Mark A. Cochrane, 171–226. Heidelberg, Germany: Springer-Praxis, 2009.
5. Einrem, Johannes. *Gjennem Grønne Briller*. Vol. 1, Samlede Verker. Bergen: A. S. Lunde, 1912 (1936).
6. Archives Nationales de Madagascar (hereafter ANM). Folder Vice Présidence 850, subfolder 'Situation des feux de brousse 1969–1971'.
7. *Express de Madagascar* (17 June 2013, http://www.lexpressmada.com/vohemar-madagascar/44084-la-moitie-d-un-village-rasee-par-les-flammes.html).

8. ANM, folder Vice Présidence 850, subfolder 'Conference international…' To my knowledge, his is the only year for which such a summary has been made. As it has a political goal, there may be some strategic conflation of bushfires with house fires.
9. Bowman, David M. J. S., Jennifer K. Balch, Paulo Artaxo, William J. Bond, Mark A. Cochrane, Carla M. D'Antonio, Ruth DeFries, Fay H. Johnston, Jon E. Keeley, Meg A. Krawchuk, Christian A. Kull, Michelle Mack, Max A. Moritz, Stephen J. Pyne, Christopher I. Roos, Andrew C. Scott, Navjot S. Sodhi, and Thomas W. Swetnam. 'The Human Dimension of Fire Regimes on Earth.' *Journal of Biogeography* 38, (2011), 2223–36; c.f. Pyne, Stephen J. *Fire: A Brief History*. Seattle: University of Washington Press, 2001; Pyne, Stephen J. 'The Human Geography of Fire: A Research Agenda.' *Progress in Human Geography* 33, no. 4 (2009), 443–6; Kull, Christian A. 'Saving Land with a Spade: Human Landscape Transformations in the Madagascar Highlands.' In *Greening the Great Red Island: Madagascar in Nature and Culture*, edited by Jeffrey C. Kaufmann, 113–35. Pretoria: Africa Institute of South Africa, 2008.
10. Significant passages of the text on the history of government attempts to manage fire are based on Chap. 7 in Kull (2004) and the references and archival sources cited therein.
11. Burney, David A., Lida Pigott Burney, Laurie R. Godfrey, William L. Jungers, Steven M. Goodman, Henry T. Wright, and A. J. Timothy Jull. 'A Chronology for Late Prehistoric Madagascar.' *Journal of Human Evolution* 47, no. 1 (2004), 25–63; Dewar, Robert E., and David Burney. 'Recent Research in the Paleoecology of the Highlands of Madagascar and Its Implications for Prehistory.' *Taloha* 12, (1994), 79–88.
12. Dewar, Robert E., Chantal Radimilahy, Henry T. Wright, Zenobia Jacobs, Gwendolyn O. Kelly, and Francesco Berna. 'Stone Tools and Foraging in Northern Madagascar Challenge Holocene Extinction Models.' *PNAS* 110, no. 31 (2013): 12583–88.
13. Bowman, David M. J. S., Jennifer K. Balch, Paulo Artaxo, William J. Bond, Mark A. Cochrane, Carla M. D'Antonio, Ruth DeFries, Fay H. Johnston, Jon E. Keeley, Meg A. Krawchuk, Christian A. Kull, Michelle Mack, Max A. Moritz, Stephen J. Pyne, Christopher I. Roos, Andrew C. Scott, Navjot S. Sodhi, and Thomas W. Swetnam. 'The Human Dimension of Fire Regimes on Earth.' *Journal of Biogeography* 38, (2011): 2223–36.
14. Bird et al. (2005).
15. Burney et al. (2004) and Dewar et al. (2013).
16. Burney et al. (2004) and Kull (2008).
17. Pyne (2001), Laris (2002) and Fowler (2013).
18. Coulaud, Daniel. *Les Zafimaniry: Un Groupe Ethnique De Madagascar À La Poursuite De La Forêt*. Antananarivo: Fanontam-Boky Malagasy, 1973;

Brand, Jürg. 'Das Agro-Ökologische System am Ostabhang Madagaskars: Ressourcen- und Nutzungsdynamik unter Brandrodung.' Dissertation, Universität Bern, 1998; Laney, Rheyna M. 'Disaggregating Induced Intensification for Land-Change Analysis: A Case Study from Madagascar.' *Annals of the Association of American Geographers* 92, no. 4 (2002), 702–26; Kull (2004).

19. Multiple interviews with farmers, 1998–2001. See Kull (2004).
20. Coulaud (1973); Humbert, H. 'Principaux Aspects de la Végétation à Madagascar. La Destruction d'une Flore Insulaire par le Feu.' *Mémoires de l'Académie Malgache* Fascicule V, (1927); Scales, Ivan R. 'Farming at the Forest Frontier: Land Use and Landscape Change in Western Madagascar, 1896–2005.' *Environment and History* 17, (2011), 499–524.
21. Abe, Yoshio. 'Ancien Mode de Subsistence Pratiqué sur les Hautes Terres Centrales de Madagascar.' *Asie du Sud-Est et Monde Insulindien* 8, no. 3–4 (1977): 79–98.
22. Burney et al. (2003, 2004).
23. Klein, Jørgen, Betrand Réau, and Mary Edwards. 'Zebu Landscapes: Conservation and Cattle in Madagascar.' In *Greening the Great Red Island: Madagascar in Nature and Culture*, edited by Jeffrey C. Kaufmann, 157–78. Pretoria: Africa Institute of South Africa, 2008.
24. Prehistoric Madagascar.' *Journal of Human Evolution* 47, no. 1 (2004): 25–63. Cori, Gilles, and Pierre Trama. *Types d'Élevage et de Vie Rurale à Madagascar*, Travaux et Documents de Géographie Tropical No. 37. Bordeaux: Centre d'Etudes de Géographie Tropicale, CNRS, 1979.
25. Kull (2004).
26. J.-P. Raison, pers. comm.; field interviews by author, August 1999.
27. Geoffroy, M. 'Les Feux de Brousse' *Bulletin Economique, Madagascar* 52, no. Mars 1931 (1931): 79–82.
28. Ranaivoarivelo, Nivo, and Pierre Milleville. 'Exploitation Pastorale des Savanes de la Région De Sakaraha (Sud-Ouest De Madagascar).' In *Sociétés Paysannes, Transitions Agraires et Dynamiques Écologiques dans le Sud-Ouest de Madagascar*, edited by Samuel Razanaka, Michel Grouzis, Pierre Milleville, Bernard Moizo and Christine Aubry, 181–97. Antananarivo: CNRE & IRD, 2001; Kaufmann, Jeffrey C., and Sylvestre Tsirahamba. 'Forests and Thorns: Conditions of Change Affecting Mahafale Pastoralists in Southwestern Madagascar.' *Conservation and Society* 4, no. 2 (2006), 231–61.
29. Kull (2004).
30. These include Andrianampoinimerina's edicts and the Code of 305 Articles, discussed in Kull (2004), where I also mention a possible exception to the lack of regulation (p. 207 n2).

31. Brown, Mervyn. *A History of Madagascar*. London: Damien Tunnacliffe, 1995.
32. Dez, Jacques. 'La Limitation des Feux de Végétation.' *Terre Malgache* 4, (1968): 97–124.
33. Journal Officiel de Madagascar (hereafter JOM), 15 Jan 1898.
34. JOM, 17 Apr 1900.
35. anon. 'Les Feux De Brousse.' *Bulletin Economique, Madagascar et Dépendences* 4, no. 1ere trimestre (1904): 25–9.
36. JOM, 30 Mar 1904. For further details, see Kull (2004, p. 208).
37. JOM, 9 Mar 1907, 12 Jun 1909, 10 May 1913, 9 Oct 1915 and 20 Nov 1920; ANM D81s and D196; Lavauden, L. 'Histoire de la Législation et de L'administration Forestière À Madagascar.' *Revue des Eaux et Forêts* 72, (1934), 949–60; Dez (1968). For details, see Kull (2004, pp. 210–15).
38. Jarosz, Lucy. 'Defining Deforestation in Madagascar.' In *Liberation Ecologies*, edited by Richard Peet and Michael Watts, 148–64. London: Routledge, 1996; Kull (2004).
39. Kull (2004).
40. Lavauden (1934), Jarosz (1996) and Kull (2004).
41. Perrier de la Bâthie, Henri. 'La Végétation Malgache.' *Annales du Musée Colonial de Marseille* Sér. 3, v. 9, (1921), 1–266 (p. 4); Humbert (1927).
42. JOM, 9 Feb. 1929.
43. JOM, 22 Nov. 1930.
44. JOM, 13 July 1937 and 20 Dec 1941; AOM fn-sg-mad 353/948; See Kull (2004, p. 222).
45. e.g. Geoffroy (1931), Lavauden (1934); Heim, Roger. 'L'État Actuel des Dévastations Forestières à Madagascar.' *Revue de Botanique Appliquée et d'Agriculture Tropicale* 15, (1935), 418–26; and Archives d'Outre Mer (Aix-en-Provence; hereafter AOM) mad-ggm-2/D/19bis/6, p. xiv. See Kull (2004, p. 220).
46. JOM, 2 July 1938, 6 Dec 1941.
47. Jarosz (1996) and Kull (2004).
48. Brown (1995), Dez (1968) and Kull (2004).
49. For instance: botanist Humbert (1949), Forest Service chief Bégué, and rangeland ecologist Bosser (1954). See: Humbert, Henri. 'La Dégradation des Sols à Madagascar.' *Mémoires de l'Institut de Recherche Scientifique de Madagascar* D1, no. 1 (1949), 33–52; Bosser, J. 'Les Pâturages Naturels de Madagascar.' *Mémoires de l'Institut Scientifique de Madagascar* Série B, V, (1954), 65–77.
50. Bergeret, Anne. 'Discours et Politiques Forestières Coloniales en Afrique et à Madagascar.' *Revue Français d'Histoire d'Outre-Mer* 79, no. 298 (1993): 23–47.
51. Kull (2004).

52. Dez (1968) and Kull (2004).
53. Unfortunately for its champions, early burns were never a perfect solution, as they were often very difficult to ignite in the rainy season, and their effects on pasture were not the same as dry season fires (Dez 1968).
54. JOM, 15 Oct 1960.
55. Sociologists: Dez (1966, 1968); rangeland specialists: Granier, P., and H. Serres. 'A Propos de L'incidence des Feux de Brousse sur l'Évolution des Pâturages.' *Bulletin de Madagascar* 16, no. 243 (1966), 789–91; Gilibert, J., P. Dubois, P. Granier, P. Grenier, and H. Serres. 'Lutte Contre les Feux Sauvages dans une Exploitation de Ranching.' *Bulletin de Madagascar* 24, no. 329 (1974), 465–80.
56. Rakatonindrina, Raymond. 'Les grands problèmes actuels de la conservation des écosystèmes: défrichement et feux de brousse.' In *Environnement et Gestion des Ressources Naturelles dans la Zone Africaine de l'Océan Indien*, edited by M. Maldague, K. Matuka, and R. Albignac, 111–22. Paris: Unesco, 1989; interviews with farmers and foresters, 1998–1999 (see Kull 2004).
57. ANM Vice-Presidence folders 831, 840, 844, 850, 851.
58. IUCN. *Comptes Rendus de la Conférence Internationale sur la Conservation de la Nature et de ses Ressources à Madagascar, Tananarive 7-11 Octobre, 1970*, Publications UICN Nouvelle Série, Document Supplémentaire No. 36. Morges, Switzerland: IUCN, 1972; Kull (1996).
59. DEF Annual Report, 1971. See Kull (2004, p. 235).
60. Aubry, 181–97. Antananarivo: CNRE & IRD, 2001. RDM. 'Rapport Du Groupe Interministériel D'etudes Sur Les Feux De Brousse.' Antananarivo: Repoblika Demokratika Malagasy, 1980.
61. Andriamampionona, A. 'Les Feux de Brousse dans la Région de l'Itasy.' Antananarivo: FAO/PNUD/MAG/88.032, 1992; JOM, 30 Sep 1972, 2 Dec 1972, 1 Nov 1975, and 27 Aug 1976; see Kull (2004, p. 236).
62. Cori and Trama (1979, p. 58).
63. Fanony (1989).
64. RDM (1980).
65. RDM (1980).
66. Rakatonindrina (1989) and Kull (2004).
67. Kull (1996, 2014) and Pollini (2011).
68. Multiple interviews, foresters and farmers, 1998–2001 (see Kull 2004, pp. 240–1).
69. JOM, 14 Oct 1996.
70. Kull (2004).
71. Gillson and Willis (2004).
72. Keeley, Jon E., Gregory H. Aplet, N. L. Christensen, S. G. Conard, E. A. Johnson, P. N. Omi, D. L. Peterson, and Thomas W. Swetnam. *Ecological*

Foundations for Fire Management in North American Forest and Shrubland Ecosystems, Portland: USDA Forest Service, 2009.

73. Rabetaliana, Hanta, Michel Randriambololona, and Peter Schachenmann. 'The Andringitra National Park in Madagascar.' *Unasylva* 50, no. 196 (1999), 25–30; Rakotoarimanana, Vonjison, and Michel Grouzis. 'Influence du Feu et du Pâturage sur la Richesse et la Diversité Floristiques d'une Savane à Heteropogon Contortus du Sud-Ouest de Madagascar.' *Candollea* 61, no. 1 (2006): 167–88.
74. The satellite data comes from NASA (earthdata.nasa.gov/data/near-real-time-data/firms), and is available via free subscription at the Conservation International website (firealerts.conservation.org/fas/home.do) (sites last accessed 25 Aug 2013).

Bibliography

Abe, Yoshio. 1977. Ancien Mode de Subsistence Pratiqué sur les Hautes Terres Centrales de Madagascar. *Asie du Sud-Est et Monde Insulindien* 8(3–4): 79–98.

Andriamampionona, A. 1992. Les Feux de Brousse dans la Région de l'Itasy. Antananarivo: FAO/PNUD/MAG/88.032.

anon. 1904. Les Feux De Brousse. *Bulletin Economique, Madagascar et Dépendences* 4, no. 1ere trimestre: 25–29.

Ballet, Jérôme, and Mahefasoa Randrianalijaona, eds. 2011. *Vulnérabilité, Insécurité Alimentaire et Environnement à Madagascar*. Paris: L'Harmattan.

Bergeret, Anne. 1993. Discours et Politiques Forestières Coloniales en Afrique et à Madagascar. *Revue Français d'Histoire d'Outre-Mer* 79(298): 23–47.

Bird, Douglas W., Rebecca Bliege Bird, and Christopher H. Parker. 2005. Aboriginal burning regimes and hunting strategies in Australia's Western Desert. *Human Ecology* 33(4): 443–464.

Bosser, J. 1954. Les Pâturages Naturels de Madagascar. *Mémoires de l'Institut Scientifique de Madagascar* Série B, V: 65–77.

Bowman, David M.J.S., Jennifer K. Balch, Paulo Artaxo, William J. Bond, Mark A. Cochrane, Carla M. D'Antonio, Ruth DeFries, Fay H. Johnston, Jon E. Keeley, Meg A. Krawchuk, Christian A. Kull, Michelle Mack, Max A. Moritz, Stephen J. Pyne, Christopher I. Roos, Andrew C. Scott, Navjot S. Sodhi, and Thomas W. Swetnam. 2011. The human dimension of fire regimes on earth. *Journal of Biogeography* 38: 2223–2236.

Brand, Jürg. 1998. Das Agro-Ökologische System am Ostabhang Madagaskars: Ressourcen- und Nutzungsdynamik unter Brandrodung. Dissertation, Universität Bern.

Brown, Mervyn. 1995. *A history of Madagascar*. London: Damien Tunnacliffe.

Burney, David A., Guy S. Robinson, and Lisa Pigott Burney. 2003. Sporormiella and the late Holocene extinctions in Madagascar. *PNAS* 100(19): 10800–10805.

Burney, David A., Lida Pigott Burney, Laurie R. Godfrey, William L. Jungers, Steven M. Goodman, Henry T. Wright, and A.J. Timothy Jull. 2004. A chronology for late prehistoric Madagascar. *Journal of Human Evolution* 47(1): 25–63.

Cori, Gilles, and Pierre Trama. 1979. *Types d'Élevage et de Vie Rurale à Madagascar*, Travaux et Documents de Géographie Tropical No. 37. Bordeaux: Centre d'Etudes de Géographie Tropicale, CNRS.

Coulaud, Daniel. 1973. *Les Zafimaniry: Un Groupe Ethnique De Madagascar À La Poursuite De La Forêt*. Antananarivo: Fanontam-Boky Malagasy.

Dewar, Robert E., and David Burney. 1994. Recent research in the paleoecology of the highlands of Madagascar and its implications for prehistory. *Taloha* 12: 79–88.

Dewar, Robert E., Chantal Radimilahy, Henry T. Wright, Zenobia Jacobs, Gwendolyn O. Kelly, and Francesco Berna. 2013. Stone tools and foraging in northern Madagascar challenge Holocene extinction models. *PNAS* 110(31): 12583–12588.

Dez, Jacques. 1966. Les Feux de Végétation: Aperçus Psycho-Sociologiques. *Bulletin de Madagascar* 247: 1211–1229.

Dez, Jacques. 1968. La Limitation des Feux de Végétation. *Terre Malgache* 4: 97–124.

Donque, Gérald. 1975. Les Cyclones Tropicaux des Mers Malgaches. *Madagascar Revue de Géographie* 27: 9–63.

Einrem, Johannes. 1912 (1936). *Gjennem Grønne Briller*. Vol. 1, Samlede Verker. Bergen: A. S. Lunde.

Eriksen, Christine. 2007. Why do they burn the 'bush'? Fire, rural livelihoods, and conservation in Zambia. *Geographical Journal* 173(3): 242–256.

Fanony, Fulgence. 1989. Un Modèle de Stratégie de Conservation de la Forêt À Madagascar. L'exemple d'Andrianampoinimerina. In *Environnement et Gestion des Ressources Naturelles dans la Zone Africaine de l'Océan Indien*, eds. Michel Maldague, Kabala Matuka, and Roland Albignac, 349–352. Paris: Unesco.

Fowler, Cynthia. 2013. *Ignition stories: Indigenous fire ecologies in the Indo-Australian Monsoon zone*. Durham: Carolina Academic Press.

Geoffroy, M. 1931. Les Feux de Brousse. *Bulletin Economique, Madagascar* 52, no. Mars 1931: 79–82.

Gilibert, J., P. Dubois, P. Granier, P. Grenier, and H. Serres. 1974. Lutte Contre les Feux Sauvages dans une Exploitation de Ranching. *Bulletin de Madagascar* 24(329): 465–480.

Gillson, Lindsey, and Katherine J. Willis. 2004. 'As earth's testimonies tell': Wilderness conservation in a changing world. *Ecology Letters* 7(10): 990–998.

Granier, P., and H. Serres. 1966. A Propos de L'incidence des Feux de Brousse sur l'Évolution des Pâturages. *Bulletin de Madagascar* 16(243): 789–791.

Heim, Roger. 1935. L'État Actuel des Dévastations Forestières à Madagascar. *Revue de Botanique Appliquée et d'Agriculture Tropicale* 15: 418–426.

Humbert, H. 1927. Principaux Aspects de la Végétation à Madagascar. La Destruction d'une Flore Insulaire par le Feu. *Mémoires de l'Académie Malgache* Fascicule V.

Humbert, Henri. 1949. La Dégradation des Sols à Madagascar. *Mémoires de l'Institut de Recherche Scientifique de Madagascar* D1(1): 33–52.

IUCN. 1972. *Comptes Rendus de la Conférence Internationale sur la Conservation de la Nature et de ses Ressources à Madagascar, Tananarive 7-11 Octobre, 1970*, Publications UICN Nouvelle Série, Document Supplémentaire No. 36. Morges, Switzerland: IUCN.

Jarosz, Lucy. 1996. Defining deforestation in Madagascar. In *Liberation ecologies*, eds. Richard Peet, and Michael Watts, 148–164. London: Routledge.

Kaufmann, Jeffrey C., and Sylvestre Tsirahamba. 2006. Forests and thorns: Conditions of change affecting Mahafale pastoralists in southwestern Madagascar. *Conservation and Society* 4(2): 231–261.

Keeley, Jon E., Gregory H. Aplet, N.L. Christensen, S.G. Conard, E.A. Johnson, P.N. Omi, D.L. Peterson, and Thomas W. Swetnam. 2009. *Ecological foundations for fire management in North American forest and shrubland ecosystems*. Portland: USDA Forest Service.

Klein, Jørgen, Betrand Réau, and Mary Edwards. 2008. Zebu landscapes: Conservation and cattle in Madagascar. In *Greening the Great Red Island: Madagascar in nature and culture*, ed. Jeffrey C. Kaufmann, 157–178. Pretoria: Africa Institute of South Africa.

Kull, Christian A. 1996. The evolution of conservation efforts in Madagascar. *International Environmental Affairs* 8(1): 50–86.

Kull, Christian A. 2004. *Isle of fire: The political ecology of landscape burning in Madagascar*. Chicago: University of Chicago Press.

Kull, Christian A. 2008. Saving land with a spade: Human landscape transformations in the Madagascar highlands. In *Greening the Great Red Island: Madagascar in nature and culture*, ed. Jeffrey C. Kaufmann, 113–135. Pretoria: Africa Institute of South Africa.

Kull, Christian A. 2014. The roots, persistence, and character of Madagascar's conservation boom. In *Conservation and environmental management in Madagascar*, ed. Ivan R. Scales. London: Earthscan.

Kull, Christian A., and Paul Laris. 2009. Fire ecology and fire politics in Mali and Madagascar. In *Tropical fire ecology: Climate change, land use, and ecosystem dynamics*, ed. Mark A. Cochrane, 171–226. Heidelberg: Springer-Praxis.

Laney, Rheyna M. 2002. Disaggregating induced intensification for land-change analysis: A case study from Madagascar. *Annals of the Association of American Geographers* 92(4): 702–726.

Laris, Paul. 2002. Burning the seasonal mosaic: Preventative burning strategies in the wooded Savanna of southern Mali. *Human Ecology* 30(2): 155–186.

Lavauden, L. 1934. Histoire de la Législation et de L'administration Forestière À Madagascar. *Revue des Eaux et Forêts* 72: 949–960.

Olson, Sherry. 1988. Environments as shock absorbers, examples from Madagascar. *Environmental Review* 12(4): 61–80.

Perrier de la Bâthie, Henri. 1921. La Végétation Malgache. *Annales du Musée Colonial de Marseille* Sér. 3, 9: 1–266.

Pollini, Jacques. 2011. The difficult reconciliation of conservation and development objectives: The case of the Malagasy environmental action plan. *Human Organization* 70(1): 74–87.

Pyne, Stephen J. 1997. *Vestal fire*. Seattle: University of Washington Press.

Pyne, Stephen J. 2001. *Fire: A brief history*. Seattle: University of Washington Press.

Pyne, Stephen J. 2009. The human geography of fire: A research agenda. *Progress in Human Geography* 33(4): 443–446.

Rabetaliana, Hanta, Michel Randriambololona, and Peter Schachenmann. 1999. The Andringitra National Park in Madagascar. *Unasylva* 50(196): 25–30.

Rakatonindrina, Raymond. 1989. Les grands problèmes actuels de la conservation des écosystèmes: défrichement et feux de brousse. In *Environnement et Gestion des Ressources Naturelles dans la Zone Africaine de l'Océan Indien*, eds. M. Maldague, K. Matuka, and R. Albignac, 111–122. Paris: Unesco.

Rakotoarimanana, Vonjison, and Michel Grouzis. 2006. Influence du Feu et du Pâturage sur la Richesse et la Diversité Floristiques d'une Savane à Heteropogon Contortus du Sud-Ouest de Madagascar. *Candollea* 61(1): 167–188.

Ranaivoarivelo, Nivo, and Pierre Milleville. 2001. Exploitation Pastorale des Savanes de la Région De Sakaraha (Sud-Ouest De Madagascar). In *Sociétés Paysannes, Transitions Agraires et Dynamiques Écologiques dans le Sud-Ouest de Madagascar*, eds. Samuel Razanaka, Michel Grouzis, Pierre Milleville, Bernard Moizo, and Christine Aubry, 181–197. Antananarivo: CNRE & IRD.

RDM. 1980. *Rapport Du Groupe Interministériel D'etudes Sur Les Feux De Brousse*. Antananarivo: Repoblika Demokratika Malagasy.

Scales, Ivan R. 2011. Farming at the forest frontier: Land use and landscape change in western Madagascar, 1896–2005. *Environment and History* 17: 499–524.

Sivaramakrishnan, K. 1996. The politics of fire and forest regeneration in Colonial Bengal. *Environment and History* 2(2): 145–194.

CHAPTER 7

Emperor Tự Đức's 'Bad Weather': Interpreting Natural Disasters in Vietnam, 1847–1883

Kathryn Dyt

Introduction

Natural disasters posed a serious threat to Vietnamese society during Tự Đức's 36-year reign from 1847 to 1883. In the nineteenth century, a string of bad weather in Vietnam including storms, floods, typhoons and droughts wreaked havoc on state infrastructure and human life, resulting in famines, epidemics and local uprisings. The idea that Tự Đức was responsible for these disasters has been perpetuated through popular folk song lyrics. Lines from a Nghệ Tĩnh folk song explicitly state, for example, that Tự Đức's ascension to the throne marked the beginning of crop failure and famine: 'Since Tự Đức's enthronement, no crop has been normal'.[1] Such blame is underpinned by the Confucian notion that environmental well-being is tied to the emperor. According to Confucian doctrine, a virtuous monarch who secures the 'Mandate of Heaven' produces good weather and bountiful crops, whereas a corrupt or illegitimate emperor loses his heavenly mandate resulting in bad weather and crop failure.

K. Dyt
Australian National University, Canberra, Australia

G. Bankoff, J. Christensen (eds.), *Natural Hazards and Peoples in the Indian Ocean World*, DOI 10.1057/978-1-349-94857-4_7

Lyrics from Nghệ Tĩnh folk songs have been uncritically recycled in historical accounts of Tự Đức's reign, perpetuating the view that the emperor was universally blamed for disasters and that he was unresponsive to the suffering of his people.[2] This view of Tự Đức has been shored up by accounts left by French missionaries during disaster events. Writing from his vicarage of West Tonkin during the great flood of 1857, Mgr Pierre Retord, for instance, chastised Tự Đức for 'leading a merry life in the depths of his palace, keeping his distance from the misery of the people'.[3] Drawing on both French and Vietnamese primary material, this chapter takes a critical look at the historical contexts within which these sources were produced and the ideological biases underpinning these texts. The Nghệ Tĩnh song texts that are frequently cited as evidence of Tự Đức's inaction in the face of disaster were collected in Nghệ An and Hà Tĩnh provinces (see Fig. 7.1) during the 1950s and 1960s when a 'new history' (sử học mới) was being forged in socialist North Vietnam.[4] As I will discuss, this historical agenda promoted a negative view of the Nguyễn dynasty (1802–1945) and circumscribed the types of materials that could be published. Furthermore, portrayals of Tự Đức within the letters of French missionaries published in the *Annales of the Propagation of the Faith* (*Annales de la Propagation de la Foi*) were shaped by a fervent Catholic world view. Priests' sensational descriptions of their experience of disasters and Tự Đức's response were bound up with their reading of calamity as a form of divine punishment and their fears and fantasies of the Orient.

In this chapter, I balance the picture of Tự Đức that emerges in much of the historiography against more positive appraisals of Tự Đức's disaster response found within Nguyễn court records. These sources suggest that Tự Đức was deeply concerned about disasters and the impact of calamity on his subjects, and they also describe a range of measures taken by the emperor to respond to disaster events. Like missionary writings and Nghệ Tĩnh folk materials, court histories were influenced by the context in which they were produced and shaped by the world view of Nguyễn intelligentsia. Moving outside of court histories, I also consider Tự Đức's reign and disaster response from the perspective of his personal poetry and edicts, which were important mediums of communication with his officials and subjects.

A detailed consideration of Tự Đức's response to disasters suggests that the cause of natural calamity was not only ascribed to the emperor's wrongdoing but also linked to a complex web of spiritual and social factors.

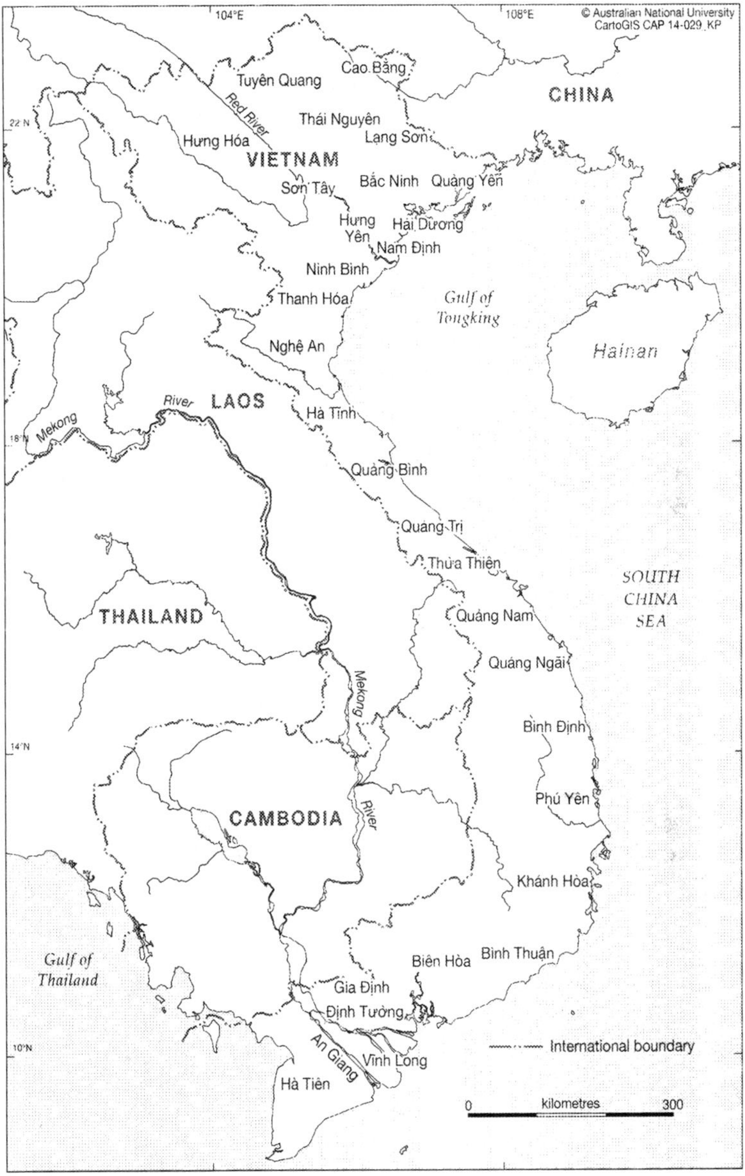

Fig. 7.1 The provinces of Vietnam during the early reign of Tự Đức (prior to the territorial changes made due to French conquest after 1862)

Moreover, disasters enabled political opportunism. As will be discussed in this chapter, calamities provided French missionaries with opportunities to proselytize and gain influence among their communities. In the Nguyễn court, disaster events led to Tự Đức inculpating disobedient officials and, conversely, calamities provided officials with the opportunity to reproach the emperor and overturn unpopular policies.

Disasters Under the Reign of Tự Đức

Descriptions of natural disasters during the reign of Tự Đức and court response to them are richly documented in the Nguyễn dynastic chronicles, the *Veritable Records* (*Đại Nam Thực Lục*). In compiling their records, officials in the Nguyễn History Bureau (Quốc Sử Quán) drew on communications between the court and the provinces.[5] The *Veritable Records* include valuable information such as the date, location and severity of disasters and document lengthy conversations between the emperor and his advisors on how best to respond to crises. Whereas the *Veritable Records* provide an account of disasters that took place throughout the country, descriptions of disasters by missionaries published in the *Annales of the Propagation of the Faith* are rooted in specific localities. Letters by foreign missionaries, who were on the ground when disasters struck, include blow-by-blow accounts of the unfolding of disasters and detailed descriptions of their impact on communities. In some cases, such letters provide a much fuller description of events that are only accorded a brief few lines in the *Veritable Records*, making them a useful source for a study of the impact of catastrophes on nineteenth-century Vietnamese society. Combining information contained in the *Veritable Records* and the *Annales*, it is possible to ascertain the incidence of certain natural disasters and the years during which they had the greatest impact (See Fig. 7.2).

As can be seen in Fig. 7.2, the most frequent and destructive bad weather of Tự Đức's reign was flood. Flooding was most common in the northern delta areas during the hottest months between July and September. The worst flooding during Tự Đức's reign occurred in the two years of 1856 and 1857 along the northern Vietnamese coastal region extending from Ninh Bình to Hải Dương. According to the *Veritable Records*, heavy downpours during these two years caused a number of dikes to break and water to cascade over rice fields built on

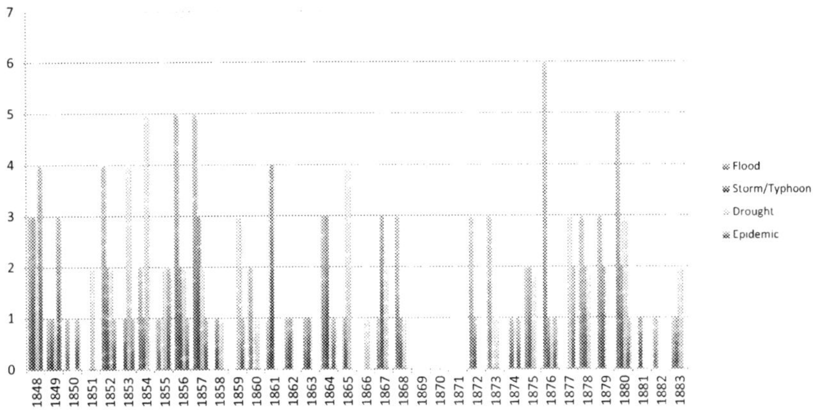

Fig. 7.2 Incidence of disaster during Tự Đức's Reign based on data compiled from the *Veritable Records* and the *Annales of the Propagation of the Faith*

the lower plains, submerging buildings, livestock and people.[6] Looking out from his Catholic parish in West Tonkin, Retord described the scene: 'in my garden, there was over six feet of water and in the fields, which are always lower than the fields in the villages, a ship of 300 tons could have navigated it like the open sea'.[7] Worst of all, the consecutive floods over two years completely destroyed the region's food supply, bringing about a terrible famine.

Flood events often occurred in tandem with storms and typhoons. In early October 1881, one of the most deadly typhoons on record, which is estimated to have reached wind speeds of up to 185 kilometres per hour, entered Vietnam just north of the Red River.[8] After devastating the coastline of Hải Phong, the typhoon moved across the delta regions of Nam Định, Hưng Yên and Hà Nội, then hurtled to the northern upper valley areas of Bắc Ninh and Thái Nguyên, decimating entire villages in its wake. Fr Paul-François Puginier, who was in the north-western province of Nam Định at the time, described the devastation wrought by the intense winds: 'homes ... collapsed one after the other and the debris was carried a great distance. Numerous trees which, hitherto, had faced all storms were uprooted or their trunks were broken down the middle'.[9] Horrific flash floods ensued, followed by tsunami-like waves of up to four metres high, destroying houses and boats and drowning numerous inhabitants.[10]

Puginer wrote wistfully: 'Woe to the villages that formed an obstacle to the passage of the waves, they were submerged and their houses were undermined'.[11] According to the *Veritable Records*, thousands of people were left homeless and communication was interrupted for almost a week.[12] Court records do not include an official death toll for the 1881 typhoon, but latter-day analysts have placed the death toll as being in the region of 30,000–300,000 people.[13]

Droughts were more common in the south and central plains, which tend to experience larger fluctuations in rainfall during the year, but they also hit northern Vietnam with devastating results. In the seventh lunar month of 1854, for example, the northern region experienced a 'great drought' (đại hạn): rice crops withered under the relentless heat, hiking up rice prices throughout the region and causing famine.[14]

Disease and epidemics were also rife during Tự Đức's reign. Tự Đức's investiture in September 1849 was marred by the onset of a deadly cholera epidemic that continued to rage in the northern provinces until well into 1853.[15] By the first lunar month of 1850, the *Veritable Records* state that 589,460 people countrywide had become victim to the epidemic.[16] The disease spread so quickly that by the end of 1849 coffins supplied by the court were already in short supply and people were forced to inter their dead directly into the earth or wash them away in rivers.[17] Writing at the end of 1849 from Huế, Mgr François Pellerin described how dead bodies had 'polluted and cluttered the waters'.[18] The *Veritable Records* describe how the disease brought large sections of the country to a complete standstill: exams were cancelled at schools in northern and central provinces, all public works were suspended, markets were closed down and even an elephant battle was called off.[19] When cholera eventually subsided, typhoid fever followed directly on its heels, persisting well into 1853.[20]

Natural disasters had a devastating impact on communities and infrastructure. Protracted epidemics such as cholera in the mid-nineteenth century decimated entire populations. Floods and typhoons arrived as sudden, catastrophic events, but the after-effects often lingered for months and even years in the form of crop failure, famine and social instability. Official histories document that the court attempted to curb such disastrous consequences though a combination of measures such as tax reduction ordinances, financial loans and rice distributions. Historical assessments of court relief efforts, however, have largely relied on descriptions left behind by the pens of French missionaries as well as commentary contained in Nghệ Tĩnh folk songs. Taking a closer look at these sources, in the next section

I consider the ways in which these sources have informed understandings of disaster management and shaped views of Tự Đức's reign more broadly.

'The Miseries of the People During the Reign of Tự Đức': The Historiography of Disaster Relief

In April 1858, Mgr Pierre Retord wrote a long letter to a member of his vicariate outlining the horrors of the previous year's flood that devastated large sections of the northern countryside. The inhabitants of the two villages surrounding Retord's Catholic parish in Western Tonkin had all but drowned, save 13 survivors. In a horrific description of the scene, Retord explains: 'everyday you can see a number of corpses floating on the river that runs through our village. We even retrieved two bodies that ended up hanging in my garden bamboo'.[21] Retord offers a grim description of the famine that resulted from the destruction of the region's food supply. 'Everywhere,' Retord writes, 'we hear about people who have starved to death – fifty in this village, a hundred in another. Dead are found everywhere, in pagodas, markets and on the roadside ... On the roads and fields, people we meet are nothing but skin and bones, they are like ghosts from another world'.[22]

Penned during the worst flood of Tự Đức's reign, Retord's negative evaluation of the imperial relief effort has informed a number of latter-day historical critiques of the court's relief effort. Writing in the midst of surrounding chaos and devastation, Retord harshly condemns the court relief effort as slow, disorganized and grossly inadequate. He explains:

> [Rice distributions] were made too early ... and now that the famine has reached its height, the granaries are completely empty. Moreover, distributions were so crowded that numerous people were crushed in their attempts to receive the most meagre rations and nine-tenths of them returned without received anything at all, despite a long wait ... Several mandarins have constructed shelters to feed the sick andhelpless, but the number of admissions is restricted to between six and ten of the leanest and most emaciated people in the village. And what dreadful places these shelters are! They are open to the elements, men and women are crammed in pell-mell, almost naked, on damp ground, breathing putrid air and devoured by vermin... Many unfortunates prefer to starve in their villages rather than seek admission in such gloomy shelters![23]

In his account, Retord is particularly critical of mandarins for exploiting the starving masses by hiking up rice prices, accepting bribes and for pocketing money from the court that was meant for labourers. According to Retord, mandarins also forced households with rice reserves to sell it, which resulted in further food loss: 'some of the people who still had rice buried it deep in the earth so that it would be out of sight from mandarins and much of this rice was spoiled and lost. Others who gave it to neighbouring houses to look after had it stolen ... rich people gave the mandarins large sums of money to fend off their searches and later they sold [their rice] for even greater profit'.[24]

In addition to blaming court mandarins for mismanaging the crisis, Retord criticizes Tự Đức for being aloof and impractical. In response to the circulation of a court edict on the 24 October 1856 in which Tự Đức outlined a series of measures for people to follow in the wake of a big flood, Retord scornfully writes: 'ideas are not enough to satisfy the hunger of the poor'.[25] Echoing Retord's criticisms of the detached, luxurious life enjoyed by the emperor within the confines of his palace during calamities, in 1849 Mgr Pellerin condemned Tự Đức's absence from public life during the cholera epidemic: 'since the beginning of the epidemic, the king has shut himself away in his palace... and has not granted an audience to anyone'.[26]

Criticism of the 'detached' emperor in missionary accounts has been echoed in historical assessments of Tự Đức's reign. As the last emperor to rule over an independent Vietnam, Tự Đức was disparaged by Marxist historians who were dominant prior to the government's Renovation policy (đổi mới) was introduced in 1986 as a tyrannical despot who 'lost the country' (mất nước) to the French.[27] While post-1986 revisionist histories have been more sympathetic to Tự Đức and the Nguyễn dynasty more generally, scholars both inside and outside Vietnam have continued to portray Tự Đức as a bookish, insular monarch, who was largely out of touch with his people, hopelessly applying outdated Chinese models to stem the tide of foreign threat.[28] In his article, 'Vietnamese Politics and Confucianism in the Nineteenth-Century', for example, Tsuboi Yoshiharu explains, 'Tự Đức lived in the heart of the palace, quite separate from the real world and without any knowledge of the institutions and the history of his own country'. Tsuboi also boldly contends that Tự Đức was a 'big head who could not see beyond the tip of his nose'.[29]

The majority of the historiography surrounding disaster relief has perpetuated a disparaging view of Tự Đức and his court based on readings of Nghệ Tĩnh folk songs and missionary accounts.[30] Retord's damning account of the imperial relief effort during the 1857 flood continues to be quoted by historians as evidence of the emperor's negligence in the face of disaster. For example, in his article 'Floods, Famine and Poverty among Northern Peasants in 1857', published in 2000 Nguyễn Văn Kiệm fills most of his paper with a Vietnamese translation of Retord's account before finally condemning the 'irresponsible' court and officials who did not 'take the opportunity to make things better for peasants'.[31] Bùi Quang Tùng's 1967 article 'The Succession of Thieu-Tri' also makes extensive use of evidence from Retord's account, marrying this with lyrics from Nghệ Tĩnh folk songs. Bùi Quang Tùng emphatically concludes, 'through a number of testimonies of Vietnamese and French writers of the time, we have stories that give us an accurate picture of the suffering of the people and the inefficiency of the royal administration in the face of appalling devastation'.[32]

The extent to which Nghệ Tĩnh folk songs and missionary accounts have influenced historical understandings of the court's disaster relief efforts and Tự Đức's reign warrants taking a closer look at these sources. The giặm folk song texts that are frequently quoted as evidence of the people's malcontent about the state's response to natural disasters were published in North Vietnam in 1963 when anti-Nguyễn sentiment was at its zenith. As outlined by Patricia Pelley, the 'new history' (lịch sử mới) being forged in socialist North Vietnam at this time sought to steer historical emphasis away from the elite ruling classes and endeavoured to 'retrieve peasants from the margins of history'.[33] Song texts from Nghệ An and Hà Tĩnh provinces in the 1950s were collected during this revolutionary drive to amass a new body of source material which would recover the authentic, uncorrupted village voice, and provide a counter-narrative to elite versions of history recorded in court chronicles. The historian Nguyễn Đổng Chi, who together with his colleague Ninh Viết Giao gathered and published two volumes of Nghệ Tĩnh folk song lyrics, was a key figure in this collection project. Nguyễn Đổng Chi edited the five-volumed tome *The Treasure of Vietnamese Folktales*, published between 1957 and 1974, which includes folktales, proverbs and poems that are presented as being 'representative' of true Vietnamese culture. As explained by Pelley, although Nguyễn Đổng Chi never defined what 'representative' meant, the term translated into the promotion of 'countless folktales celebrat[ing] the cunning of villagers and their ability to thwart the designs of officials, who were typically depicted as embodiments of arrogance'.[34] In line with the anti-Nguyễn stance of new socialist history, the collection of Nghệ Tĩnh folk songs published by Nguyễn Đổng Chi and Ninh Viết Giao conveniently voice the populace's discontent with Tự Đức. The editors of the volume preface the song text, 'The miseries of the people under the reign of Tự Đức' with a short introduction that draws attention to the verses that condemn Tự Đức's court most harshly. They write:

> [The song] aims directly at the emperor and mandarins who turned a blind eye to the people's immense suffering and it suggests that they were not worthy of protecting the people and the nation. [The song includes phrases like] 'The emperor and his mandarins do not reach [the people]' [and] 'The emperor does not think about the people'.[35]

The extent to which these folk song texts present an accurate picture of people's feelings from Nghệ Tĩnh in the mid-nineteenth century is questionable, not only because of the political agenda which spurred the production of the text but also because of the oral transmission of these folk songs over the course of a century. Further, while Nguyễn Đổng Chi and Ninh Viết Giao's two volumes of Nghệ Tĩnh folk songs overwhelmingly include lyrics that are critical of corrupt Nguyễn officials, a more positive appraisal of the court's relief efforts can be found in the last few verses of the song text, 'Hey Tự Đức!':

> Hardships brought about by unfavourable weather
>
> There is nothing anyone can do.
>
> Everywhere we go,
>
> We see dead trees and withered leaves.
>
> Our clever and resourceful emperor,
>
> His treasury is full of money,
>
> And he is using it
>
> To give aid and relief to his people.
>
> One person after another,
>
> Everyone will receive aid.
>
> So let us hurry![36]

Lyrics like this one, which describe a 'clever and resourceful' Tự Đức distributing aid amongst the people, have not previously been considered within the historiography and furnish us with a new perspective on the emperor's efforts to respond to disasters. Rather than blaming the emperor for calamities, the poem portrays bad weather as a random natural occurrence. While it is not possible to conclude from such evidence that relief efforts were effective, it does bring to light the hazards of relying solely on Nghệ Tĩnh folk songs as evidence of the Vietnamese experience of relief efforts and the allocation of blame in connection with disaster events.

Tự Đức's lack of concern for his people during disasters is also challenged by evidence within Nguyễn dynastic chronicles. While the effects of flood and other natural disasters were devastating and the measures taken were inadequate to alleviate their effects, numerous passages in the *Veritable Records* describe Tự Đức's efforts to expedite the court's response to disaster and to stamp out corruption amongst his officials. Tự Đức's reflections on relief efforts during a drought in the fourth lunar month of 1854 suggest that he was acutely aware of the problems of misconduct and profiteering among officials:

> Day in day out [officials] are chipping away at the people and lining their pockets ... These evil officials just sit there and callously watch the suffering of the people, they cause loses to state revenue and act like the rulers from the Qin dynasty did when they watched the Viet people suffer without showing any concern.[37]

Tự Đức sought to address problems by issuing warnings that he would 'look closely into corrupt practices' and harshly punish offending parties.[38] He also tried to facilitate a more rapid response to crisis. In 1860, three years after the devastating flood of 1857, for example, Tự Đức drafted new laws for disaster response, stipulating the precise level of aid to be allocated per household for a range of disaster events.[39] Reflecting on aid relief efforts at the time, Tự Đức bemoaned: 'the people need aid, but local officials have been slow in responding: they wait to make a report to the court before giving aid to the people. The situation is an emergency! It is like when people are drowning, you need to pull them out immediately!'[40] Tự Đức ordered the Ministry of Finance (Bộ Hộ) to distribute a copy of emergency procedures to local areas and he implored officials to 'give aid at the same time as submitting a report in order to save time'.[41] In addition to implementing new laws for disaster relief, in 1863 Tự Đức issued laws that awarded honorific titles to wealthy households who offered donations during crises.[42] Tự Đức also established a system of rewards for local officials who were able to prevent famine during natural disasters through drawing on the contents of their own granaries.[43]

The implementation of new laws for disaster relief as well as the numerous edicts issued for tax relief ordinances and rice handouts during Tự Đức's reign suggest that effort was made by the court to assist people suffering the effects of disaster. While Tự Đức's criticism of officials has some points of convergence with Retord's damning remarks about the inadequacies of disaster relief efforts, the *Veritable Records* indicate that far from being detached and unconcerned with the plight of his people, Tự Đức was extremely vexed about the unfolding of disasters. In the following section, I discuss how missionary portrayals of Tự Đức and the relief efforts were bound up with their feelings of helplessness and their Catholic faith before elaborating further on Tự Đức's responses to bad weather.

Biblical Interpretations of Disasters in Missionary Letters

Many foreign missionaries living in nineteenth-century Vietnam interpreted natural calamities as divine retribution for their repressive treatment under an anti-Catholic regime. Soon after Tự Đức's ascension to the throne in 1848, the Nguyễn court moved swiftly to stamp out heterodox creeds and put a stop to all mission activity in the country. The court issued a decree that included in its twelfth clause severe penalties for missionaries found to be spreading the Catholic faith. The decree stipulated that foreign missionaries be immediately executed following their arrest and interrogation, and that local priests and converts were to be branded with tattoos, which would alienate them from non-Christian communities if they refused to recant at their trial.[44] Anti-Catholic edicts continued to be issued by the Nguyễn court throughout the late 1840s and 1850s, outlining punishments of fines, torture and death for those who were complicit in the spread of the alien faith.[45] During the Franco-Spanish invasion between 1858 and 1862, persecution of Catholics reached a violent apogee as numerous vicariates were destroyed and assets were confiscated, leaving many missions penniless.[46]

In correspondence sent back home, missionaries related natural disasters to political events. When the cholera epidemic struck in 1849, Mgr Pellerin was quick to explain the plague as God's retribution for Tự Đức's persecution of Catholics: 'there is no doubt that this is a punishment exercised by God on a nation that has declared war on Him, and perhaps a punishment for the edict of last year in particular'.[47] Six years later in 1855, when persecution of Catholics had again intensified, Pellerin attributed 'all the ills which plague the country' to Tự Đức's violent policies and refusal to accept Christianity as the true faith:

> For a long time, the harvest has not been sufficient and this year it has been very bad. Although the harvest managed to withstand the drought, locusts devoured it and rice has never been so expensive. The people are starving, bands of thieves add to the horrors of famine ... This poor kingdom is crumbling under the weight of its transgressions. Christianity would relieve these ruins and yet it rejects the hand that could save it.[48]

In her analysis of the *Annales of the Propagation of the Faith*, Nola Cooke suggests that most foreign priests represented their life in Vietnam as an 'apostolic existence in which Divine Providence intervened visibly and repeatedly on their behalf'.[49] Indeed, the arrival of a continuous stream of calamities heightened the sense missionaries had of living in biblical times or at the cusp of the apocalypse. The stubborn Egyptian Pharaoh of the Old Testament who would not release the Israelite people despite the stream of plagues and calamities sent by God found a fitting analogy in the figure of Tự Đức. When anti-Catholic fervour was at his height and a series of calamities, including 'an army of caterpillars', typhoon, flood, drought and cholera, followed one after another, the missionary Sohier made this link explicit:

> The Lord has inflicted terrible punishment on this unhappy country to punish them for so many crimes, but our *Annamite Pharaohs* [italics mine], rather than open their eyes and look within, do nothing but provoke more anger from heaven.[50]

In another letter in December the same year, Sohier reiterates the connection between the obstinate Pharaoh of the Bible and Tự Đức and his court mandarins: 'our new Pharaoh has not become any wiser, but continues to make war on God and his worshippers'.[51]

Like biblical accounts of the Israelites who were protected from the angel of death sent to kill every first born in the kingdom, missionaries believed that they and their Christian flocks were divinely protected from the natural disasters that befell the kingdom.[52] Writing in 1849, at the time of the devastating cholera epidemic, Pellerin explains:

> Everyone agrees that the Christians are being protected. I do not know exactly the number of our dead, but it is much less, proportionally, than the pagans. Once the epidemic was declared, I wrote a circular ordering public prayers to Mary. These pious exercises have been carried out everywhere. As a result the illness diminished and it even disappeared completely in some places as we started the prayers.[53]

Two years later when cholera was still raging in the north, Retord similarly observed: 'the idolaters die in greater numbers, either because they are more commonly attacked, or because they abandon their patients immediately without giving them any help'.[54] In addition to proclaiming divine protection, Retord offers a more sinister reason for proportionally higher death tolls among non-Christian populations. A number of missionary letters describe how non-Christian populations abandoned or buried members of their families alive during natural disasters.[55] Retord's specific reference to the demonic practice of 'burying children alive' paints an image of depraved and wild people in need of proper moral guidance from Christianity.[56]

Missionaries not only believed that they were divinely protected during disasters, they also perceived that disasters were Heaven-sent for the purpose of their liberation. Writing in 1852, the missionary Galy calls the epidemic of 1849 a 'powerful liberator', as it enabled him to go out in the community without fear of being hunted down.[57] Wistfully describing his busy activities during the plague 'flying from one house to another' in order to administer the Last Rights, Galy regretfully writes: 'Cholera had given me a kind of freedom, its disappearance returned me to my solitary life'.[58] The second part of the editorial piece, 'Notes on the Mission in Cochinchina and Tonkin', similarly describes how the cholera epidemic enabled priests to freely attend to the sick and dying, while the emperor remained 'deep within his walled palace, shut off from the suffering outside, and hardly daring to breathe the tainted air exhaled from his royal city':[59]

> They [i.e. the priests] are free; they are masters by the grace of cholera. See them at work. Day and night, at the bedside of the pagans abandoned by their brothers, collecting abandoned orphans, turning their homes into hospitals to console the pain that is without remedy, preparing the ground for corpses that clutter the place ... When the danger was past, the persecutors who had fled reappeared and it was the Christians turn to hide.[60]

The juxtaposition between an inert and isolated Tự Đức 'deep within his walled palace' and the charity of missionaries on the ground served an important purpose. Not only did it reinforce notions of the evil nature of Tự Đức and his court against the virtue of those belonging to the 'true' faith, it also provided fodder for proselytizing efforts. Missionaries, it seems, purposefully drew attention to Tự Đức's absence from public life as a strategy to win over converts. One piece describing missionary activities during the cholera epidemic suggests that the rhetoric of abandonment was widely propagated: 'the dismayed people were without help. Only one class of people did not abandon them: the Christians'.[61] Missionary letters regularly detail the charity that they administered to their communities during calamities, as well as the miracles that resulted from their prayers to the Christian God. Retord describes the large number of conversions to Christianity during the plague of 1850: 'what convinced them of the truth of the religion was having the plague among them'.[62]

The fact that Tự Đức usually remained within the confines of his palace during disasters thus provided opportunities for French missionaries to foster loyalty among the local people. But while missionaries equated Tự Đức's absence with a lack of concern, Vietnamese court sources tell a different story. In the next section, I draw on the perspectives provided in official histories to reflect on Tự Đức's isolation and consider the range of measures taken by the court to fight calamity.

'Gazing up at Clouds': Tự Đức's Response to Disasters

The *Veritable Records* indicate that Tự Đức did indeed confine himself to his palace during disasters. In stark contrast to the image painted by missionaries, however, in the *Veritable Records* Tự Đức's isolation appears as a self-imposed penance to restore heavenly favour and a necessary condition for the performance of an array of rituals, fasting and emotional pleading. For example, while French priests ridiculed Tự Đức for being inert and 'shut off from the suffering outside' during the great cholera epidemic of the mid-nineteenth century, court records report that Tự Đức was involved in 'vegetarian fasting, praying and reading scriptures so that heaven would be moved'.[63] Tự Đức's strict personal austerity during disaster events—which included, among other things, rationing the number of courses he ate at each meal and refusing to listen to music—are well documented within court histories.[64]

Tự Đức's emotional outcries and concern for the suffering of his subjects are recorded in detail within the *Veritable Records*. These are often linked to somatic symptoms such as crying, sweating or an inability to eat or sleep. For example, during the cholera epidemic of 1849 Tự Đức exclaimed, 'I feel great sympathy, and I worry late into the night and cannot sleep' and later he added 'I thought about the people affected by the epidemic and I could not rest'.[65] When floods hit the northern regions in the seventh lunar month of 1856, Tự Đức beseeched his officials to 'accomplish all works that I have ordered. I humbly request this so that I can eat and sleep normally'.[66] In the aftermath of the deadly typhoon of 1881 Tự Đức expressed that he was painfully aware of, and engaged with, his people's suffering: 'How could I sit by and watch and not feel sympathy?'[67]

The reporting of Tự Đức's suffering and concern for his subjects in the *Veritable Records* should be critically evaluated. Officials in the court History Bureau (Quốc Sử Quán) were schooled in Confucian thought and this ideological orientation undoubtedly influenced the way in which events were reported. The emperor's empathetic outpourings and emotional pleas to heaven would likely have been perceived by court intelligentsia as a fitting response from the emperor as the Son of Heaven who was entrusted with the responsibility for the welfare of his subjects and the proper functioning of the natural world during crises. In line with their views of effective governance, they may well have prioritized passages relating the emperor's sincere attempts to restore clement weather for inclusion in the records. Yet the *Veritable Records* also documents tense conversations between court officials as well as blunt criticism of the emperor himself, contesting the view that the historian's sole agenda was to project a seamless image of the court through time. As noted by Alexander Woodside, the honest and fulsome recording of events pertaining to court life was considered a hallmark of responsible governance. He writes: 'communications ... were faithfully recorded. The emperor was supposed to reveal his worth not be suppressing them but by publishing them, answering them, acting upon them and thus transcending them'.[68]

While court historians no doubt wanted to portray the emperor's responses to disasters in a positive light, it seems unlikely that Tự Đức's expressions of sympathy and concern for his people were purely fabrications dreamt up by officials. Moving outside official court histories, Tự

Đức own poems provide further evidence that the emperor was deeply tormented by bad weather. Indeed, a vast number of the poems that Tự Đức wrote during his reign relate to his concern for his subjects during disasters. In his poem 'Gazing up at the Clouds', for instance, Tự Đức divulged that he had been anxious about the elements for many years. In the poem, he also conveys his fear of famine and his perpetual sadness in thinking about his subjects:

> If there has long been rain, then I desire sunshine;
>
> If there has long been sunshine, then I desire rain.
>
> I've had thirteen years of worrying,
>
> And half of them are about clear skies and cloudy days.
>
> ...
>
> Peasants are the foundation of the state;
>
> I hope their troubles are not great.
>
> Yet now we have experienced ten days of summer heat,
>
> A light shower is unable to provide a nourishing soaking.
>
> Now my only fear is of recurring famine;
>
> Awake and asleep I am beset by despondency.
>
> In the evening I gaze upon the clouds as they wander;
>
> In the morning I gaze upon the clouds as they arise from the hills.
>
> When the clouds disperse I gaze ever more anxiously;
>
> When the clouds gather I am beside myself with joy.[69]

Tự Đức's poetry was not just a medium for private insular self-reflection; it also enabled the emperor to project his emotions beyond himself. Evidence suggests that Tự Đức used poems to communicate his worry and empathy with his officials and subjects. For example, during an ongoing drought in the capital and surrounding districts in early 1853, Tự Đức wrote two poems 'expressing his deep feelings and wish for rain'.[70] He ordered that the poems be engraved and distributed to all his literary and military mandarins 'from the lowest to the highest level' so that 'everyone had a copy to read'.[71]

Edicts were another means through which Tự Đức transmitted his feelings and concern for his people during crises. In an 'Edict in Favour of the Christians', reportedly distributed by Tự Đức in 1864 and translated into French in the thirty-seventh volume of the *Annales*, Tự Đức discusses a series of natural disasters including a cholera epidemic and he appeals to his subjects to understand and 'believe' the feelings in his heart:

> I fear that you are not well informed of my feelings, which is why I want to tell you what I have in my heart. Maybe you will believe me … I have only one thought: the happiness of my people.[72]

Such passages call into question the view that Tự Đức was detached, both physically and emotionally, from the people during disasters. Rather, they highlight Tự Đức's attempts to widely convey his concern for his people's suffering.

While Tự Đức's personal strictures and emotional pleading underscore the link between the reigning monarch and the weather, the causes of disaster were not only linked to the correct behaviour of the emperor. Rather, Tự Đức sought to elicit the collaboration of a pantheon of spirits to help mitigate disaster. At the first sign of disharmony in the weather Tự Đức usually sent out one of his officials to conduct a 'calling for wind and rain' ritual, known as cầu đảo, at a local temple.[73] For example, in November 1849, when the death tolls from cholera in the provinces of Vĩnh Long and Quảng Bình reached nearly 67,000, Tự Đức sent out an entourage of three of his most trusted men to appeal to the protective tutelary deities of the capital of Huế for assistance.[74] During disasters, court officials petitioned a range of spirits drawn from Daoist belief systems, animism and Buddhism. Spirits who had connection to the elements were particularly valued for their ability to restore environmental harmony. If spirits did not comply with their requests through prayer and ritual offerings, sometimes officials resorted to coercing spirits to restore clement weather through beating and exposure.[75] The numerous ways in which the court called on the spirits during disasters suggest that they were regarded as important agents of environmental change.

Alongside prayers and rituals, Tự Đức ordered that judicial processes be expedited and he also circulated injunctions for people to cease frivolity and restrict their movement.[76] During the flood of 1856, for example, Tự Đức issued the following plea:

> I ask all offices that verdicts do not take too long, that prisoners held on minor charges be released and that the palace stop all unnecessary work ... Movement should be made carefully, people should not go out and seek amusement, and all hunting should be stopped. With these measures, we should be able to completely suppress the calamities and bring good omens in all things.[77]

The practice of freeing prisoners and speeding up procedures for those awaiting trial has its roots in the Chinese idea that detaining the innocent created too much 'yang' in the atmosphere, thus impeding the circulation of 'yin' which created rain.[78] In a nineteenth-century Vietnamese context, it seems that the practice of releasing prisoners became disconnected from a singular association with rain bringing. Rather, releasing prisoners, like the suspension of work, leisure and hunting, became part of a general strategy aimed at restoring heavenly harmony across a wide range of disaster events.

As the cholera epidemic lingered and a wave of other natural calamities swept through the kingdom, Tự Đức also considered the role played by officials in his administration in causing disaster. The implication of Tự Đức's officials in environmental well-being is illustrated in an entry included in the *Veritable Records* in the eighth lunar month of 1849. Here, in addition to questioning his own moral integrity as a causative factor, Tự Đức also reflects on the morality of the people and his officials:

> Lately the weather has not been harmonious and the people have contracted many infections ... Inside my palace I have cried out to heaven with all my heart, I have prayed that things will rapidly become wholesome ... I do not know what crimes the people have committed or what error I have committed to bring about such a calamity, but I am unable to bring back the favourable harmony of heaven. What do I need to do to achieve this? Or is it the case that my high-ranking officials are not fully loyal and reverent, and so the harmony of heaven cannot be restored?[79]

Tự Đức's accusations directed at his administrative officials became increasingly loud and forceful as the disasters piled on top of one another. In the midst of ongoing epidemic, drought and crop failure stretching from the central province of Quảng Nam to the southern province of Hà Tiên in 1854, Tự Đức explained to his officials that they were responsible for good governance:

> I need my officials to be my eyes, ears and hands to assist me. That is the reason why I appoint my officials to govern the provinces and districts and to oversee each local area ... I am concerned about affairs in the border areas because it is difficult to know what is going on there and that is why I send envoys out to inspect ... Your job is not a small matter ... so you had better find a way not to disappoint me. This task weighs heavily on you![80]

As the passage continues, the tone becomes increasingly accusatory. Tự Đức probes, 'could it be that my officials are not doing their job and have been robbing the people blind?'.[81] In another instance, Tự Đức directed criticism at his officials for withholding crucial information from him. Following a violent storm in the provinces of Bình Thuận and Khánh Hòa and a spell of 'cold rain' around the capital in the second lunar month of 1852, Tự Đức summoned a meeting with his high-ranking mandarins and complained:

> You only know how to remain silent and think that this is good. You do not report anything to me about the life of the people and the affairs of the state. There is no way that... I can bring back the favour of heaven without your help.[82]

In an effort to refract some of Tự Đức's criticism and the responsibility for environmental calamity, officials started doing some finger pointing of their own. In the midst of a long-standing drought in the sixth lunar month of 1857, a group of Tự Đức's trusted court officials submitted a report to the throne in which they inculpated officials in outlying southern and northern provinces:

> For months there has been very little rain around the capital and many provinces have been experiencing disease and epidemic. The people on the ground, are they good people or not? I have heard rumours that the chief of Định Tường province, Nguyễn Tường Vĩnh, and the treasurer of Bình Định province, Nguyễn Hữu Độ have been embezzling state money. The treasurer and provincial judge of Khánh Hòa province, Nguyễn Quýnh and Lê Cán, have accused each other of accepting large bribes ... Respectfully order that two righteous mandarins go out and examine the north and south.[83]

This passage echoes the anxiety about officials in border regions evoked earlier by Tự Đức. The recently conquered southern plains and the outlying northern provinces, which were a hotbed for frustrated literati and scholars loyal to the previous Lê dynasty, had long been viewed with mistrust by the Huế administration.[84] It is possible to view the indictment of officials located in these areas not only as strategic—that is, as a means of circumventing imperial blame—but also as opportunistic. In other words, environmental calamity enabled questioning and scrutiny of mistrusted officials on the periphery.

The political opportunism enabled by bad weather has been touched upon by Mark Elvin. In his discussion of 'moral meteorology' in late imperial China, he contends that the Emperor Yongzheng (1723–1736) employed discourses of official responsibility for the weather as 'a weapon of psychological terror'.[85] Elvin does not consider, however, the way in which mandarins may have also manipulated weather ideology and capitalized on disaster events as opportunities to speak out against the emperor and mend problems within the administration. A passage included in the *Veritable Records* in 1849 includes a strikingly candid remark about high-ranking officials 'using the epidemic as a pretext' (lấy cớ là dich lệ) to make five requests to the throne. Their list of requests included a plea that the descendants of Mỹ Đường, who was banished from the royal court following charges of incest with his mother, be reinstated as members of the royal family with the relevant stipend.[86] They also requested a pardon for the crimes of Lê Văn Duyệt, who had been posthumously tried and found guilty of treason by Emperor Ming Mạng (r. 1820–1841) for defending Catholic missionaries.[87] In the second lunar month of 1852, the director of the Imperial Censorate, Hoàng Thu, responded to Tự Đức's accusations that mandarins had been withholding information from the throne with some damning criticism of his own:

> When you came to the throne, you tended to the needs of the people and wished to bring back harmony. Now you go off to gallop on horses in the back garden and shoot birds at Tĩnh Tâm lake. You only play and you do not nourish a comfortable climate... Recently, Mai Anh Tuấn and Phạm Phú Thứ came to report some matters, but they were punished... We must open the way for people to speak. Tôn Thất Đắc and Nguyễn Văn Nhị were put in the pillory for committing small mistakes. We must forgive trifling errors such as these, so that the emperor and people can be shown to trust each other.[88]

In response, Tự Đức criticized Hoàng Thu for his 'vain, empty words' but nonetheless, ordered that an edict be circulated to encourage all grievances to be brought to forward to the throne for consideration. Such exchanges highlight how disasters provided officials with opportunities to openly air grievances, to converse frankly with the emperor to the point of criticism and to seek resolutions to problems within the court administration.

Conclusion

The view that Tự Đức was ineffective during disasters has largely arisen from a selective reading of historical evidence contained in Nghệ Tĩnh folk song lyrics and in the letters of French missionaries. The perception of Tự Đức presented in these sources has reinforced historiographical representations of Tự Đức as an incompetent and insular monarch who was remote from his people. Through considering the prejudices and political agendas underpinning French and Nghệ Tĩnh materials, and bringing to light new evidence from Vietnamese court materials and Tự Đức's own poetry, this chapter has taken a fresh look at court responses to disaster in the mid-to-late nineteenth century.

In different ways, both Nghệ Tĩnh folk songs and the letters of French missionaries ascribed responsibility for natural disasters to Tự Đức and blamed him for not doing enough during crises. Missionaries interpreted natural disaster in biblical terms as a form of divine retribution for the court's persecution of Catholics and scorned Tự Đức for shutting himself in his palace when disaster struck. Nghệ Tĩnh folk song lyrics connect natural calamities to Tự Đức's moral shortcomings and charge that he did 'not think of the people' and their suffering. In this chapter, I have suggested that these sources tell us more about the political composition of the era and the cultural contexts within which they were produced, than they do about Tự Đức's management of crises. French missionary depictions of disasters and the court relief effort are imbued with a fervent Catholic world view and their own terrifying experience of persecution. French missionaries emphasized Tự Đức's confinement to his palace during crises in order to foster loyalties and establish their own connections with local people. The disdain for Tự Đức expressed in giặm folk songs from Nghệ Tĩnh, on the other hand, were well suited to North Vietnam's socialist agenda to construct a 'new history' that disparaged the feudalism of the Nguyễn dynasty. Even though Nghệ Tĩnh folk song lyrics are typically critical of relief efforts, an overlooked song, 'Hey Tự Đức!', contains a more positive appraisal, calling into question the universal nature of discontent.

A close reading of court records relating to disaster complicates a simplistic understanding of Tự Đức as aloof and out of touch. Missionary sources interpret the emperor's absence as indifference, whereas Vietnamese chronicles present Tự Đức's propensity to conduct routines of self-reflection and moral cultivation within his palace as an appropriate response aimed at winning back heavenly favour. While Tự Đức's practical relief efforts were clearly inadequate to effectively mitigate all of the effects of the numerous large-scale disasters during his reign, court records show that he nonetheless tried to implement a range of measures to fight disasters, including ritual responses and the distribution of aid. The poems that Tự Đức wrote and distributed during his reign make explicit his deep concern about bad weather and his empathy for people's suffering.

The strict Confucian view that the emperor was responsible for disasters downplays the role of human and cosmological collaboration in the maintenance of environmental harmony. In this chapter, however, I have considered how the causes for disaster were not only linked to the correct behaviour of the emperor. Rather, a range of factors, including discontented spirits and corrupt officials, were linked to balancing heavenly harmony and natural calamities. Vietnamese chronicles document how disasters provided opportunities for the emperor to point fingers at disobedient officials and, in turn, for officials to reproach the emperor. Consideration of the court's responses to 'bad weather' therefore provides insights into the mechanisms of Nguyễn governance and enables a re-evaluation of Tự Đức's reign.

Notes

1. 'Providence in Times of Trouble' [Cơ Trời Tao Loạn] in Nguyễn Đống Chi and Ninh Viết Giao. *Hát Giặm Nghệ Tĩnh, Vol. 2.* (Hà Nội: Khoa học Xã Hội, 1963, pp. 24-26. Another folk song in the same volume entitled 'Tự Đức's Reign' [Là Cái Đời Tự Đức] concludes with the lines, 'Using my brush, I note these few verses to leave to posterity the memories of the miseries of the reign of Tự Đức' (Ibid., pp. 20–24). All translations from Vietnamese and French are my own unless stated otherwise. I would like to extend my thanks to Phạm Thu Thúy for her assistance with Vietnamese translation.

2. Nguyễn Thế Anh. *Kinh Tế và Xã Hội Việt Nam dưới các Vua Triều Nguyễn* [*The Economy and Society of Vietnam under Nguyễn Kings*].Saigon: Lửa Thiêng 1971, pp. 136–137; Phạm Cao Dương. *Vietnamese Peasants under French Domination, 1861–1945*. Berkeley: Center for South and Southeast Asia Studies 1985, p. 4; Bùi Quang Tùng. 'La Succession de Thieu-Tri', *Bulletin de la Société des Études Indochinoises*, 42, 1967, pp. 23–106. David Marr also cites Nghệ Tĩnh folk lyrics broader analysis of Tự Đức's reign, which includes discussion of the intellectual and cultural roots of Vietnamese nationalism in the mid-nineteenth century. See David Marr. *Vietnamese Anticolonialism: 1885–1925* (Berkeley: University of California 1980, p. 25.
3. *Annales de la Propagation de la Foi* (hereafter *APF*), vol., 31 (1859), p. 28.
4. The giặm folk songs spread over both Hà Tĩnh and Nghệ An provinces, and I follow the common usage of referring to them as 'Nghệ Tĩnh' folk songs here.
5. The *Veritable Records* quote extensively from the *Châu Bản* (*Vermillion Records*), a collection of reports from local areas that were viewed by the emperor.
6. *Đại Nam Thực Lục* [*Veritable Records of the Great South*](*hereafter ĐNTL*), vol. 7. Hà Nội: Nhà Xuất Bản Giáo Dục, 2007, p. 458.
7. *APF*, vol. 30 (1858), p. 210.
8. David Longshore. *Encyclopaedia of Hurricanes, Typhoons, and Cyclones.* New York: Facts on File 1998, p. 95.
9. *APF*, vol. 54 (1882), p. 157.
10. Ibid.
11. Ibid.
12. *ĐNTL*, vol. 8, pp. 489–490. Puginier attributes this delay in communication to the fact that people in the region were consumed with the task of finding shelter, see *APF*, vol. 54, p. 159.
13. Longshore (1998), p. 95.
14. *ĐNTL*, vol. 7, pp. 325, 327 and 328.
15. There are numerous references to cholera's impact in the northern provinces over a number of years. See, *ĐNTL*, vol.7, pp. 244, 287.
16. *ĐNTL*, vol. 7, p. 155.
17. *ĐNTL*, vol. 7 p. 134; *APF*, 22 (1850), p. 394.
18. *APF*, 22 (1850), p. 394.
19. An edict was sent out to cancel all exams at schools in Thừa Thiên (Huế), Nghệ An, Thanh Hóa and Hà Nội, *ĐNTL*, vol. 7, p. 134. See also *ĐNTL*, vol. 7, p. 140.
20. Writing in August 1853, Berger reflects on these sequential epidemics: 'Cholera in 1851 took away nearly ten thousand neophytes and typhoid

fever has thinned down the ranks last year and this year again', *APF*, 27 (1855), p. 23.

21. *APF*, vol., 30 (1858), p. 211.
22. *APF*, vol. 31 (1859), p. 27.
23. Ibid., pp. 31–32.
24. Ibid., p. 32.
25. Ibid., p. 30.
26. *APF*, vol. 22 (1850), p. 394. A similar observation is made by Galy three years later: 'to escape the deadly influence of the epidemic, the king rendered himself inaccessible in his palace', *APF*, vol. 25 (1853), pp. 24–25.
27. Trần Văn Giàu's study of 'The development of Vietnamese thought from the nineteenth century to the August Revolution' is exemplary here. In particular, Trần Văn Giàu disparages Tự Đức for his strict and inflexible adherence to Confucian doctrine that weakened the country against foreign threat. Trần Văn Giàu. *Sự phát triển của tư tưởng ở Việt Nam từ thế kỷ XIX đến Cách Mạng Tháng Tám* [*The Development of Thought in Vietnam from the Nineteenth Century to the August Revolution*]. Hà Nội: Khoa Học Xã Hội, 1973.
28. Bruce Lockart discusses the continuous reworkings of the Nguyễn dynasty, including more recent positive portrayals in his article 'Re-Assessing the Nguyễn Dynasty', *Crossroads* 15(1) (2001), 9–53. For discussions of Tự Đức's personality see Tsuboi, Yoshiharu. *Politique et Confucianisme Vietnamien au 19 siècle: Le cas de L'Empereur Tự Đức (1847–1883)*. Tokyo: Jochi Daigaku Ajia Bunka Kenkyu, 1988, Hồ Bạch Thảo 'Tìm hiểu về con người vua Tự Đức' [Understanding the person Emperor Tự Đức], *Tập Chí nghiên cứu và phát triển*, 5(64), (2007), Jamieson, Neil. *Understanding Vietnam*. Berkeley: University of California Press 1993, pp. 45–46; Chapuis *The Last Emperors of Vietnam: From Tự Đức to Bao Dai*. Westport, CT: Greenwood Publishing Group 2000, p. 11.
29. Tsuboi (1988) p. 11.
30. It should be noted that Đỗ Bang has provided a more positive appraisal of state relief efforts drawing on information contained in the *Veritable Records* in his article, 'Tình Hình Bão Lụt ở Thanh Hóa Hồi Thế Kỷ' [The Situation of Storms and Floods in Thanh Hóa during the Nineteenth Century] *Nghiên Cứu Lịch Sử* [*Historical Research*], 4(323), (2000), pp. 79–82.
31. Nguyễn Văn Kiệm. 'Nạn, Lụt và Tình Trạng Khốn Cùng của Nông Dân Bắc Kỳ Trong Năm 1857 [Floods, Famine and the Poverty of Northern Peasants in 1857]', *Nghiên Cứu Lịch Sử* [*Journal of Historical Research*], vol. 5, no. 312, (2000), p. 85.
32. Bùi Quang Tùng (1967), p. 69.

33. P. Pelley, Patricia. *Postcolonial Vietnam: New Histories of the National Past* (Durham and London: Duke University Press 2000, p. 132.
34. Ibid., p. 136.
35. Nguyễn Đổng Chi (1962), p. 26.
36. Nguyễn Đổng Chi and Ninh Viết Giao. *Hát Giặm Nghệ Tĩnh*, vol. 1 Hà Nội: Khoa Học Xã Hội, 1963, pp. 46–47.
37. *ĐNTL*, vol. 7, p. 309. The passage makes reference to the Chinese Qin dynasty (221 to 206 BC), which expanded its borders south to present-day Vietnam.
38. See for example, *ĐNTL*, vol. 7, p. 310.
39. According to these laws, aid was to be applied across all 'disasters' (tai ương) including fires, storms, flood and epidemic. According to these new regulations, aid was to be allocated per household (rather than per person as in previous years) and depended on whether the degree of damage was 'major' (hạng nặng) or 'minor' (hạng nhẹ). In the case of major damage, households were to be given two quan and one phương of rice, whereas those suffering from minor damage were to be given half this amount. *Khâm Định Đại Nam Hội Điển Sự Lệ Tục Biên* [*Official Compendium of Institutions and Usages of Imperial Vietnam*], Huế: Nhà Xuất Bản Giáo Dục, 2004, vol. 3, p. 183.
40. *ĐNTL*, vol. 7, pp. 660–661.
41. Ibid.
42. Van Nguyễn-Marshall. *In Search of Moral Authority: The Discourse on Poverty, Poor Relief, and Charity in French Colonial Vietnam.* New York: Peter Lang, 2008, p. 25. Nguyễn-Marshall explains that a donation of 2000 quan was to be rewarded with a promotion to the ninth rank, whereas a donation of 900 quan qualified for 14 years' exemption from paying taxes and military duty.
43. Van Nguyễn-Marshall (2008), p. 23, Nguyễn Thế Anh (1971), p. 141. Village granaries that were supervised and maintained by a well-regarded community figure, proliferated under the reign of Tự Đức, and the court rewarded areas that were able to prevent famine through drawing on the contents of its own storehouse. Following the flood of 1856, the province chief of Hà Tĩnh, Lâm Duy Thiếp, sent a report to the throne noting that 7000 phương of rice had been distributed from the village granary, and requested a reward for his efforts (*ĐNTL*, vol. 7, p. 460).
44. Ramsay, Jacob. *Mandarins and Martyrs: The Church and the Nguyễn Dynasty in Early Nineteenth-Century Vietnam.* Stanford: Stanford University Press, 2008, pp. 68–91.
45. Ibid., p. 138.

46. Cooke, Nola. 'Early Nineteenth-Century Vietnamese Catholics and Others in the Pages of the Annales De La Propagation De La Foi', *Journal of Southeast Asian Studies*, 35, no. 2, 2004, p. 274.
47. *APF*, vol. 22 (1850), p. 394.
48. *APF*, vol. 28 (1856), p. 122.
49. Cooke (2004), p. 267.
50. *APF*, vol. 32 (1860), pp. 347–348.
51. *APF*, vol. 34 (1862), p. 48.
52. In her article, Nola Cooke (2004) also notes that deaths were also believed to be fewer among Catholic populations during the cholera epidemic of 1820, p. 276.
53. *APF*, vol. 22 (1850), p. 393.
54. *APF*, vol. 23 (1851), p. 286.
55. In 1849, Pellerin wrote: 'Men still alive have been thrown out of houses, then hastily buried or brought to the water. There are a few who have returned and have been able to get up, and are still full of life', APF, vol. 22, (1850), p. 394. Writing a year later, Retord says of the non-Christian Vietnamese population: 'they bury even more before they breathe their last', *APF*, vol. 23 (1851), p. 286. Retord offers a similar observation during the flood of 1857. He wrote: 'There are villagers who, unable to feed their children, bury them alive', *APF*, vol. 37 (1859), p. 27.
56. Ibid.
57. *APF*, vol. 25 (1853) p. 24.
58. Ibid., pp. 29–30.
59. *APF*, vol. 27 (1855), p. 347.
60. Ibid.
61. Ibid., p. 347.
62. *APF*, vol. 23, pp. 288–289.
63. *ĐNTL*, vol. 7, p. 153.
64. See, for example, DNTL, vol. 7, pp. 139–140; pp. 153–157.
65. *ĐNTL*, vol. 7, p. 139, p.149.
66. *ĐNTL*, vol. 7, p. 458.
67. *ĐNTL*, vol. 8, p. 490.
68. Woodside, Alexander. *Vietnam and the Chinese Model: A Comparative Study of Nguyen and Ch'ing Civil Government in the First Half of the Nineteenth Century*. Cambridge: Harvard University Press, 1971, p. 324.
69. *Tự Đức Cơ Dư Tự Tỉnh Thi Tạp* (嗣德機餘自省詩集), Han-Nom Institute Manuscript No: A.1541. I am grateful to Dr Nathan Woolley and Dr Li Tana at the Australian National University for helping me translate this poem.
70. *ĐNTL*, vol. 7, p. 266.
71. Ibid. p. 267.

72. *APF*, vol. 37 (1865), pp. 331–332.
73. For more information about 'cầu đảo' rituals, see Dyt, Kathryn. 'Calling for Wind and Rain Rituals: Environment, Emotion and Governance in Nguyễn Vietnam, 1802–1883', *Journal of Vietnamese Studies*, vol. 10, no. 2, 2015, pp.1–42.
74. *ĐNTL*, vol. 7, p. 148.
75. To find out more about the types of spirits petitioned, see Dyt (2015) pp. 12–16. The coercion of spirits through beating and exposure is discussed in pp. 19–26.
76. *ĐNTL*, vol. 7 (1847–1873), p. 138, pp. 459–460.
77. Ibid., p. 458.
78. Elvin, Mark. 'Who Was Responsible for the Weather? Moral Meteorology in Late Imperial China', *Orsis* 13, 1998, p. 214. Elvin notes the practices of freeing prisoners in the later Han dynasty (25–220) as a strategy to increase the 'yin principle' and hence bring rain in times of drought.
79. Ibid., p. 138.
80. *ĐNTL*, vol. 7, p. 308.
81. Ibid.
82. *ĐNTL*, vol. 7, p. 240.
83. Ibid., pp. 506–507.
84. See Woodside (1971), p. 133; Nola Cooke, Nola. (1995) 'The Composition of the Nineteenth-Century Political Elite of Pre-colonial Vietnam (1802–1883)', *Modern Asian Studies*, 29, 4, 1995, p. 764.
85. Elvin (1998), p. 226.
86. Mỹ Đường was the eldest son of Prince Cảnh, the son of Emperor Gia Long (r. 1802–1820).
87. For more on Lê Văn Duyệt see Taylor, Keith. *A History of the Vietnamese*. Cambridge: Cambridge University Press, 2013, pp. 418–425.
88. ĐNTL, vol. 7, pp. 240–241.

Bibliography

Annales De La Propagation de la Foi [*Annales of the Propagation of the Faith*].

Bùi Quang Tùng. 1967. La Succession de Thieu-Tri. *Bulletin de la Société des Études Indochinoises* 42: 23–106.

Chapuis, Oscar. 2000. *The last emperors of Vietnam: From Tu Duc to Bao Dai*. Westport: Greenwood Publishing Group.

Cooke, Nola. 1995. The composition of the nineteenth-century political elite of pre-colonial Nguyễn Vietnam (1802–1883). *Modern Asian Studies* 29(4): 741–764.

Cooke, Nola. 2004. Early nineteenth-century Vietnamese Catholics and others in the pages of the Annales De La Propagation De La Foi. *Journal of Southeast Asian Studies* 35(2): 261–285.

Đại Nam Thực Lục [*Veritable Records of the Great South*]. 2007. Hà Nội: Nhà Xuất Bản Giáo Dục.

Đỗ Bang. 2002. Tình Hình Bão Lụt ở Thanh Hóa Hồi Thế Kỷ [The situation of storms and floods in Thanh Hóa during the nineteenth century]. *Nghiên Cứu Lịch Sử* [*Historical Research*], 4(323): 79–82.

Dyt, Kathryn. 2015. 'Calling for wind and rain' rituals: Environment, emotion and governance in Nguyễn Vietnam, 1802–1883. *Journal of Vietnamese Studies* 10(2): 1–42.

Elvin, Mark. 1998. Who was responsible for the Weather? Moral meteorology in late imperial China. *Orsis* 13: 213–237.

Hồ Bạch Thảo. 2007. Tìm Hiểu về Con Người Vua Tự Đức [Understanding Tự Đức: The emperor and the man]. *Tập Chí Nghiên Cứu và Phát Triển* 5(64): 37–52.

Jamieson, Neil. 1993. *Understanding Vietnam*. Berkeley: University of California Press.

Khâm Định Đại Nam Hội Điên Sự Lê Tục Biên [Official compendium of institutions and usages of imperial Vietnam].2004. Huế: Nhà Xuất Bản Giáo Dục. vol. 3.

Lockart, Bruce. 2001. Re-assessing the Nguyễn Dynasty. *Crossroads* 15(1): 9–53.

Longshore, David. 1998. *Encyclopedia of hurricanes, typhoons, and cyclones*. New York: Facts on File.

Marr, David. 1980. *Vietnamese anticolonialism: 1885–1925*. Berkeley: University of California.

Nguyễn Đổng Chi, and Ninh Viết Giao. 1962. *Hát Giặm Nghệ Tĩnh, Tập 2*. Hà Nội: Khoa Học Xã Hội.

Nguyễn Đổng Chi, and Ninh Viết Giao. 1963. *Hát Giặm Nghệ Tĩnh, Tập 1*. Hà Nội: Khoa Học Xã Hội.

Nguyễn-Marshall, Van. 2008. *In search of moral authority: The discourse on poverty, poor relief, and charity in French colonial Vietnam*. New York: Peter Lang.

Nguyễn Thế Anh. 1971. *Kinh Tế và Xã Hội Việt Nam Dưới Các Vua Triều Nguyễn* [*The economy and society of Vietnam under the Nguyễn Kings*]. Saigon: Lửa Thiêng.

Nguyễn Thế Anh. 1985. The Vietnamese monarchy under French colonial rule 1884–1945. *Modern Asian Studies* 19(1): 147–162.

Nguyễn Văn Kiệm. 2000. Nạn, Lụt và Trình Trạng Khốn Cùng của Nông Dân Bắc Kỳ Trong Năm 1857 [Floods, famine and the poverty of northern peasants in 1857]. *Nghiên Cứu Lịch Sử [Journal of Historical Research]* 5(312): 85–90.

Pelley, Patricia M. 2002. *Postcolonial Vietnam: New histories of the national past*. Durham and London: Duke University Press.

Phạm Cao Dương. 1985. *Vietnamese peasants under French domination, 1861–1945*. Berkeley: Center for South and Southeast Asia Studies.

Phan Huy Le. 1993. *The traditional village in Vietnam*. Hanoi: The Gioi.

Ramsay, Jacob. 2008. *Mandarins and martyrs: The Church and the Nguyễn Dynasty in early nineteenth-century Vietnam*. Stanford: Stanford University Press.

Smith, Ralph B. 1974. Politics and society in Vietnam during the early Nguyễn period (1802–62). *Journal of the Royal Asiatic Society*, 2: 153-169.

Snyder-Reinke, Jeffrey. 2009. *Dry spells: State rainmaking and local governance in late imperial China*. Cambridge: Harvard University Press.

Taylor, Keith W. 2013. *A history of the Vietnamese*. Cambridge: Cambridge University Press.

Trần Văn Giàu. 1973. *Sự Phát Triển của Tư Tưởng ở Việt Nam từ Thế Kỷ XIX đến Cách Mạng Tháng Tám* [The development of thought in Vietnam from the nineteenth century to the August Revolution].Hà Nội: Khoa Học Xã Hội.

Tsuboi, Yoshiharu. 1988. *Politique et Confucianisme Vietnamien au 19 siècle: Le cas de L'Empereur Tự Đức (1847–1883)*. Tokyo: Jochi Daigaku Ajia Bunka Kenkyu.

Woodside, Alexander. 1971. *Vietnam and the Chinese model: A comparative study of Nguyễn and Ch'ing civil government in the first half of the nineteenth century*. Cambridge: Harvard University Press.

CHAPTER 8

Storm over San Isidro: Repeated 'disasters' and Civic Community Culture in the Nineteenth-Century Philippines

Greg Bankoff

'Thanks to the measures taken and the cooperation of all the people concerned, I have the pleasure to inform you that as of this moment there has not been any report of serious injury to persons in the *cabecera* [seat of local government] despite it being the point where the floodwaters were deepest.'[1] So wrote the Spanish civil governor of Nueva Ecija, Joaquin Reyes, to his superiors in Manila, the capital of the colonial Philippines, about the typhoon and flood that struck the town of San Isidro on the night of 19–20 September 1887. By Filipino standards, the event was not particularly noteworthy, not like the great storm of 1882, though it did cause considerable flooding in at least three provinces of northern Luzon. Rather its historical value lies in the measures taken prior to, during and after that night. This small provincial community, defined here in terms of both municipal governance and local capacity, responded rather

This chapter is adapted from an article first published as 'Storm over San Isidro: "Civic Community" and Disaster Risk Reduction in the Nineteenth Century Philippines,' *Journal of Historical Sociology* 25, 3 (2012): 331–51.

G. Bankoff
Department of History, University of Hull, Hull, UK

G. Bankoff, J. Christensen (eds.), *Natural Hazards and Peoples in the Indian Ocean World*, DOI 10.1057/978-1-349-94857-4_8

well to preserve life and protect property as indicated by the governor's report. The community and its leaders seemed to know what was needed to reduce townspeople's vulnerability and enhance local resilience in such situations. It was also widely recognised that the capacity for action was vested in the community's own resources. Using a detailed archival account of this typhoon-induced flood, this chapter examines how hazard was dealt with in one late nineteenth-century provincial town in the Philippines. Andrew Maskrey warns us that it is easy to romanticise 'the virtues of traditional techniques and methods' and to confuse contemporary community-based mitigation practices with the all too often inadequate, small, makeshift, self-help measures of the past.[2] The flood of 1887 and the account of the actions taken by the community in San Isidro challenge any such romanticised notions of the past. It also hints at the origins of the vibrant civil society that is a characteristic of contemporary Filipino society.

Repeated 'disasters'

Sociologists and psychologists have long been intrigued by the idea that people who are frequently exposed to hazard adapt their behaviour to accommodate risk and that these patterns can become embedded to some extent in cultures over time. Interest, however, has not been with the one-in-one hundred year 'big one' as with the smaller scale, more frequent hazards that can be regularly anticipated. The constant recurrence of such events, they argue, can give rise to alternative normative structures that replace the routine social norms with ones more in accord to a time of crisis: That is, the very frequency of hazard can engender cultural adaptation. The degree of cultural adaptation can vary enormously from specific forms of risk handling to counter a repeated threat, to a geographically bounded subcultural accommodation to a single risk, to a more wholesale societal adjustment to one or more hazards. None of these categories are mutually exclusive in that risk handling may form an aspect of both subcultural and societal adaptations, and one or more subculture may be embedded within the larger society.[3]

Harry Moore is credited with first coining the term 'disaster subculture' to describe 'those adjustments, actual and potential, social, psychological and physical, which are used by residents of such areas in their efforts to cope with disasters which have struck or which tradition indicates may strike in the future.'[4] He was writing about how people who lived in coastal communities behaved as Hurricane Carla hit Texas in September 1961. Moore concluded by explaining people's actions as pertaining to a 'hurricane subculture.' A few years later, William Anderson similarly identified a 'flood disaster subculture' in Cincinnati to designate a social–physical environment of repeated inundations along the Ohio River Valley.[5] Moore never went on to fully formulate his ideas; this was left to a research paper by Dennis Wenger and Jack Weller written nearly a decade later. Based on research carried out in a number of communities across the USA regularly affected by flooding and hurricanes, they sought to uncover the existence of these disaster subcultures through indices that showed 'the perpetuation of successful patterns of adaptation to the disaster context through socialization.'[6] Rather than viewing disaster subcultures as special arrangements apart from the general culture of the community that are resorted to in times of need—more in the way Moore argued, they viewed the subcultural traits that evolved out of disaster experiences as being fully integrated into the dominant non-disaster culture. Moreover, these traits constituted a residue of learning that is both passed down from one generation to the next and are transmissible to new community members.[7] Drawing on their study of these Midwestern North American communities and the variations in the extent to which they possessed distinctive cultural characteristics, Wenger and Weller were able to identify three factors that appear to be crucial to the emergence of a disaster subculture within a community: that a hazard has to be repetitive, of such a nature as to allow a period of forewarning, and that its consequences inflict significant damage to human and material resources.[8]

In some places, moreover, the disaster subculture may not be confined to only a single geographical area but more generally permeate the entire society. The peoples of the Philippines share a distinctive and common cultural range of adaptation to disasters.[9] The threat of what would otherwise be regarded as regional or local subcultures is so frequent and widespread in these instances that they have come to constitute part of the wider culture. The sheer variety and magnitude of typhoons, floods and flood-related threats in the archipelago, as so graphically illustrated recently by Typhoon Haiyan (Yolanda) have influenced social and material culture on a truly archipelagic scale. The repeated nature of typhoons and floods exposes people to risk and also motivates them to develop risk-handling strategies drawn on their past experiences to make themselves more resilient in the future.[10] Typhoons and subsequent flooding satisfy the three factors identified by Wenger and Weller as crucial to the emergence of a disaster subculture within a community: they occur at frequent if irregular intervals; people living in all but the southernmost part of the archipelago know they are at risk; and the consequences of such events inflict significant losses in terms of mortality and material damage.

Typhoons are by no means the only natural hazard to afflict the Philippines but they are, perhaps, the most regularly destructive.[11] Each year about 20 severe typhoons, equivalent to a quarter of all such events in the world cross over or near the islands.[12] Tropical cyclones can occur at any time but are more frequent between July and November.[13] Some provinces are also more vulnerable than others, with northern Luzon receiving the highest number. There are two types of typhoons: the severe (aka destructive) typhoons that are 'one of the greatest natural calamities that may occur in any place,' and the lower intensity but much more frequent ones that are still potentially life threatening but are responsible for much of the rain that makes the climate of the islands so conducive to agriculture.[14] Colloquially referred to as *bagyos*, there are a surprising number of historical sources on typhoons testifying to what happened to people in the past and how communities fared in such events.[15] Local histories indicate that in some parts of Luzon people might have expected to live through a major or destructive typhoon every four to five years during the eighteenth and nineteenth centuries, storms that were 'so violent that objects weighing more than 100 kilos were easily carried away by the blasts of the storm.'[16] Each year, however, most of the inhabitants of the islands experienced high winds that frequently destroyed their light nipa palm and bamboo constructed houses, flattened their crops, stripped the trees of leaves and even their bark.[17]

Along with the typhoons came heavy rain falling over very short periods of time. While tropical cyclones make a significant contribution to total precipitation in the islands, they are often also responsible for the frequency of flooding.[18] A 'furious typhoon' struck the province of Capiz on 19 November 1835 and was 'accompanied by a dense and cold rain that lasted from daybreak till seven o'clock at night without diminishing in intensity.'[19] Floods have historically been the source of much privation and suffering in the Philippines and were more to be feared than the high winds. 'Almost always during typhoons,' an official noted in 1854, 'the floods are more terrible and destructive than the winds of the storm.'[20] Again, local accounts testify to the frequency of such hazards. A list drawn up from sources found in the Archive of the Manila Observatory constitutes a record of major floods that occurred between 1691 and 1900.[21] While almost certainly incomplete, it does provide an indication of the primary causes, geographical predisposition and even the frequency of inundations in specific areas. Local histories, too, regularly refer to floods in association with the passage of typhoons, citing them as responsible for over 50 % of recorded incidences.[22]

Wind and rain, storm and water, typhoon and flood were what communities expected. Document after document repeats the same litany of destruction bemoaning both the frequency of the event and the scale of the damages wrought. 'These phenomena are so unfortunately frequent in these latitudes,' writes the provincial governor of Camarines Sur in October 1882, 'some occurring so unexpectedly that it is almost impossible to take the precautionary measures needed at those places where it might matter most.'[23] Streams turned into raging torrents in moments and rivers formed walls of water that swept all before them submerging roads and sweeping away wooden bridges, isolating communities, disrupting commerce and trade and transforming the lowlands into an aqueous world where boats were the only sure means of transport.

The principal flood-prone areas in the archipelago and those hardest hit were the ancient channels of river systems filled with Quaternary alluvial deposits. Many of these sites had been made more vulnerable to flood through the already extensive deforestation of surrounding slopes in the more populous provinces of Luzon. Perhaps as much as 50 % of the forest cover in those areas of the Philippines under Spanish control had disappeared by the end of the nineteenth century, diminishing the absorption level of soils and increasing the magnitude of floodwaters.[24] An 1877 report on forest clearance in Tayabas warned that the general loss of forest cover along river banks and creeks had 'deprived the towns and surrounding areas of an efficient defence against floods.' It further forbade the cutting of any more timber along riverbanks or in any other place where trees 'exercise a beneficial influence on the hydrological system and serve to protect against floods.'[25] However, the growing demand for wood to satisfy both commercial and private purposes and the straightening of rivers to facilitate the transport of logs largely rendered such ordinances ineffective and the flow of floodwaters only increased.[26] These lowland areas were also among the flattest, most fertile and easiest to irrigate landscapes and so were the richest centres of agriculture and the site of the most intensive human settlement.

Storm over San Isidro

The typhoon that struck central Luzon on 19–20 September 1887 was by no means exceptional even if it caused considerable flooding in Nueva Ecija and in the neighbouring provinces of Nueva Vizcaya and Pampanga (See Fig. 8.1).[27] It was not considered sufficiently noteworthy to warrant a mention in either Fr. Miguel Selga's catalogue of such events or the lists complied by the Manila Observatory.[28] Yet the governor of Nueva Vizcaya still reported a great quantity of water that inundated the whole province 'from North to South,' sweeping away all the bridges and roads, carrying off animals and tree trunks and putting the telegraph and mail out of action and leaving the province 'in complete isolation' from the rest of the country.[29] His counterpart in Pampanga noted how what had been but 'yesterday fertile fields and roads' were 'today converted into one immense lake.'[30] However, the historical value of this particular typhoon lies not so much in the scale of damages it inflicted as in the detailed record of its passage.

The document consists of a 23-page report written by Joaquin Reyes, the Civil Governor of the province to the Director-general of Civil Administration. Governance in the Spanish Philippines left much to the discretion of its provincial and municipal representatives. A governor-general in Manila presided over an administrative structure composed of a royal court (*real audiencia*) at its apex that acted as both the highest court of appeal and as an advisory council. Provinces were administered by a combination of civil and military governors known respectively as *alcaldes mayores* and *corregidores*, the former designated by rank according to the importance of the appointment. All these officials were Spaniards, either born in the Philippines (*Filipinos*) or posted there from Spain (*Peninsulares*). Local government was administered by an extensive indigenous bureaucracy of town mayors (*gobernadorcillos*), village headmen (*cabezas de barangay*), police chiefs (*alguaciles mayores*) and a host of minor officials.[31] As Civil Governor of Nueva Ecija, Reyes was charged with providing his superiors with an account of his activities and of the measures he took prior to, during and after the typhoon and ensuing flood of September 1887. His account furnishes a unique record of how the repeated nature of storms gave rise to a form of civic community culture in the islands.

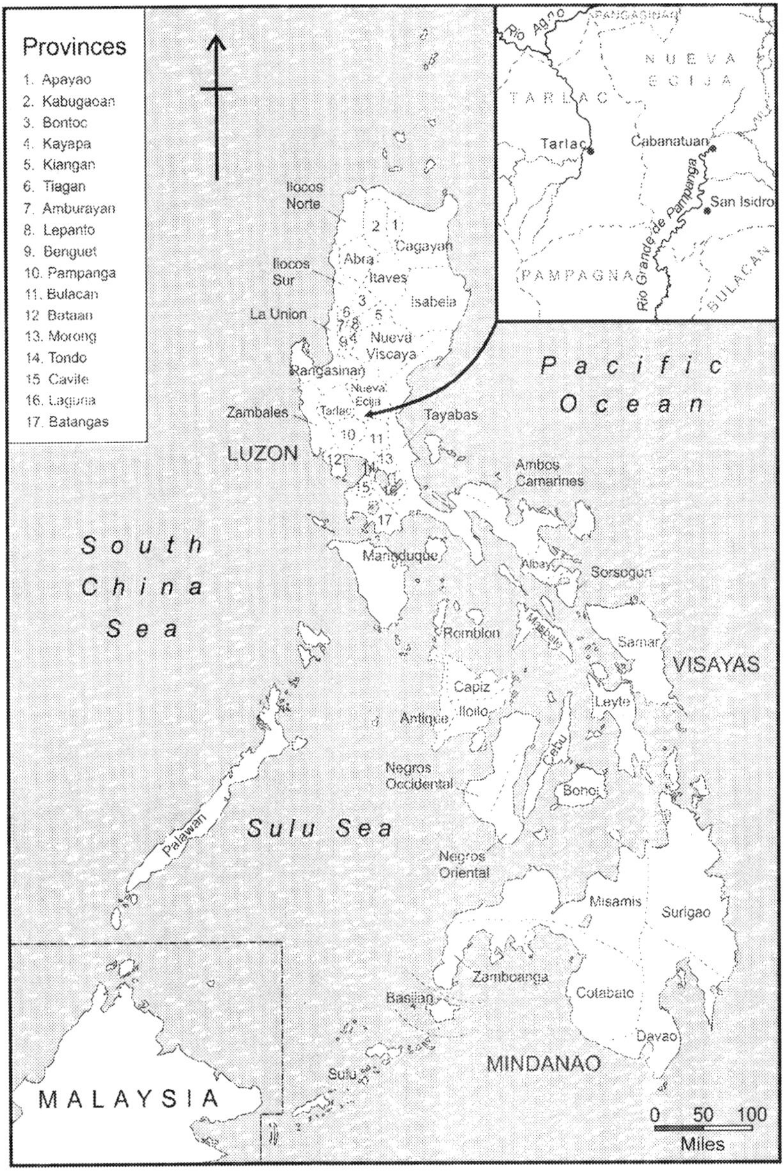

Fig. 8.1 The Philippines showing a detail of San Isidro and surrounding areas c.1887

Civic Community

A civic community is one rich in social capital. Social capital refers to those 'norms and networks that enable people to act collectively.'[32] It both 'binds' people together through self-help as well as provides a 'bridge' to help others get ahead. The more social capital a community 'possesses,' so the argument goes, the more its inhabitants are civically engaged. Alexis De Tocqueville in his political ethnography of the USA, *Democracy in America*, famously attributed the vibrancy of early Republican America to the fact that its citizens were forever forming associations and the not unrelated phenomenon that extra-local government seemed barely present.[33] If De Tocqueville celebrates the birth of civic engagement in the USA, Robert Putnam charts its decline, arguing in his academic bestseller, *Bowling Alone*, that there has been a generational change since the 1960s with younger age cohorts less inclined to participate in community life and politics.[34] Much more controversially, his argument has been extended to include most other Western democracies.[35] Less detailed analysis, however, has so far been carried out on non-Western 'democratic' societies and the nature of civic engagement at the local and regional level.[36] Social relations in these instances usually have been either cast as 'obstacles' to development and dismissed by modernisation theorists or romanticised in terms of the poor's inherent solidarity from a communitarian view. In the case of the Philippines, worse was yet to follow. Governance in that country has been variously labelled kleptocracy, a government led by thieves, narco-democracy, the rule of criminal overloads in collusion with bent politicians, or the abode of crony capitalism, where the political and economic spoils of office are shared between a select group of family, friends and key supporters.[37] In a now infamous article, James Fallow went so far as to credit this state of affairs as the product of 'a damaged culture.'[38]

According to Putnam, the measure of a civic community culture can be gauged by membership in communitarian associations of all descriptions. In an earlier work, he based his assessment of civic engagement on the different historical development within a nation, choosing as examples a comparison of regional governance in Italy. In this, as in much of his later work, he was concerned with understanding why civic engagement was so evident in some regions and was less pronounced in others. Putnam is interested in civic community only as a precursor to effective participatory governance. Its absence or decline, he argues, is responsible for the 'backwardness' of southern Italy and for the waning of democratic participation in the USA after the Second World War.[39] 'The civic context,' he writes, 'matters for the way institutions work.'[40] Putnam identifies four aspects that he claims characterise civic community: active engagement in public affairs; a strong sense of solidarity, trust and tolerance among community members; equality that binds people together through horizontal relations of reciprocity; and a commitment to cooperation expressed through the vibrancy of local social and political associations.[41] In his analysis, however, he fails to consider the role that the regular occurrence of hazards, disaster as a 'frequent life experience' might have in engendering a civic community culture.[42] The civic community in question here is not so much one of engagement, solidarity, equality and cooperation as identified by Putnam as a prelude to stable and democratic systems of government, as one of mutual dependence in the face of extreme adversity: not so much about the benefits of shared interests as about the dangers of going it alone under perilous conditions.[43] Without forcing the record unduly, there is evidence of all four of Putnam's attributes in the detailed account of the storm over San Isidro in 1887.[44]

Civic Engagement

Putnam defined civic engagement as an active concern with broader public needs over those of the individual, a steady recognition and pursuit of the public good. Certainly, measures were taken in advance of the storm to better prepare San Isidro for its passage and to safeguard the lives and property of the community. Local officials received advance warning of the impending typhoon by telegraph from the Jesuit-run Manila Observatory.[45] In this sense, of course, the community was no longer completely 'alone' but had recently been integrated into a wider state network of information dissemination and expanded services that included the establishment of a meteorological observatory in 1865. These synergistic developments allowed the first typhoon warning to be made on 7 July 1879, all too accurately predicting the path of an approaching storm. The timely circulation of this alert, however, depended upon the concurrent development and expansion of the telegraphic service, the first line of which was opened in 1872. The last of these stations, the one at San Isidro, was only established on 7 July 1887 or barely two months before the typhoon in question struck Nueva Ecija.[46]

As a result of these developments, Governor Reyes was notified by telegraph that a typhoon was approaching, a forecast he was able to verify by consulting the *cabecera*'s new barometer and noting the falling needle. Realising the storm was imminent, he took precautionary measures. He ordered the evacuation of certain houses regularly exposed to flooding, forbade any further attempts to wade across the river and had boats and rafts firmly secured.[47] He also instructed the province's *gobernadorcillos* to take all necessary steps to protect people, livestock, crops and buildings within their jurisdictions.[48] By 11 p.m. on 19 September, strong gusts of wind and heavy rain were lashing San Isidro. The governor instructed his principal officials, the doctor and members of the local police force (*cuadrilleros*) to search the houses previously ordered abandoned to ensure that their occupants had decamped to designated evacuation centres at the *casa gobierno*, *casa tribunal* and *casa parroquial* (the seat of administration, the court house and the priest's residence respectively) or some other substantial building constructed of stone that might offer more security against floodwaters.[49] The river burst its banks at 1 a.m. on 20 September. Any subsequent crossing of the river became extremely hazardous due to the strength of the current and the great number of trees borne by the floodwaters. Large parts of the town were soon inundated to a depth of at least two metres.[50]

Solidarity

Members of a civic community are more than simply publically minded and are helpful, respectful, tolerant and concerned for one another's welfare. Soon after the storm struck San Isidro, the town's rescue measures were put to the test and their sense of trust and solidarity towards one another was made manifest. The people appointed 'to patrol the *cabecera*' reported hearing cries for help from the middle of the rushing waters. The 'fully equipped' rescue boat manned by *cuadrilleros* was immediately launched but lost track of the voices in the dark, wind and rain. At considerable risk to their own lives, the searchers persisted until early the next morning when they were rewarded by the sight of two men, four women and three children clinging to the debris of their house that had been washed more than a league away. They were quickly taken to safety. Another less dramatic rescue was carried out by two other boats dispatched to fetch farmers who had been stranded in their fields by the speed of the rising waters and were in imminent danger of drowning. Vigilance was redoubled as the whole plain was transformed into one large 'lake' but the force of water and particularly 'the multitude of trees' dragged through the town temporarily made any further rescue attempts too hazardous to consider and effectively divided the town in two.[51]

Equality

While notions of equal rights and obligations for all are more difficult to discern from such an account, the care taken to safeguard each member of the community regardless of their position is a noticeable feature of events that night. News began to reach Reyes about other members of the community in trouble and of the measures taken to aid them. The *alcaide* (warden) reported that the prison was flooded and that he had temporarily transferred its inmates to one of the storehouses used by the Tobacco Monopoly.[52] By 3 a.m., floodwaters were lapping at the *casa gobierno*, the 'nerve centre' of emergency management operations, necessitating the precautionary relocation of numerous people seeking shelter there because 'they had already had their houses destroyed.' This threat to the seat of administration posed problems of another kind, requiring the hasty removal of all official documentation to prevent their spoilage in the floodwaters.[53] Across the now divided town, families were also in dire need of assistance but effectively cut off and thrown back on their own resources. Communications were not restored until early on the morning of 21 September and then only after repeated efforts and at great personal risk to the lives of those who made the crossing in boats. Emergency personnel appear to have been mainly local government employees, members of the municipal police force and agents of the Tobacco Monopoly and *Hacienda* (Treasury Department) many of whom clearly exceeded their duties to effect these operations. While overall coordination of activities remained in the hands of the provincial governor, the captain of cuadrilleros aided by a lieutenant from a neighbouring settlement took charge of actual rescue attempts.[54]

Cooperation

There is no reason to doubt the veracity of Reyes's account of the flood but one might wonder whether his explanation for the lack of casualties in San Isidro was the whole story. After all, it is a 'court history': He is the governor; it is an official document; and it is written to be read by his superiors. In other words, the report might highlight those activities successfully carried out under his direction and downplay those ones equally meritorious carried out by others—even if only inadvertently. Reading between the lines of his text suggests that the community may have also played an important role in this event even if its members' contribution is less visible and more difficult to discern in the official account.

Putnam argues that a dense network of associations both embodies and contributes to creating a cohesive local community bound together by a shared sense of collective responsibility. And there is just such a long history of formal and informal associations at the local level in the Philippines committed to individual and community welfare that enhance people's capacity to withstand the magnitude and frequency of misfortune.[55] The first evidence of mutual aid associations—religious fraternities known as *cofradías* common across the Hispanic world—dates back to at least 1594. Religious orders promoted the establishment of these lay organisations to deepen the faith of converts and to foster a sense of Christian community.[56] Although they were primarily religious, such associations also had important charitable roles in caring for the sick and dying, providing funds for illness and bereavement and aiding 'the unfortunate and needy people in general.'[57] These forms of associations were still very active in the nineteenth century. Another form of communitarian spirit can be discerned in the organisation of extra-familial work that had its roots in customary practice at the village level. Aid in the field or for personal matters like the moving or building of a house was rendered on the expectation that it would be returned in kind. Need or sometimes lot determined the order in which help was received, the notion of taking one's turn suggested by the Tagalog term used to identify such practices, *turnuhan* derived from the Spanish word *turno* meaning 'a turn.'[58] Mary Hollnsteiner credits the Spanish missionary, Fr Colin, with describing such arrangements as early as 1663 but these forms of reciprocity were very still much in evidence in the nineteenth century.[59]

This form of community labour was resorted to also in times of emergency. In his report on the activities of communitarian associations, the American 'Thomasite' schoolteacher, Harvey Edgar Hostetter, observed that it 'was a custom here' to build a special house 'which might be occupied by anyone whose residence would be destroyed by a typhoon.'[60] The communal construction of dams to protect *barrios* (settlements) from floods was apparently common practice in Antique. In Batangas, community help might be organised by a popular or influential person as a charity measure to help some person or family that had suffered loss through 'flood, fire or some similar disaster.' The purpose of cooperative associations on Bohol was to assist families with illness or death, paying for the burial service when someone died insolvent.[61] The similarities between some of these activities and those of the *cofradías* suggest that there was a measure of overlap between local religious associations and *turnuhans*. They served as expressions of common solidarity through faith and as vehicles to render mutual assistance. Filipinos rarely establish rigid epistemological divisions between the spiritual and human constructions of nature.[62] Not unexpectedly, there are few signs of these associations in Reyes' account of the storm over San Isidro. But when the raging floodwaters of the Rio Grande de Pampanga cut off the farther side of the town from the *cabecera*, the governor noted that the families in those areas worst affected were later found to be safe and well-provided for 'as they had been offered refuge by others.'[63] Such might be accounted for as nothing more than good neighbourliness in the circumstances or of the safety net conferred by extended families. However, it also hints at another world at work, one where people helped one another and where reciprocity was not just a code but an aspect of civic community.[64]

Storm, Flood and Cultural Memory

The flood peaked at 7 a.m. on 21 September and by midday water levels were falling rapidly. By midnight only the lowest parts of the town were still inundated. Reyes embarked on a tour of inspection of the worst-affected areas, something he had not been able to do since early the previous morning.[65] On the following day, he set off for neighbouring towns to survey the damage there. He also petitioned the Governor-general for permission to cut down trees to provide the timber necessary to carry out essential repairs, noting that floodwaters had apparently rendered inoperable many bridges in the province.[66] Outside of San Isidro, people had not fared so well. There were at least five fatalities in Cabanatuan. In Jaen, a three-year-old had drowned when the raft on which she was sheltering with her parents collided with a tree. In Gapan, a tree fell on a house killing a young woman and badly injuring her two uncles. Nor, of course, did San Isidro escape altogether unscathed, though losses here were confined to property. The *casa gobierno* was badly damaged, eight houses were washed away in the floodwaters and another 16 flattened by the wind, 'an infinite number' of animals were lost and countless trees blown down. Moreover, some 2000 *cavans* of *palay* (unhusked rice), 200 cavans of husked rice, 600 *fardos* of tobacco, 120 *pitones* of sugar and 200 *ollas* of honey were spoilt by the floodwaters.[67] But there was no loss of life 'despite it being the point where the floodwaters were deepest.' The governor concludes his report on this positive note, claiming credit for the lack of serious injuries in the *cabecera* and attributing this outcome to the effective emergency operations carried out under his direction.[68]

The storm over San Isidro passed, as all such tropical storms must, and the floodwater receded into memory soon to be replaced by still other storms and still other floods. This long history of coping with repeated storm and flood stretching back through the generations has generated a sense of community self-reliance and belief in their own efficacy. As Andrew Maskrey argues 'only local people know their own needs and therefore only they can define their own priorities for mitigation, within a given context.'[69] Such sentiments are rooted in a cultural tradition based on the expectation of mutual assistance freely given and consistently repaid. Responsibility is vested in the local community and is not dependent on external assistance for anything more than some general sense of forewarning. Disasters, at least in the first instance, are always local affairs. The record of the storm demonstrates how effective risk handling ultimately depends on grass-roots resilience—municipal governance and local capacity or what Putnam calls civic community. When it comes to confronting an extreme event and mitigating its life-threatening and property-damaging aspects, every community is thrown back on its resources. The outside world can do little more than warn of an impending danger and, perhaps, provide some material and financial aid to assist the community to rebuild afterwards.

If civil society with its social capital is largely seen as the products of a contemporary world with its complex class structures, economic diversity and political pluralism, people in the past had just as much need of non-state structures to help meet the challenges and vicissitudes of everyday life. If it is premature to talk about civil society prior to the twentieth century, perhaps a more appropriate term is one that is more socially and geographically bounded and that limits the concept to a more localised area or civic community. Civic community constitutes an essential aspect of self-reliant communities, ones that are able to withstand and rebound from extreme events and perhaps even learn from them to make themselves more resilient in the future. Repeated experience of disasters can engender adaptations that become an integral aspect of local and even national culture. These cultural adaptations are of no mean value when the very scale of modern societies and the uncertainties of climate change make everyone increasingly vulnerable to just such an event as the storm over San Isidro.

Notes

1. Joaquin Reyes, Civil-governor of Nueva Ecija to Director-general of Civil Administration, Manila, 1 October 1887, *Philippine National Archive* (hereafter PNA), Corte de Maderas, Bundle 6.
2. Andrew Maskrey, *Disaster Mitigation: A Community Based Approach* (Oxford: Oxfam Professional, 1989): 88.
3. Greg Bankoff, 'Living with Hazard: Disaster Subcultures, Disaster Cultures and Risk Societies,' in Gerrit J. Schenk (ed.) *Disasters, Risks and Cultures: A Comparative and Transcultural Survey of Historical Disaster Experiences Between Asia and Europe* (Heidelberg: Springer, forthcoming 2016).
4. Harry Estill Moore, *And the Winds Blew* (Austin, Texas: Hogg Foundation for Mental Health, 1964), 195.
5. William A. Anderson, *Some Observations on a Disaster Subculture: The Organisational Response of Cincinnati, Ohio, to the 1964 Flood, Research Note 6* (Columbus, Ohio: The Disaster Research Centre, Ohio State University, 1965).
6. Dennis E. Wenger and Jack M. Weller, *Disaster Subcultures: The Cultural Residues of Community Disasters, Preliminary Paper No. 9* (Columbus, Ohio: Disaster Research Centre, Ohio State University, 1973), 1.
7. Ibid., 2–5.
8. Ibid., 9.
9. Greg Bankoff, *Cultures of Disaster: Society and Natural Hazard in the Philippines* (London: Routledge, 2003).
10. Ben Wisner, Piers Blaikie, Terry Cannon, and Ian Davis (eds), *At Risk: Natural Hazards, People's Vulnerability and Disasters*, London: Routledge, 2004; Daniel Lorenz, 'The Diversity of Resilience: Contributions from a Social Science Perspective', *Natural Hazards*, published online 2010, http://dx.doi.org/10.1007/s11069-010-9654-y [accessed 4 December 2012].
11. On earthquakes, see Bailey Willis, 'Philippine Earthquakes and Structure,' *Bulletin of the Seismological Society of America* 34, no. 2 (1944): 69–81 and Maria Leonila Bautista and Kazuo Oike, 'Estimation of the Epicenters and Magnitudes of Philippine Historical Earthquakes,' *Technophysics* 317 (2000): 137–69. On weather-related phenomena, see Greg Bankoff, 'Storms of History: Society and Weather in the Philippines 1565–1930,' in

Peter Boomgaard (ed.), *Water in Maritime Southeast Asian Societies, Past and Present* (Leiden: KITLV, 2007), 153–83.

12. Nicholas Brown, Leoncio A. Amadore and Emmanuel C. Torrente, 'Philippine Country Study,' in *Disaster Mitigation in Asia and the Pacific* (Manila: Asian Development Bank, 1991), 196. The number of typhoons is calculated on the Philippine Area of Responsibility (PAR) that includes a rectangular area of ocean with the Philippine Islands at the centre, Palau at the eastern edge, Taiwan in the north and Sabah in the south. Only a proportion of tropical cyclones and typhoons actually make landfall somewhere in the archipelago. Bernardo Soriano, 'Tropical Cyclone Statistics in the Bicol Region,' *AngTagamasid* 20, no. 4 (1992): 12.
13. PAGASA (Philippine Atmospheric, Geophysical and Astronomical Services Administration) uses a three now four level warning system based on the wind speed of storms expected within 12–8 hours: Signal Number 1, 30–60 kph (kilometres per hour); Signal Number 2, 60–100 kph; Signal Number 3, 100–185 kph; and Signal 4, over 185 kph—the latter being first raised on 27 October 1991 when Typhoon Trining hit Northern Luzon. Emelina Almarios (ed.), *Disasters: The Philippine Experience* (Quezon City: Citizens' Disaster Response Center, 1992), 41.
14. José Coronas, 'The Climate and Weather of the Philippines, 1903–1918,' in *Census of the Philippine Islands: 1918*, Vol. I (Manila: Bureau of Printing, 1920), 446.
15. A detailed discussion of the historical sources on the nomenclature of typhoons is given in footnote 40 of Miguel Selga, 'Relacion Inedita del P. Francisco Ignacio Alzina S.J. sobre los Baguios, Mareas y Terremotos de Filipinas,' *Revista de la Sociedad Astronomica de España y América* 18, no. 118 (1928): 43. The primary historical source on typhoons in the Philippines that provides a record of such events between 1565 and 1863 is Miguel Selga, 'Primer Catálogo de Baguios Filipinos,' *Revista de la Sociedad Astronomica de España y América* 19, no. 125 (1928): 15; 19, no. 127 (1928): 42; 19, no. 128 (1928): 63–4; 19, no. 130 (1928); 94–5; 19, no. 131 (1928): 110–11; 19, no. 133 (1928): 142–3; 20, no. 138 (1929): 73–9; 20, no. 139 (1929): 89–96; 20, no. 142 (1929): 141–4; 21, no. 143 (1930): 7–13; 21, no. 144 (1930); 25–32; 21, no. 145 (1930): 46–7. The Manila Observatory compiled data on typhoons for the second half of the nineteenth century that was published in 1899 and later reproduced in the *First Report of the Philippine Commission to the President*, Vol. IV (Washington, DC: Bureau of Printing, 1901): 290–344. A reliable list of such events only really exists from 1948 onwards. Vicente Manalo, Vic Leaño and Ernesto Verceles, 'Frequency of Tropical Cyclones, by Intensity, Crossing the Philippines,' *Ang Tagamasid* 23, no. 2 (1995): 8.

16. 'List of Typhoons,' *Archive of the Manila Observatory* (*hereafter AMO*), Box 9-35.
17. 'List of Typhoons.'
18. A study undertaken by PAGASA attributes 38 % of annual average rainfall from 1951–1997 to the passage of tropical cyclones. Bureau of Agricultural Statistics, *Documentation on the Impacts of and Responses to Extreme Climate Events: Food and Agricultural Sector*, (Quezon City: Bureau of Agricultural Statistics, 2001), 9.
19. Miguel Duran to Governor and Captain General, Panitan, 20 November 1835, *AMO*, Box 4-11.
20. Miguel Selga, 'El Baguio del 4 de Octubre de 1854 en Pangasinan,' *AMO*, Box 10-36/10.
21. The list, simply entitled 'Floods in the Philippines 1691–1911' is anonymous and does not seem to have been composed by Miguel Selga but makes frequent reference to his works and so presumably post-dates him.
22. 'Floods in the Philippines 1691–1911,' *AMO*, Box 10-37.
23. Alcalde-mayor de Camarines Sur to Governor-general, Nueva Caceres, 30 October 1882, *PNA*, Calamidades Publicas, Baguios y Huracanes, Bundle 1.
24. Greg Bankoff, 'One Island Too Many: Reappraising the Extent of Deforestation in the Philippines Prior to 1946,' *Journal of Historical Geography* 33, no. 2 (2007): 314–34.
25. Inspección general de Montes, 'Summary of Regulations,' Manila, 19 October 1877, *PNA*, Corte de Maderas, Bundle 3.
26. Inspección general de Montes, 'Summary of regulations.'
27. Nueva Ecija and Nueva Vizcaya were sites of pioneering settlement along a shifting inland frontier and Pampanga the location of extensive sugarcane estates. Marshall McLennan, *The Central Luzon Plain Land and Society on the Inland Frontier* (Quezon City: Alemar-Phoenix Publishing House, 1980); John Larkin, *Sugar and the Origins of Modern Philippine Society*, (Berkeley: University of California Press, 1993).
28. See footnote 15 for a fuller discussion of this catalogue. Selga, 'Primer Catálogo de Baguios Filipinos;' 'List of typhoons;' 'Floods in the Philippines 1691–1911.' Floods are reported from the Visayas and from northwestern Luzon for 1887. The historical record for Iloilo mentions strong *baguios* and accompanying floods that buffeted the towns of Alimodian on 25–26 November and Arevalao on 1 December. A big flood likewise affected Bontoc and Ilocos Norte. Selga's catalogue lists a severe typhoon on 27 May–3 June which caused the Abra River to flood, destroyed several bridges in Pangasinan and considerably damaged the tobacco plantations of the Cagayan Valley.

29. 'Espediente sobre Corte de Maderas que Solicita el Gobernador Civil de Nueva Vizcaya por Espacio de un Año, a Causa del los Destrozos Ocasionados por el Baguio Ocurrido en 19 de Septiembre de 1887,' *PNA*, Corte de Maderas, Bundle 6.
30. Francisco Gomez to Governor-general, Bacalor, 25 September 1887, *PNA*, Calamidades Publicas, Baguios y Huracanes, Bundle 2.
31. An extensive discussion of the administrative structure of the Spanish Philippines can be found in Elidoro Robles, *The Philippines in the Nineteenth Century* (Quezon City: Malaya Books, 1969) and Greg Bankoff, *Crime, Society and the State in the Nineteenth Century* (Quezon City: Ateneo de Manila University Press, 1996).
32. Michael Woolcock and Deepa Narayan, 'Social Capital: Implications for Development Theory, Research, and Policy,' *World Bank Research Observer* 15, no. 2 (2000): 226.
33. Alexis De Tocqueville, *Democracy in America* (Garden City, NY: Doubleday, Anchor Books, 1969 [1835-1840]).
34. Robert Putnam, *Bowling Alone: The Collapse and Revival of American Community* (New York: Simon & Schuster, 2000).
35. Dietlind Stolle and Marc Hooghe, 'Review Article: Inaccurate, Exceptional, One Sided or Irrelevant? The Debate About the Alleged Decline of Social Capital and Civic Engagement in Western Societies,' *British Journal of Political Science* 35 (2004): 149–67.
36. See for instance, Evan Schofer and Marion Fourcade-Gourinchas, 'The Structural Context of Civic Engagement: Voluntary Association Membership in Comparative Perspective,' *American Sociological Review* 66, no. 6 (2001): 806–28. Despite its purported perspective most of the countries compared are European or North American.
37. Paul Hutchcroft, *Booty Capitalism: The Politics of Banking in the Philippines* (Ateneo de Manila University Press, Quezon City, 1998); UPAC (United People Against Crime) 'Changing the Culture of Criminality and Corruption,' *Philippine Graphic* 11, no. 23 (2000): 40; D. Arroyo, 'Government by Plunder Can No Longer Last,' *Philippine Daily Inquirer* (5 March 2001): C3.
38. James Fallows, 'A Damaged Culture,' *The Atlantic Monthly* (November 1989): 49–58.
39. Putnam et al., *Making Democracy Work*; Putnam, *Bowling Alone*. Needless to say, many scholars have not agreed with Putnam's conclusions. For an overview of such criticisms, see Stolle and Hooghe, 'Review Article.'
40. Robert Putnam with Robert Leonardi and Raffaella Y. Nanetti, *Making Democracy Work: Civic Traditions in Modern Italy* (Princeton: Princeton University Press, 1993).
41. Putnam et al., *Making Democracy Work*, 87–91.

42. Bankoff, *Cultures of Disaster*.
43. Greg Bankoff, 'The Dangers of Going It Alone: Social Capital and the Origins of Community Resilience in the Philippines,' *Continuity and Change* 22, no. 2 (2007): 327–55.
44. There is some ambiguity as to the community referred to in this report which is actually never named but just referred to as the *cabecera*. The seat of provincial government in Nueva Ecija, however, has shifted on numerous occasions from Baler (now in Quezon), to Bongabon, Cabanatuan, San Isidro, back to Cabanatuan in 1912 and is currently Palayan City.
45. The Jesuits created the first meteorological service able to monitor atmospheric disturbances in Eastern Asia. Under the energetic directorship of Fr. Federico Faura (1840–1897), the Manila Observatory began collating and publishing daily observations on barometric pressures, thermometer readings, wind force and direction and rainfall as monthly bulletins. Marcial Solá, *Report of the Philippine Weather Bureau 1902 Part II Meteorological Service of the Philippine Islands Report of its Establishment and Development Under the Spanish Government and Its Reorganization under the Government of the United States 1865–1902* (Manila: Bureau of Printing, 1903), 8–10; Angel Hidalgo, *El P. Federico Faura, S.J. y el Observatorio de Manila* (Quezon City: Observatorio de Manila, 1974), 16–7.
46. In 1878, the director of civil administration mandated that all weather observations be made available to the Jesuit fathers and that the Observatory was to be linked to the telegraphic network. The colonial state took an even more active role in 1884, funding the establishment of 13 secondary meteorological stations throughout Luzon that were progressively equipped and made operational between 1885 and 1887. Solá, *Report of the Philippine Weather Bureau* 1902, 8–10, 12–16. Communities had never been completely without advance notice of hazards even prior to meteorological forecasts. There are numerous references to the unusual behaviour of animals. José Marina to Governor-general, Cottabato, 12 November 1894, *PNA*, Calamidades Publicas, Baguios y Huracanes, Bundle 1. See also the account in D. Santiago Patero, *Reglas Practicas para Librarse de los Huracanes en el Archipiélago Filipino* (Madrid: Imprenta de Miguel Ginesta, 1873).
47. The river in question is the Rio Grande de Pampanga that converges with the Rio Chico at a site close to San Isidro and frequently threatens the town with inundation. Joaquín Rajal y Larre, *Memoria Acerca de la Provincia de Nueva Ecija en Filipinas* (Madrid: Establecimiento Tipográfico de Fortanet, 1889), 295.
48. Joaquin Reyes, Civil-governor of Nueva Ecija to Director-general of Civil Administration.

49. Joaquin Reyes, Civil-governor of Nueva Ecija to Director-general of Civil Administration. On various village and local police forces, see Bankoff, *Crime, Society and the State*, 129–54.
50. Joaquin Reyes, Governor-governor of the Philippines, 26 September 1887, *PNA*, Corte de Maderas, Bundle 6.
51. Joaquin Reyes, Civil-governor of Nueva Ecija to Director-general of Civil Administration.
52. On the Tobacco Monopoly and its role in Philippine history and finances, see Edilberto de Jesús, *The Tobacco Monopoly in the Philippines: Bureaucratic Enterprise and Social Change, 1766–1880* (Quezon City: Ateneo de Manila University Press, 1980).
53. This is a repeated and often commented upon consequence of flooding and may go far to explain the destruction of local government correspondence in many communities. The ingenuous measures taken to preserve official documentation, such as sealing them in fired clay pots is deserving of an article in its own right. Francisco de Ripoll to Director-general de Hacienda, Manila, 13 January 1873, *PNA*, Calamidades Publicas, Baguios y Huracanes, Bundle 1; José Martin to Intendente-general de Hacienda Pública de Filipinas, Tacloban, 20 November 1879, *PNA*, Calamidades Publicas, Baguios y Huracanes, Bundle 1.
54. Joaquin Reyes, Civil-governor of Nueva Ecija to Director-general of Civil Administration.
55. Emmanuel Luna, 'Rethinking Community Development in the Philippines: 'Indigenizing' and Regaining Grounds,' in Virginia Miralao (ed.), *Philippine Social Sciences in the Life of the Nation*, Vol. I, (Quezon City: Philippine Social Sciences Center, 1999), 315–16; Bankoff, 'The Dangers of Going It Alone.'
56. M. Caridad Barrion, *Religious Life of the Laity in Eighteenth Century Philippines: As Reflected in the Decrees of the Council of Manila of 1771 and the Synod of Calasiao of 1773* (Maniia: University of Santo Tomas Press, 1961); Setsuho Ikehata, 'Popular Catholicism in the Nineteenth-century Philippines: The Case of the Cofradía de San José,' in *Reading Southeast Asia: Translation of Contemporary Japanese Scholarship on Southeast Asia*, Vol. I (Ithaca: Southeast Asia Program, Cornel University, 1990), 111–12. On their existence in the rest of the Hispanic world, see María del Pilar Martínez López-Cano, Gisela von Wobeser, Juan Guillermo Muñoz Correa, *Cofradías, Capellanías y Obras Pías en la América Colonial* (Mexico City: Universidad Nacional Autónoma de México, 1998); and Ramón Franquesa Artés, *Las Cofradías en España: Papel Económico y Cambios Estructurales* (Girona: Museu de la Pesca, 2005).
57. 'Estatutos que observa los Hermanos y Hermanas de la Cofradía de Smo. Niño Jesús,' *Archive of the Archdiocese of Manila (hereafter AAM)*, 40A1,

Folder 8, Ternate 1866; 'Estatutos de la Cofradía de Jesús Nazareno,' *AAM*, 40A1, Folder 9, Cofradías, 1891–1897; 'Articles of Incorporation of 'The Crusade',' *AAM*, 40A1, Folder 11, Cofradías, 1914–1935.

58. Julian Balmaceda, "'Turnuhan' as Practised in Various Provinces,' *Philippine Agricultural Review* 20, no. 4 (1927): 381–421.
59. Mary Hollnsteiner, 'Reciprocity in the Lowland Philippines,' in Frank Lynch, *Four Readings on Philippine Values* (Quezon City: Ateneo de Manila University Press, 1968), 28.
60. Balmaceda, 'Turnuhan,' 387. Hostetter was stationed in Panay (1914–1917).
61. Balmaceda, 'Turnuhan,' 386, 389–90, 394.
62. Greg Bankoff, 'In the Eye of the Storm: The Social Construction of the Forces of Nature and the Climatic and Seismic Construction of God in the Philippines,' *Journal of Southeast Asian Studies* 35, no. 1 (2004): 91–111.
63. Joaquin Reyes, Civil-governor of Nueva Ecija to Director-general of Civil Administration.
64. Bankoff, 'The Dangers of Going It Alone.' Such culturally specific coping practices have been identified by the Filipino anthropologist Felipe Landa Jocano and others as *bayanihan* or toiling on another's behalf and of assuming another's burdens, *pakikipagkapwa* or 'being one with the other or with others,' and *pakikisama* or sanction against breaking ranks with the group. All these are attempts to express a sense of shared community, often defined in operational terms as neighbourhoods, which will guarantee support for its members especially during times of personal travail or common hardship. Felipe L. Jocano, *Growing Up in a Philippine Barrio* (New York: Holt, Rinehart and Winston, 1969), 96–98; Felipe L. Jocano, *Slum as a Way of Life: A Study of Coping Behaviour in an Urban Environment* (Quezon City: University of the Philippines Press, 1975), 166–87; Felipe L. Jocano, *Working with Filipinos: A Cross-Cultural Encounter* (Manila: PUNLAD Research House,1999), 66.
65. Joaquin Reyes, Civil-governor of Nueva Ecija to Director-general of Civil Administration.
66. Joaquin Reyes, Governor-governor of the Philippines. The governor of the neighbouring province of Nueva Vizcaya also sought unrestricted authorisation to cut timber for one year to repair damages to public buildings and works and to allow people to rebuild their homes. 'Espediente sobre corto de maderas que solicita el Gobernador Civil de Nueva Vizcaya.' Such requests were commonplace under these circumstances.
67. A *cavan* is a measurement equivalent to about 44 kilograms, a *fardo* (or pack consisting of 40 bundles) weighed between 18 lbs and 60 lbs depending on its grade, first to fourth class, a *piton* describes the mound-shaped body into which raw unprocessed sugar was dried, and *ollas* are jars or pans.

68. Joaquin Reyes, Civil-governor of Nueva Ecija to Director-general of Civil Administration.
69. Maskrey, *Disaster Mitigation*, 87.

Bibliography

AAM.Estatutos de la Cofradía de Jesús Nazareno.Archive of the Archdiocese of Manila, 40A1, Folder 9, Cofradías, 1891–1897.

AAM. Articles of incorporation of "the crusade". Archive of the Archdiocese of Manila, 40A1, Folder 11, Cofradías, 1914–1935.

Almario, Emelina, ed. 1992. *Disasters: The Philippine experience.* Quezon City: Citizens' Disaster Response Center.

AMO. 'List of typhoons. Archive of the Manila Observatory, Box 9-35.

AMO. Floods in the Philippines 1691–1911. Archive of the Manila Observatory, Box 10-37.

AMO. Miguel Duran to governor and captain general, Panitan, 20 November 1835. Archive of the Manila Observatory, Box4-11.

AMO. Miguel Selga, 'El baguio del 4 de Octubre de 1854 en Pangasinan.' Archive of the Manila Observatory, Box 10-36/10.

Anderson, William A. 1965.Some observations on a disaster subculture: The organisational response of Cincinnati, Ohio, to the 1964 flood. Research Note 6. Columbus: The Disaster Research Centre, Ohio State University.

Arroyo, D. 2001. Government by plunder can no longer last. *Philippine Daily Inquirer* (5 March): C3.

Artés, Ramón Franquesa. 2005. *Las Cofradías en España: Papel Económico y Cambios Estructurales.* Girona: Museu de la Pesca.

Balmaceda, Julian. 1927. Turnuhan' as practised in various provinces. *Philippine Agricultural Review* 20(4): 381–421.

Bankoff, Greg. 1996. *Crime, society and the state in the nineteenth century.* Quezon City: Ateneo de Manila University Press.

Bankoff, Greg. 2003. *Cultures of disaster: Society and natural hazard in the Philippines.* London: Routledge.

Bankoff, Greg. 2004. In the eye of the storm: The social construction of the forces of nature and the climatic and seismic construction of god in the Philippines. *Journal of Southeast Asian Studies* 35(1): 91–111.

Bankoff, Greg. 2007a. Storms of history: Society and weather in the Philippines 1565–1930. In *Water in maritime Southeast Asian societies, past and present*, ed. Peter Boomgaard, 153–183. Leiden: KITLV.

Bankoff, Greg. 2007b. The dangers of going it alone: Social capital and the origins of community resilience in the Philippines. *Continuity and Change* 22(2): 327–355.

Bankoff, Greg. 2007c. One island too many: Reappraising the extent of deforestation in the Philippines prior to 1946. *Journal of Historical Geography* 33(2): 314–334.

Bankoff, Greg. 2016. Living with hazard: Disaster subcultures, disaster cultures and risk societies.In *Disasters,risks and cultures: A comparative and transcultural survey of historical disaster experiences between Asia and Europe*.ed. Gerrit J. Schenk. Heidelberg: Springer, forthcoming 2016.

Barrion, M. Caridad. 1961. *Religious life of the laity in eighteenth century Philippines: As reflected in the decrees of the Council of Manila of 1771 and the Synod of Calasiao of 1773*. Manila: University of Santo Tomas Press.

Bautista, Maria Leonila, and Kazuo Oike. 2000. Estimation of the epicenters and magnitudes of Philippine historical earthquakes. *Technophysics* 317: 137–169.

Brown, Nicholas, Leoncio A. Amadore, and Emmanuel C. Torrente. 1991. Philippine country study. In *Disaster mitigation in Asia and the Pacific*, 193–253. Manila: Asian Development Bank.

Bureau of Agricultural Statistics. 2001. *Documentation on the impacts of and responses to extreme climate events: Food and agricultural sector*. Bureau of Agricultural Statistics: Quezon City.

Calamidades Publicas. 1873. Francisco de Ripoll to Director-general de Hacienda, Manila, 13 January 1873. *Philippine National Archive*, Calamidades Publicas, Baguios y Huracanes, Bundle1.

Calamidades Publicas. 1879. José Martin to Intendente-general de Hacienda Pública de Filipinas, Tacloban, 20 November 1879. *Philippine National Archive*, Calamidades Publicas, Baguios y Huracanes, Bundle1.

Calamidades Publicas. 1882. Alcalde-mayor de Camarines Sur to Governor-general, Nueva Caceres, 30 October 1882. *Philippine National Archive*, Calamidades Publicas, Baguios y Huracanes, Bundle1.

Calamidades Publicas. 1887. Francisco Gomez to Governor-general, Bacalor, 25 September 1887. *Philippine National Archive*, Calamidades Publicas, Baguios y Huracanes, Bundle2.

Calamidades Publicas. 1894. José Marina to Governor-general, Cottabato, 12 November 1894. *Philippine National Archive*, Calamidades Publicas, Baguios y Huracanes, Bundle1.

Coronas, José. 1920. The climate and weather of the Philippines, 1903–1918. In *Census of the Philippine islands: 1918*, vol I, 291–467. Manila: Bureau of Printing.

Corte de Maderas. 1877. Inspección general de Montes, Summary of regulations, Manila, 19 October 1877. *Philippine National Archive*, Corte de Maderas, Bundle3.

Corte de Maderas. 1887a. Joaquin Reyes, Civil-governor of Nueva Ecija to Director-general of Civil Administration, Manila, 1 October 1887. *Philippine National Archive*, Corte de Maderas, Bundle6.

Corte de Maderas. 1887b. Espediente sobre corte de maderas que solicita el Gobernador Civil de Nueva Vizcaya por espacio de un año, a causa de los destrozos ocasionados por el baguio ocurido en 19 de septiembre de 1887. *Philippine National Archive*, Corte de Maderas, Bundle6.

Corte de Maderas. 1887c. Joaquin Reyes, Governor-governor of the Philippines, 26 September 1887. *Philippine National Archive*, Corte de Maderas, Bundle6.

Fallows, James. 1989. A damaged culture.*The Atlantic Monthly*, November: 49–58.

Hidalgo, Angel. 1974. *El P. Federico Faura, S.J. y el Observatorio de Manila*. Quezon City: Observatorio de Manila.

Hollnsteiner, Mary. 1968. Reciprocity in the lowland Philippines. In *Four readings on Philippine values*, ed. Frank Lynch, 22–49. Quezon City: Ateneo de Manila University Press.

Hutchcroft, Paul. 1998. *Booty capitalism: The politics of banking in the Philippines*. Quezon City: Ateneo de Manila University Press.

Ikehata, Setsuho 'Popular Catholicism in the nineteenth-century Philippines: The case of the Cofradía de San José,' in *Reading Southeast Asia: Translation of contemporary Japanese scholarship on Southeast Asia*, Vol. I, (Ithaca: Southeast Asia Program, Cornel University, 1990), 109-188.

de Jesús, Edilberto. 1980. *The tobacco monopoly in the Philippines: Bureaucratic enterprise and social change, 1766–1880*. Quezon City: Ateneo de Manila University Press.

Jocano, Felipe L. 1969. *Growing up in a Philippine barrio*. New York: Holt, Rinehart and Winston.

Jocano, Felipe L. 1975. *Slum as a way of life: A study of coping behaviour in an urban environment*. Quezon City: University of the Philippines Press.

Jocano, Felipe L. 1999. *Working with Filipinos: A cross-cultural encounter*. Manila: PUNLAD Research House.

Larkin, John. 1993. *Sugar and the origins of modern Philippine society*. Berkeley: University of California Press.

López-Cano, María del Pilar Martínez, Gisela von Wobeser, and Juan Guillermo Muñoz Correa. 1998. *Cofradías, Capellanías y Obras Pías en la América Colonial*. Mexico City: Universidad Nacional Autónoma de México.

Lorenz, Daniel. 2013. The diversity of resilience: Contributions from a social science perspective. *Natural Hazards* 67(1): 7–24.

Luna, Emmanuel. 1999. Rethinking community development in the Philippines: "indigenizing" and regaining grounds. In *Philippine social sciences in the life of the nation*, vol 1, ed. Virginia Miralao, 315–343. Quezon City: Philippine Social Sciences Center.

Manalo, Vicente, Vic Leaño, and Ernesto Verceles. 1995. Frequency of tropical cyclones, by intensity, crossing the Philippines. *Ang Tagamasid* 23(2): 7–8.

Maskrey, Andrew. 1989. *Disaster mitigation: A community based approach*. Oxford: Oxfam Professional.

McLennan, Marshall. 1980. *The central Luzon plain land and society on the inland frontier.* Quezon City: Alemar-Phoenix Publishing House.

Moore, Harry Estill. 1964. *And the winds blew.* Austin: Hogg Foundation for Mental Health.

Patero, D. Santiago. 1873. *Reglas Practicas para Librarse de los Huracanes en el Archipiélago Filipino.* Madrid: Imprenta de Miguel Ginesta.

Putnam, Robert. 2000. *Bowling alone: The collapse and revival of American community.* New York: Simon & Schuster.

Putnam, Robert, Robert Leonardi, and Raffaella Y. Nanetti. 1993. *Making democracy work: Civic traditions in modern Italy.* Princeton: Princeton University Press.

Rajal y Larre, Joaquín. 1889. *Memoria Acerca de la Provincia de Nueva Ecija en Filipinas.* Madrid: Establecimiento Tipográfico de Fortanet.

Robles, Elidoro. 1969. *The Philippines in the nineteenth century.* Quezon City: Malaya Books.

Schofer, Evan, and Marion Fourcade-Gourinchas. 2001. The structural context of civic engagement: Voluntary association membership in comparative perspective. *American Sociological Review* 66(6): 806–828.

Selga, Miguel. 1928. Relación Inédita del P. Francisco Ignacio Alzina S.J. sobre los Baguios, Mareas y Terremotos de Filipinas. *Revista de la Sociedad Astronómica de España y América* 18, no. 118: 18(117): 13–21, 18(118): 39–43, 18(119): 61–62, 18(120): 75–77, 18(121): 89–91, 18(122): 103–105, 18(123): 123–127, 18(124): 139–140.

Selga, Miguel. 'Primer Catálogo de Baguios Filipinos,' *Revista de la Sociedad Astronómica de España y América* 19, no. 125 (1928): 15; 19, no. 127 (1928): 42; 19, no. 128 (1928): 63–64; 19, no. 130 (1928); 94–95; 19, no. 131 (1928): 110–111; 19, no. 133 (1928): 142–143; 20, no. 138 (1929): 73–79; 20, no. 139 (1929): 89–96; 20, no. 142 (1929): 141–144; 21, no. 143 (1930): 7–13; 21, no. 144 (1930); 25–32; 21, no. 145 (1930): 46–47.

Solá, Marcial. 1903. *Report of the Philippine Weather Bureau 1902 Part II meteorological service of the Philippine islands report of its establishment and development under the Spanish Government and its reorganization under the government of the United States 1865–1902.* Manila: Bureau of Printing.

Soriano, Bernardo. 1992. Tropical cyclone statistics in the Bicol region. *Ang Tagamasid* 20(4): 12–15.

Stolle, Dietlind, and Marc Hooghe. 2004. Review article: Inaccurate, exceptional, one sided or irrelevant? The debate about the alleged decline of social capital and civic engagement in western societies. *British Journal of Political Science* 35: 149–167.

Tocqueville, Alexis De. 1969[1835–1840]. *Democracy in America.* Garden City: Doubleday, Anchor Books.

UPAC (United People Against Crime). 2000. Changing the culture of criminality and corruption. *Philippine Graphic* 11(23): 40.

Wenger, Dennis E., and Jack M. Weller. 1973. *Disaster subcultures: The cultural residues of community disasters, preliminary paper no. 9.* Columbus: Disaster Research Centre, Ohio State University.

Willis, Bailey. 1944. Philippine earthquakes and structure. *Bulletin of the Seismological Society of America* 34(2): 69–81.

Wisner, Ben, Piers Blaikie, Terry Cannon, and Ian Davis, eds. 2004. *At risk: Natural hazards,people's vulnerability and disasters.* London: Routledge.

Woolcock, Michael, and Deepa Narayan. 2000. Social capital: Implications for development theory, research, and policy. *World Bank Research Observer* 15(2): 225–249.

CHAPTER 9

Disaster Management and Colonialism in the Indonesian Archipelago, 1840–1920

Alicia Schrikker

The Indonesian archipelago is geophysically one of the more dynamic regions in the Indian Ocean, with recurring earthquakes and over 100 active volcanoes. Recent catastrophes, such as the 2004 Indian Ocean tsunami, the earthquakes in Yogyakarta and Padang in 2006 and 2009, respectively, and the Merapi eruption of 2010, have resulted in a growing disaster consciousness in Indonesian society, and disaster management and mitigation rank high on the agenda of the central and local governments.[1] This contribution traces the early history of disaster management in the archipelago, by looking at indigenous and colonial perceptions of and responses to disasters in the nineteenth and early twentieth centuries.

Our knowledge of Indonesia's historical experience with disasters is limited. Natural disasters in Indonesia in the past were the terrain of geologists rather than historians. This is even the case with iconic nineteenth-century disasters such as the Tambora and Krakatoa eruptions of 1815 and 1883. Interest from the field of environmental history has been limited as environmental historians, taking a *longue durée* perspective, argue that the average disaster did not have a lasting impact on Indonesia's demography. Disasters have thus been written off as irrelevant events.[2] But in some non-environmentally oriented studies, historic disasters do appear as mean-

A. Schrikker
Institute for History, Leiden University, Leiden, The Netherlands

G. Bankoff, J. Christensen (eds.), *Natural Hazards and Peoples in the Indian Ocean World*, DOI 10.1057/978-1-349-94857-4_9

ingful events in political and cultural terms. For example, major events in Javanese political history have been related to disasters. Peter Carey argues that tectonic unrest on Java and the surrounding areas in the early nineteenth century contributed to the support that Pangeran Dipanegara gained for his claim to the throne of Mataram and in his battle against the Dutch during the Java War. No surprise then that the eruption of Gunung Merapi in 1822 received much attention in the prince's autobiographic *Babad Dipanegara*.[3] The eruption added charisma to Dipanegara's self-representation as the just king (*ratu adil*), fitting Javanese understanding of cyclic kingship. This cyclic symbolism relates to more universal interpretations of disaster invoking change, marking the end of a bad era and heralding a new start.[4]

The social construction of disasters, or the way in which disasters are interpreted and experienced, is important because, in the words of Greg Bankoff, 'they affect people's behavior and so they ultimately govern their actions.'[5] One of the leading questions in this article concerns the extent to which colonial rule led to meaningful interventions in existing modes of political and cultural understanding of disasters in the archipelago. The focus in this article will be on responses to disasters in their immediate aftermath, during recovery and reconstruction. Anthropologist Susannah Hoffman has illustrated in her analysis of the Oakland firestorm of 1992 how disaster experience, in particular the meaning victims give to disaster, varies even within one single community.[6] Inspired by the work of Bankoff and Hoffman, this article problematizes the colonial presence in the archipelago. How did disaster management work in a colonial setting, where different belief systems, stakes and knowledge come together? Do we observe change over time? After all, it was during the nineteenth century that colonial rule underwent major territorial and political transformations. From about the mid-nineteenth century, the textual production on disasters in the archipelago expanded greatly, and together with the few remaining indigenous accounts, these colonial texts provide us with a layered view of disaster experiences in the archipelago. While these written sources form the basis of this research, the analysis draws from insights in the field of disaster studies.

This chapter discusses three cases: a volcanic eruption and earthquake on the island of Ternate in Eastern Indonesia in 1840; a volcanic eruption, earthquake and tidal wave on the island of Sangihe Besar, north of Menado, in 1856; and the eruption of the Gunung Ḳelud in 1919 in central Java. Each case is analysed through the following set of questions: what action was taken during and after the disaster hit? Who was involved? Was there any meaningful cooperation between colonial and local authorities? How were disasters interpreted and experienced by political authorities? What knowledge was available and how was it used? The decision to study the first two cases (Ternate and Sangihe Besar) is source driven, as the availability of indigenous and colonial accounts allows us to analyse both local and colonial perspectives on these disasters. Together they form a snapshot of mid-nineteenth-century responses to disaster in colonial Indonesia, which will be related to other experiences throughout the period and throughout the archipelago. Developments in colonial disaster responses are examined against the backdrop of the ongoing colonial state formation process and expanding scientific interest in the archipelago. The Kelud disaster of 1919 is singled out as a third case because it marked the beginning of a new, prevention-oriented, colonial policy towards volcanic hazard in Indonesia.

Ternate 1840: Religious Politics of Risk and Blame

The volcanic island of Ternate is part of the Moluccas in Eastern Indonesia. In the sixteenth and seventeenth centuries the Islamic Sultanate was one of the major political players in the region. Traders were attracted to the island because of its central position in the clove trade. In the course of the seventeenth century, the Dutch East India Company (VOC) took control of the island and the sultanate was subordinated to Dutch rule. In the nineteenth century, when colonial rule became more formalized, the Dutch turned Ternate into an administrative unit, with a *resident* seated in Fort Oranje. Cocoa, coffee and cotton plantations were set up, owned by Europeans and worked by slaves. This prosperous plantation economy was short-lived and came to an end after the disaster of 1840.[7]

No lives were lost during the volcanic eruption and earthquake of 1840, but it proved destructive nonetheless. The Dutch *resident*, D.F.H. Helbach, noted that all stone buildings had collapsed and many fields and trees were damaged by stones and ashes. He estimated that a total sum of at least 900,000 guilders was needed for reconstruction of the town and the economy.[8] When Alfred Russel Wallace spent some time on the island 18 years later, he noted that the town had not yet recovered from the earthquake.[9] Thanks to the efforts of the Dutch colonial administrator–scholar Frederik S.A. de Clercq[10], we have at hand the translation of a Ternatese account of the 1840 disaster, drawn up by one of Sultan

Muhammad Zain's (r. 1823–59) clerks.[11] De Clercq worked on the island in the 1880s and wrote a scholarly treatise on the Ternatese language and culture. The disaster text is one among a collection of Ternatese texts from the palace that he copied and translated. The account is a curious composition: after the first few paragraphs the text becomes rather disorganized, and De Clercq suggests that the apparently unrelated paragraphs simply seem to follow the consecutive experiences. This makes the text a chaotic, but also very direct, account, providing us with a unique princely perspective on the disaster.[12]

The eruption of the Gunung Gamalama and the subsequent earthquake took place during the busy festive season of 1840: the eruption lasted for three days between the 2nd and the 5th of February and heralded the Chinese New Year. The Haj celebrations coincided with the earthquake that shook the island on the 14th of that month and that was followed by three weeks of severe aftershocks, lasting until the 9th of March. We know from the account of *resident* Helbach that Fort Oranje was severely damaged and that the 500 families living in town lost their

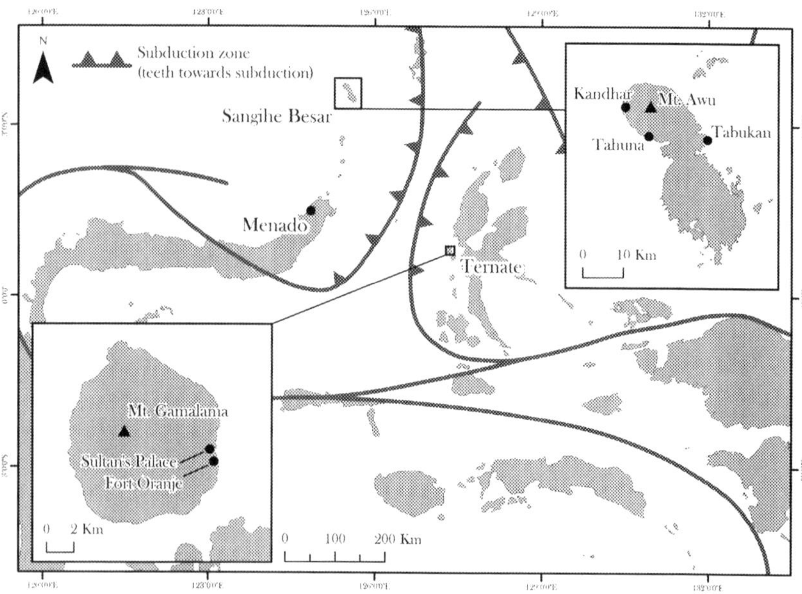

Fig. 9.1 Ternate (*left*) and Sangihe Besar (*right*), Eastern Indonesia

stone houses. They were ordered to take refuge in the shipsheds on the shore, and, when the aftershocks persisted, the families moved into the boats that were kept in the shallow water. The village *gabah* [palmleaf] houses were built on poles and proved more resilient to the aftershocks, but, as we will see below, after some time the villagers, too, took refuge on the beach and on the water. In material terms, the town community of traders, free burghers and slaves suffered the most, and it was this colonial community that depended on help from Ternate's sultan, rather than the other way around (Fig. 9.1).[13]

In contrast to the rather dull and brief colonial report, the Ternatese account gives us a fascinating impression of the earthquake as a political–religious experience. Throughout the text the behaviour of the sultan's subjects is closely monitored and judged morally. Furthermore, the text informs us of the sultan's cooperation with local religious and political powerholders, including the Dutch *resident* in Fort Oranje. This moral–religious framework in which the sultan perceived the disaster is discussed first, followed by an analysis of his political considerations. The section concludes with an impression of the communication between the sultan and the *resident*.

Religion and Disaster Causation

The sultan was very keen on keeping records of the religious behaviour of his men during the quake and aftershocks. At first, after the eruption of Gamalama, all men were instructed to observe regular prayers in the mosque, and to visit and pray at the *kramats* [tombs] of past religious leaders and former sultans. Ten days later, when the earthquake hit in the early morning during prayer time, we learn exactly who fled from the mosque and who stayed to complete the prayers.[14] When aftershocks continued to afflict the island, the sultan decided to reserve an open area on the beach for praying, as he saw it as his task to facilitate dutiful religious performance. According to the sultan, pious praying was essential to ward off further disaster, and therefore he kept strict note of men who got distracted from prayer and of those who fled during the more severe shocks. The *haj* celebrations were cancelled because of the quake, but the sultan made sure that alms were given to the victims nonetheless, and throughout the three weeks of aftershocks he repeatedly referred to his observation of this ritual charity.[15]

The relationship between morality, religion and disaster is a *topos* that runs throughout the text, and it is epitomized through the powerful story of Buniu. The fate of this lady symbolizes the strong religious component of the sultan's understanding of the disaster. Buniu was the daughter-in-law of the exiled sultan of Palembang (South Sumatra) and rumour had it that she had committed adultery with her father-in-law.[16] The sultan believed that it was essential to punish her for this immoral sexual behaviour in order to avert further disaster. As she was staying in the part of the town that fell under direct Dutch rule, the sultan asked the Dutch *resident* to apprehend her, which he did. Next, the account vividly describes how Buniu was put on a boat, screaming with terror while holding her crying daughter tightly in her arms. The writer gloats when reporting how her expectation to be drowned offshore had filled her with fear. Death by drowning was not to be her punishment as the sultan only exiled her from the island. Buniu was not the only one to be exiled; a second person to be removed from the island was a certain Tidorese man who had committed incest with his own daughter, still a child.[17] When the man appeared untraceable, the sultan concluded that this person must have fled the island already. The sultan's interpretation of the earthquake as divine punishment for corruption relates directly to Koranic representations of earthquakes and clearly connects to more general Islamic traditions of understanding earthquakes.[18]

Political Understanding of Risk

The sultan's anxiety to cleanse the island of immoral subjects and to stimulate religious diligence should be understood politically too. As was the case with other leaders in similar circumstances elsewhere in the world, the sultan's political interpretation of the disaster related directly to the stability of his rule: a disaster could damage his political legitimacy in the eyes of his subjects and rivals, as it could be understood as a supernatural sign that his rule was unlawful and had to come to an end. This development was something the sultan wanted to prevent at all costs, and so the political interpretation, in conjunction with his religious interpretation, was decisive for the way he responded to this disaster. Ternatese politics were defined mainly in relation to a long-standing rivalry with the sultan of the neighbouring island of Tidore.[19] No surprise then that the Sultan of Ternate kept a keen eye on his neighbour during the disaster and that he seemed pleased when he was informed that the sultanate of Tidore had also experienced the earthquake. Sometime after the initial earthquake, the Sultan of Tidore paid a visit to the Dutch *resident* and sent food, men and materiel to help the Dutch feed refugees and clear some of the damage and repair the roads. This act was noted with irritation by Mohammad Zain, who was frustrated that the Sultan of Tidore had not treated him and his people with similar respect.[20]

The sultan's management of the movement of his subjects during the disaster and the weeks of insecurity that followed the aftershocks further illustrate his political anxieties. Throughout the period a central concern of the sultan was to keep the inhabitants on the island, which he made a matter of honour:

> At this moment, you people, nobles and *bobatos* [...], are in doubt as to whether to stay or to leave, because of the situation in the village; but I, the Sultan, and my wife and children will stay in the palace until we die. We will not leave.[21]

But despite his precautions, we read repeatedly of attempts made by people to flee by *prahu* [boat]. The sultan was of course quick to condemn the behaviour of those who left as immoral, an immorality that was further illustrated by the fact that these men were stealing boats or that they were abandoning their families. There was one instance where a large group of women and children attempted to leave on a boat by themselves, without male supervision. They failed in their escape. Still, it was unfortunate for the sultan that, the longer the aftershocks persisted, the more people succeeded in fleeing the island. The sultan felt threatened and tried to regulate these actions by allowing the villagers to stay in their boats in the shallow water near the shore. It is written that he even guarded the property of those who stayed on shore or in the shallow water to underline his benevolence. About a week after the earthquake, panic among the villagers mounted when rumours circulated about sunken islands and beaches nearby. This panic led to a further upsurge in attempts to flee the island, including by nobles and men of higher status. Their names, and thus their disloyalty, were noted: 'There were many who fled, but a few of the nobles were clever enough to return during the night, which explains why their names do not appear in this book.'[22]

The departure of his subjects could damage his political charisma and undermine his political position in relation to Tidore and, to a lesser extent, to the colonial government. In fact, the sultan feared that those people leaving the island would choose to settle under Tidore, which, in contrast to the rumours mentioned above, had hardly suffered from the earthquake, and this development had to be prevented at all costs. It was unthinkable that Tidore would benefit in this way from Ternate's misfortune.

Colonial Cooperation and Divergent Understandings

To complicate matters for the sultan, it was at this point that the Dutch *resident* and leading European civilians too started to make preparations to abandon the island. These plans were discussed openly with the sultan and reflected upon in both the Dutch and the Ternatese accounts. The cross-cultural communication between the two parties reveals surprisingly close cooperation during the month of repeated hazards. This cooperation was paradoxically based on very divergent understandings of the disaster.

Throughout these weeks of chaos, the sultan kept regular contact with the Dutch *resident*. At first they together investigated the lava flow and the damage to the fields after Gunung Gamalama had calmed down. When, after the first quake, a cannon was fired from the fort, the sultan was informed that this had been done to keep everyone alert and to prevent people from going back inside their houses to sleep because of the risk of stone houses collapsing during the aftershocks. Some of the military officers of Fort Oranje had adopted these measures based on their experience during an earthquake in Ambon a couple of years before.[23] Communications with the *resident* involved such practical matters, but also dealt with the management of the people who had moved down to the beach, including the distribution of food. One has to remember that the colonial community depended on the help of the sultan rather than the other way around. After about ten days of aftershocks, the *resident* started to look for options to evacuate the people of the town of Ternate, preferably to somewhere on the coast of relatively nearby Halmahera. *Resident* Helbach insisted that Mohammad Zain should join them. The sultan of course refused, but he offered help in the form of manpower, materiel and food for the expedition to Halmahera.[24]

During one of their conversations about the evacuation, one of the Dutch officials read Mohammad Zain an account of the disaster that had hit the island in the 1770s that was kept among the VOC papers in his office. At that time an eruption of Gamalama was followed by a combination of an earthquake and a typhoon. In the Ternatese text this encounter is reflected upon in a singular way: Mohammad Zain remarked that the sultan reigning in the 1770s withstood three disasters and had stayed on. The story strengthened Mohammad Zain in his conviction to remain on the island. In his view, the moral of the story seems to have been that good sultans survive disasters. But it looks as if the Dutch official read him this account for a very different reason. The VOC official in charge of Fort Oranje in 1773–74 had left quite a panic-ridden account, in which he vividly described how frightened they had been that the island would sink, due to what he called 'unusual vertical shocks'.[25] Helbach and his colleagues seemed to have interpreted the narrative of the 1773–74 disaster as a warning of what might come and thought it best to evacuate the island. Just like the villagers, the Dutch *resident* feared that the island might sink into the sea—hence his insistence on the evacuation plan. Helbach had read the text to Mohammad Zain in his attempt to persuade the sultan to join him. It is important to realize that neither the Dutch *resident* nor the sultan, not to mention the villagers, would have been as knowledgeable of seismic patterns and exact risks in the region as we are today. Memories of previous disasters, be they represented through rumours or in VOC texts, would have played a role in the risk assessment of all stakeholders.

While the Dutch *resident* occupied himself searching for a suitable place to which to evacuate the town's population, the sultan anxiously continued to observe the panic among his subjects. Eventually by 9 March the aftershocks ceased and the account ends abruptly. In all, we are left with an ambiguous image of the disaster management of Sultan Mohammad Zain. On the one hand, we see him helping his people where he can through religious diligence but also more practically through charitable acts and material assistance. On the other hand, we see how he actively limits the movement of his subjects. It is as if he were waging a personal battle with nature: giving up his island and his people was simply not an option for him, as that would imply giving up his political authority. To the sultan, the disaster was in essence a political–religious experience that directly influenced his actions at the moment disaster struck. The sultan and the Dutch *resident* kept close contact over the four weeks, and helped each other: the Dutch delivered Buniu, and the sultan provided goods for the expedition to Halmahera. What is striking is that, despite these cordial relations, they had little impact on each other's actions, and it is clear that the two men understood and experienced the disaster in very different terms. This finding is in line with Hoffman's argument about divergent disaster understandings.[26] The question is whether this apparent incommensurability of disaster interpretations hampered disaster management on the ground, a point that will be further elaborated upon in the case of Sangihe Besar below.

For the Dutch, there was much less at stake in political terms, and this factor is reflected in their action. The region had become economically weak, and the decision to evacuate the town can be seen as symbolic of the lack of inclination on the part of the Dutch to further invest in the island. The Dutch reaction was determined by the waning economic and strategic importance of the island, and it revealed the limits of colonial commitment. In a vast and diverse region such as the Indonesian archipelago, where alien rule was in the process of establishing itself, disasters can form a tool to measure colonial commitment and local ties with the colonial centre.[27] Dutch indifference also explains the sorry state in which Wallace found the Dutch town some 18 years later. In fact, some historians argue that the disaster of 1840 gave a final blow to the island's colonial plantation economy.[28] Although the two powers on the island differed fundamentally in their political understanding of the disaster, this difference should not be taken for a clash between modern and traditional world views, but rather related to the political and economic stakes of the two parties. At the same time, there might have been more parallels in the religious interpretation than emerged from the texts at hand. After all, the biblical representations of earthquakes as divine punishment and apocalyptic sign are not so very different from Koranic understanding.[29] Unfortunately, the Dutch official accounts are rather silent on this point. However, we do know that decades later the Protestant inhabitants of the island still commemorated the disaster every year during a special church service.[30] The role of religion in Dutch disaster response comes more to the fore in the following case study from Sangihe Besar.

Awu 1856: Cross-Cultural Cooperation at the Colonial Frontier

The second example takes us to Sangihe, a range of islands north of Menado bridging the sea to Mindanao (Fig. 9.1). The eruption of Mount Awu on Sangihe Besar in 1856 proved extremely disastrous and was accompanied by hot mudstreams [lahar] from the lake of Awu, seaquakes and tidal waves. Menado was the nearest Dutch administrative post in the area, and the Dutch *resident* described the disaster vividly in an account that was published soon after the event in various colonial journals in the Netherlands and the Indies:[31]

> On the second of March, between 7 and 8, we heard an extreme thunder that went beyond our imagination, which announced the eruption and filled the Sangirese with fear. Simultaneously, the glowing lava flowed down in various directions with a terrible violence, and took along whatever came across its destructive path and it heated up the sea . Hot sources opened up and spit out floods of boiling water and the hot mudstreams dragged along everything that had thus far been saved from the fire. Enormous waves, the result of a seaquake, opened up the coast with thundering roars and the sea turned against the land with a wild surge, as if she wanted to master the fire...[32]

Nearly 3000 people died, between one-fifth and one-tenth of the island's population.[33] Many were uprooted and unable to return to their homes because most of the land had been made unsuitable for cultivation due to the damage caused by volcanic ash, mud and stones. The National Archives of Indonesia (ANRI) keeps a bundle of papers from the residency of Menado that contains the administrative resonance of this disaster. It includes letters from the rajahs of Tahuna, Tabukan and Kandhar (Kendahe), the three polities on the island of Great Sangihe. These, together with the papers written by the *resident* of Menado, allow us a close look at the process of recovery and reconstruction of the region in the two years after the disaster. Besides these, sources of a more literary and public nature are found in Dutch colonial journals, including the *resident*'s account quoted above and a lengthy poem that was written for fundraising purposes.[34] Again, the focus here is on response, cooperation and interpretation of colonial and indigenous powerholders.

We know from the papers of the residency that action was taken in the first weeks after the disaster; one of the naval vessels that cruised the region in search of pirates visited Sangihe Besar, unloaded emergency shipments of rice and reported to the *resident* in Menado about the situation on the island. From the correspondence with the local indigenous powerholders, it becomes clear that direct disaster mitigation had a strong political component. The chaotic movement of people troubled the local rajahs. This concern was made explicit by the Rajah of Tahuna in the most severely hit region who asked the *resident* for help to get his people back. He complained that they had fled the island and had submitted themselves to neighbouring rajahs. The *resident* answered decisively when he wrote: 'for now, it is best to leave the people where they have fled to, after all, his entire land is destroyed and there is not even enough food for those people who did stay behind.' He urged the rajah to get started with clearing the fields and preparing them for the cultivation of fruit trees and other crops and in the meantime accept the temporary lease of his people to other lords. In this way, he would become able to sustain his people again in the future.[35]

What happened to the Rajah of Tahuna is exactly what the Sultan of Ternate had feared 16 years earlier: disasters were perceived as moments when the social and political order was challenged and new alliances could be made. In an environment where people rather than land forms the basis of power, the management of people becomes crucial, leading political leaders to take calculated risks to the detriment of their subjects' welfare. But then the *resident* takes a curious turn in his reply to the Rajah of Tahuna when he remarks that, 'if the Rajah still has people who are unable to maintain themselves, they are certainly welcome to come and work on the public works in Menado or in private service.'[36] And thus we see the *resident* envisaging the Sangirese refugees in a role that under normal circumstances would be fulfilled only by the local serfs, convicts and slaves of Menado. In a way, we see here the local Dutch *resident* adapting himself to the local custom of managing refugees; studies of slavery and bondage connect the prevalence of these phenomena to, among other factors, the relatively frequent occurrence of natural disasters.[37] As elsewhere in the archipelago, debt could be a reason to end up in a subordinate bondage position. After a disaster, refugees could quickly find themselves in such a position, when they submitted to a new lord in search of protection and basic food provisions. Such protection was necessary because refugees were vulnerable to raiding bands who could market their victims as slaves at commercial centres like Jolo, Makassar or Batavia.[38]

While it is not clear how many people left the island for good, the Dutch assisted the remaining inhabitants and their rajahs in recovering and reconstructing their villages and land. Two months after the disaster, the *resident* launched a local fundraising campaign. In a circular written in Dutch and Malay, he called upon the inhabitants of the Menado residency to help. He asked people not only to donate money, but also to give in kind, because clothing and foodstuffs were what was needed most.[39] The language used in the text was objective in tone, indicating that it was aimed to appeal to the Muslims and Christians in the region alike.

The administration of this campaign, largely written in Malay, surprisingly reveals this as a cross-cultural event; we find people from a variety of backgrounds and from many of the villages of Menado engaging in this operation. People were keen to give blue-and-white linen for *kabayas*, headscarfs and rice, but also practical tools like needles and thread were collected. One person in the coastal settlement of Amourang donated his striped pants. Others like *kepala kampung* (village chief) Iman Gaawanko and a certain Baboe Abdullah gave money. Most frequently a sum between 80 cents and 3 guilders was given. All in all, 444 guilders and 85 cents were collected, and together with the donated goods, it was calculated that the resultant sum was sufficient to help 842 people. The distribution of the goods and food to the refugees involved close cooperation between the *resident* and the rajahs of Tabukan, Tahuna and Kandhar. Thanks to this meticulous administration we can trace the striped pants all the way to the moment when they were handed to one of the victims out in Kandhar.[40]

A second phase in fundraising started in June, and activities first spread throughout the archipelago and later expanded to the Netherlands. Over the next year another 10,000 guilders were raised, which allowed the *resident* to send more practical goods for the reconstruction to Sangihe Besar, like spades, nails and hammers. Finally, after about two years, goods for spiritual recovery were sent as well, in the form of a shipment of 1000 hymn books and 200 copies of the Book of Matthew. These goods hint at the religious frame through which Dutch colonial society perceived the disaster, despite its formal, objective representation in the circular that was distributed in the area around Menado.[41]

Acknowledging Blame: A Dutch Protestant Perspective on Awu's Victims

In the analysis of the Ternate earthquake, the question was raised as to what extent Dutch interpretations of disaster were informed by religion. The case of Awu provides us with ample material to reflect on Dutch Protestant interpretations of disaster. During the second phase of fundraising, the attention of the wider members of the colonial community in the Dutch Empire was drawn to the disaster by a small body of texts, consisting of the account of the *resident* and a poem written by an anonymous author. These texts contained vivid descriptions of the situation on Sangihe Besar, clearly meant to evoke a searing image of the misery on the island, and thus motivating readers to donate generously. At the same time, these texts were shaped by a clear Protestant Christian motive, another important unexpected parallel between the experience of the Awu eruption and that of Ternate. The following story illustrates the Protestant frame in which the Awu disaster was understood in the Dutch colonial world.

Not long before the eruption, 'a Spaniard' had visited the island and climbed Awu. The local population believed that this man had 'extinguished' the volcano and that they were safe now forever. For this reason the islanders had not recognized the early signs of the volcano becoming active again. This story was told to illustrate the dire need of these people for a proper Protestant mission. It was pointed out that the Sangirese had after all converted to Christianity during VOC times, but that their spiritual needs had been shamefully neglected over the previous decades. In this way the Dutch blamed themselves too for this disaster. This argument worked well in the colonial world and in calls for fundraising, as witnessed by the considerable inflow of goods and money over the next two years.

The Dutch, in their typically paternalistic way, acknowledged partial blame for the spiritual neglect of these people, and thus their part in the disaster. Ultimately, however, they reasoned, the problem lay with the inhabitants of Sangihe Besar for not being true Christians. Had they been faithful Protestants, this disaster would not have befallen them as they would not have been led astray by heathen superstition. The implication of this line of thought is not so different from the response of Ternate's ruler, who, too, stressed religious observance as a preventive measure against disaster. The fact that the state appeared secular in outlook, as represented in the circular discussed above, obscures the fact that religion was (and is) one of the most important frames in which disasters are understood. In his work on cultural representations of floods in Germany, Franz Mauelshagen has argued that it is impossible to draw a strict distinction in that respect between early modern and modern societies.[42]

As we have seen, spiritual reconstruction of the Sangirese started in the last phase of the relief campaign when Bible texts and hymn books were included in the relief package. Furthermore, in the years after the disaster the Protestant mission on the island prospered. Thanks to the active involvement of these missionaries in setting up coconut plantations, their presence even resulted in a relative economic boom in the decades to come. Still, these missionary endeavours could not prevent major tragedy when on 17 June 1892 Awu erupted again.[43]

Politics, Humanitarianism and Religion in the Colonial Realm

The disasters in Ternate and Sangihe Besar provide us with a glimpse of mid-nineteenth-century disaster response. In both cases the immediate response was characterized by cross-cultural interaction and cooperation between indigenous and colonial powerholders. Their views of the disasters differed to a great extent, though, and depended on specific political contexts. For the local rulers one of the most important challenges in the aftermath of disaster was the containment of refugees. For the Sultan of Ternate and the Rajah of Tahuna, political considerations seemed to outweigh humanitarian concerns in their management of refugees. The Dutch officials were less occupied with this post-disaster migration; on the contrary, the *resident* of Menado gladly tapped into it. In other areas, where Dutch rule was more direct we detect similar anxiety among Dutch administrators.[44] Later when the colonial state expanded and the monitoring of population became a more central concern of the colonial government, disaster-related migration certainly became an issue for the colonial government. In the late nineteenth century, relief works formed an important form of government aid and bore comparison to the indigenous system of post-disaster servitude that we saw at work in Menado. In the early twentieth century, refugee camps, ringed by wire fences, emerged at disaster sites, such as after the Merapi eruption of 1930. Despite their humanitarian outlook, the relief works and refugee camps have to be understood as means of containing migration or restoring order in a post-disaster chaos that had the potential to disrupt administrative control of the region and threaten the legitimacy of the colonial state.[45]

Governmental response to disaster remained by and large ad hoc throughout the period and was characterized by minimal action, rather than by active intervention. The Dutch state itself was for the larger part of the nineteenth century still a minimal state and started to develop slowly towards a more interventionist state only from the last quarter of the nineteenth century. Only by 1900 would the Dutch manifest developmentalist attitudes towards the archipelago and their Caribbean colonies. The so-called ethical policy, a Dutch version of the civilizing mission, was announced by Queen Wilhemina in her opening speech for the parliamentary year of 1901–02. This new policy was geared towards infrastructure, agriculture, water management, health and education. It was a highly utilitarian programme, but, except through water management and irrigation projects, disaster prevention was not given priority.[46] Mutual interdependence and cooperation on the spot between colonial and indigenous powerholders continued to characterize colonial disaster management as it had done in the mid-nineteenth century. Expenditures on long-term relief and reconstruction by the colonial government were minimal, and relief and reconstruction were, as in the Netherlands, considered to be part of a civic sphere, rather than a governmental responsibility.

The case of Sangihe showed how through fundraising activities and pamphlets a disaster in the archipelago was appropriated by a Dutch audience far away. Nor was this case unique in this respect. Rather, the Awu disaster represents an early stage of a process of gradual integration of the East Indies into a Dutch humanitarian culture that was taking place over the course of the nineteenth century. In the years after Awu, the central Java floods of 1861 and the Yogyakarta earthquake of 1867 received similar attention in the Netherlands and the Dutch colonial world.[47] It was a culture with strong civic and Christian roots that over the course of the nineteenth and early twentieth centuries became more and more institutionalized and authorized by government. The integration of Dutch and Indonesian disaster experiences through aid led to reciprocal campaigns and exchange of funds, as was the case, for example, with the concurrent floods in Java and in the Dutch province of Overijssel in the 1860s.[48] In the 1920s, this reciprocity was epitomized by the institutional unification of the Oranjekruis that coordinated fundraising after disasters in the Netherlands and the Smeroe fonds, a centralized fundraising body in colonial Indonesia that was established in 1909 after the eruption of Gunung Semeru on Java.[49] Throughout the colonial period disasters entered Dutch and colonial households through these civic fundraising campaigns that, thanks to technological developments, produced more and more elaborate brochures, photobooks, postcards and eventually even films.

A close reading of these humanitarian actions shows that there was more at stake than simple intra-empire solidarity. Certainly, disasters in remote places like Awu contributed to an imagined unity between the Netherlands and the Dutch Indies, but, at the same time, such events reinforced contrasts between the two. One of the recurring issues in Dutch and colonial newspapers in the aftermath of disasters was the question of whether funds collected by Christians should benefit Islamic victims as well, as the latter were expected to rely on their own charity networks. Over time, the direct religious connotation was replaced by one that was more racial and discussions revolved around the question whether gifts from Dutchmen and Europeans should benefit the indigenous population as well as colonists. In 1883, after the Krakatoa eruption, the result of this public discussion led some members of the European community in Batavia to set up a separate fund for European victims. Other members of the same community, however, decried this action. Segregation was after all a given in this colonial society as is reflected in the arguments used by both sides. The central Krakatoa committee, when rendering account of their operations in a public report, neatly showed how much was spent on each of the communities respectively.[50] The representation of this disaster, or more specifically of the victims, led to discussions about solidarity and apartheid, two irreconcilable ideals that underline very effectively the ambiguity of Dutch–Indonesian colonial society. It would be worthwhile to investigate the extent to which postcolonial humanitarian aid structures in the Netherlands built on this ambiguous colonial humanitarian tradition.[51]

Colonial Science and Political Limitations: Tunnels for Gunung Kelud

So far this contribution has focused on political and religious aspects of disaster response, emphasizing parallels, contrasts and continuities in indigenous and colonial experiences. There is one field where one might expect that the colonial presence would have resulted in meaningful intervention and that is in the field of knowledge and science. It is surprising, however, that, in contrast to the prompt action taken by Dutch colonial society after disasters, science made a slow start. In his study on colonial botany, Andrew Goss discusses the particularly utilitarian nature of colonial science in the archipelago. The questions colonial scientists were concerned with were not necessarily always the most relevant for local society.[52] This argument applies also to the development of scientific fields such as geology and volcanology that were to become important for the prevention of seismic disasters.

Scientific interest in the archipelago grew from the 1820s, but really only took off by the middle of the century—around the time of the disasters in Ternate and Sangihe.[53] Colonial science in this period was characterized by a functional interest in minerals and the drive for a comprehensive understanding of the region's natural history. From 1850 onwards, the Society for Natural Science of the Dutch East Indies published its own journal.[54] One of the early contributors was the scientist J. Hageman, who regularly reported on the archipelago's volcanic eruptions and earthquakes. Occasionally he and his colleagues included lists and descriptions of historic earthquakes and volcanic eruptions around the archipelago. This descriptive approach to the seismic and volcanic activities in the archipelago continued to dominate colonial science until the eruption of the Gunung Kelud volcano in 1919. The scientists revealed no structural interest in prevention, unlike their counterparts in Italy and Japan in the last decades of the nineteenth century.[55]

The Kelud volcano in the Kediri residency was and is an active and extremely dangerous volcano (Fig. 9.2). Like Awu it is a cone-shaped strato volcano, filled with a lake. In addition to the dangers of lava, hot ash and stones, these volcanoes pose a threat of flooding the region with boiling water, resulting in fatal *lahars* or mudstreams, as occurred in 1848, 1875[56], 1901 and 1919, the last producing the most destructive and fatal mudstreams of all. In the course of the nineteenth century, the slopes of the volcano had become populated, and its fertile grounds were

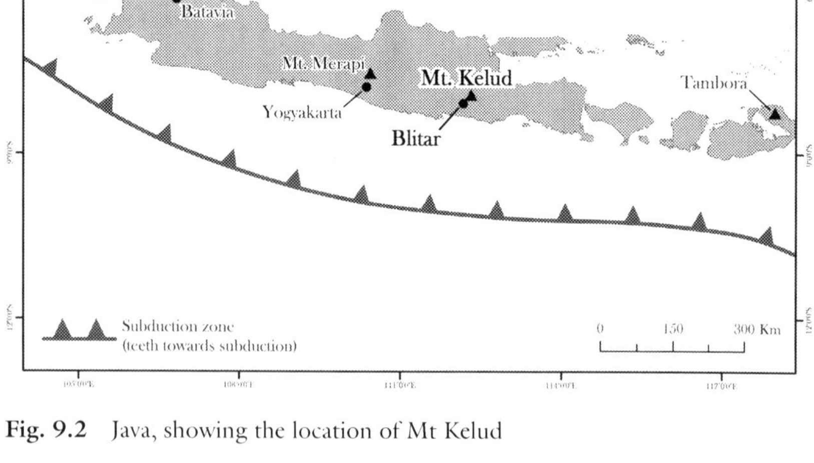

Fig. 9.2 Java, showing the location of Mt Kelud

cultivated by Dutch tobacco planters and Javanese peasants alike. After the eruption of 1901, planters and farmers were not driven away from the environs of the volcano by their cumulative experience with the *lahars*. Instead they stayed on and attempted to limit the risks of residing close to the volcano.

The planters started setting up prevention projects: they designed hide-outs on higher ground to provide refuge from the *lahar* and trenches and dams to control the mudflow and to protect the town of Blitar. Furthermore, one of the engineers came up with an ambitious and imaginative plan to tap all the water from the lake—without water an eruption would be much less of a threat. This initial design of what was later to be called the Keludworks was highly utilitarian, as the plan was to use the water as fertilizer for the plantations' fields. In that respect it is not difficult to relate this scheme to the Dutch ethical policy. No matter how promising the plan sounded, its execution, however, suffered major setbacks. Lobbying from the planters and engineers produced no results, much to their frustration.[57]

By the 1910s, some scientists came to realize that the lack of structural monitoring of the archipelago's most dangerous volcanoes had become a true embarrassment. A number of Dutch scientists advocated that they should follow the example of countries with similar risks like Italy and Japan.[58] Apparently, it was difficult to convince politicians in volcano-free Netherlands of the necessity of such prevention projects, as no funds were made available. This situation lasted until Kelud erupted again in 1919, causing 5000 fatalities, destroying the town of Blitar and leaving a much larger number of people landless and destitute.[59]

In one respect, the Kelud eruption came at an auspicious moment: parliament in the Netherlands had just voted for the most ambitious disaster prevention project the country had ever witnessed. The closing off of the Zuiderzee by a giant dike (aptly called *Afsluitdijk*) between the provinces of North Holland and Friesland was meant to prevent the regular flooding of the northern parts of the country.[60] Perhaps the government was in a mood to consider prevention projects immediately after the Kelud disaster because it agreed to set up a volcano-monitoring service (*vulkaanbewaking*) and, in addition, approved the tunnel project to tap the water from Kelud's crater lake. The budget for the project was set at a maximum of two million guilders. This quite unorthodox project turned out a success. By the time it was completed in 1926, it had drained about 80 million cubic metres of water from the lake and continued to keep the water level low. In 1951 and 1965 it prevented disaster, and it is still in use and has been improved upon by Indonesian government scientists.[61]

As a result of the Kelud eruption of the 1920s, science entered Dutch colonial disaster policies and ideas about prevention started to dominate, but never entirely took over, attitudes. With the creation of the volcanological survey, a modern scientific framework emerged in which disasters continue to be understood by postcolonial Indonesian governments. The Kelud project, however, proved to be an exception, and it should be pointed out that it very closely mirrored the interests of the European community and that there was a strong link to developments in the Netherlands. It is doubtful whether government would have agreed on such an expensive project if the Kelud disaster had not occurred exactly at a time when prevention stood so high on the Dutch agenda. Perhaps, too, it helped that the Kelud project, like the *afsluitdijk*, was essentially a water-related prevention project—something the Dutch, with their long history of floods, could appreciate. Indeed, the Kelud works formed a one-time-only project. As one member of the volcanological service critically pointed out, when assessing the risks posed by Mt Awu in the 1920s: 'to make the region safe, the Awu would need a tunnel drainage project similar to the Kelud, but the government would never be prepared to pay for this.'[62] Instead the Sangirese had to rely on Christianity as prevention or move away from the volcano. Indeed, it was relocation that was most often advocated by the volcanologists as the most effective way of preventing disaster, and, in a number of cases, relocation was carried out by government.[63]

The Politics of Disaster in Colonial Indonesia

This article should be seen as a first attempt to analyse responses to seismic and volcanic disasters in the Indonesian archipelago in a colonial context. More structural research is needed to get a comprehensive view of the history of disaster management and disaster cultures in Indonesia. These three cases, when analysed together, provide more than simple anecdotes of disaster.

In the first two cases, disaster response was characterized by cross-cultural communication and interdependence, but it was very clear from the case of Ternate that such cooperation should not be mistaken for shared experience. The two parties understood the disaster in their own political and religious terms. In Ternate and Sangihe, religion informed humanitarian action and provided rulers with a causal explanation for the disaster. Through spiritual interference rulers attempted to avert further disaster; in the case of Ternate the sultan did this during the earthquake through charity, intensive prayer and punishment of immoral behaviour. In Sangihe, this was done by colonial society afterwards, when Protestant missions became actively involved on the island. The textual production on disasters from the colonial world bears witness to the way in which catastrophes caused by natural hazards in the archipelago became well-publicized experiences throughout the empire, adding yet a new layer to the already differentiated disaster experience.

The examples from Ternate and Sangihe stress the extent to which powerholders were troubled by the movement of people in the aftermath of a disaster. Risk was calculated by these powerholders not merely in relation to people's safety, but also in relation to the very specific political contexts in which disasters took place. In Ternate and Sangihe, these calculations led to contradictory responses by the colonial and local powerholders. Over time this context changed a great deal: the expansion and integration of colonial rule led to an extension of the role of the colonial government in disaster management. Under colonial influence the management of refugees seemingly underwent a major transformation: the colonial government tried to regulate post-disaster migration first through relief works and later through encampment. However, despite the modern outlook of the military-run refugee camps of the early twentieth century, the guiding principle here remained control over people's movements, just as had been the case previously. Furthermore, disaster victims became integrated in a Dutch world of charity and humanitarianism.

Risk assessment was largely experience based, and so villagers acted on rumours and the Dutch fell back on their archived memory. Despite the late development of volcanology and seismology, incidental innovation in relation to prevention took place in the colonial period, and from the 1920s onward science started to inform formal government responses to volcanic and seismic disasters. This led to structural volcanic monitoring and active relocation of inhabitants away from 'danger zones'. These two approaches to prevention were to have a long afterlife in postcolonial Indonesia. The Kelud project of the 1920s is quite a spectacular example of scientific, technology-based intervention. It was a project that was, in fact, less exotic to the Dutch than it seems because, in principle, the Kelud tunnels were a flood prevention project. As water management is the field in which Dutch science stands out, this was no coincidence. In fact, the project was executed at a time when flood prevention ranked high on the Dutch political agenda, and it is likely that this mindset greatly informed the political and financial support for the Kelud tunnels at the time. Colonial disaster management in Indonesia developed not only in response to local practices but also in dialogue with developments in the Netherlands.

Notes

1. See, for example, President Yudhoyono's speech 'Disasters are "greatest threats to our national security"' at the 5th Asian Ministerial Conference on Disaster Risk Reduction 2012: http://www.unisdr.org/archive/29378 (last visited 15 March 2014).
2. Peter Boomgaard, D. Henley and F. Colombijn (1998), *Paper landscapes, explorations in the environmental history of Indonesia* (Leiden: KITLV Press). This is a general trend in environmental history observed by Juneja and Mauelshagen (2007), 'Disasters and pre-industrial societies: historiographic trends and comparative perspectives' in: *The Medieval History Journal* 10, 1–31.
3. Peter Carey (2007), *The power of prophecy: Prince Dipanagara and the end of an old order in Java, 1785–1855*, (Leiden: KITLV) 515–17.
4. Susannah Hoffman (2002), 'The monster and the mother' in: Susannah Hoffman and Anthony Oliver-Smith, *Catastrophe and culture. The anthropology of disaster* (Santa Fe, NM: School of American Research Press) 113–43. For examples of cyclic symbolism from Africa: Astrirf Meier (2007) 'Natural disasters? Droughts and epidemics in pre-colonial Sudanic Africa', *The Medieval History Journal* 10, 209–36.
5. Greg Bankoff (2004), 'In the eye of the storm: the social construction of the forces of nature and the climatic and seismic construction of God in the Philippines' in: *Journal of Southeast Asian Studies*, 35 (1), 91–111. 110
6. Susannah Hoffman, 'The monster and the mother', 113–42.
7. Ulbe Bosma and Remco Raben (2003), *De oude Indische Wereld, 1500–1920* (Amsterdam: Bakker),160 (Translated into English in 2008 as *Being Dutch in the Indies* Singapore: NUS Press) and F.S.A. de Clercq (1890), *Bijdragen tot de kennis der residentie Ternate* (Leiden: Brill), 174.
8. F.S.A. de Clercq (1890), *Bijdragen tot de kennis der residentie Ternate* (Leiden: Brill) 338–46.
9. A.R. Wallace (2000), *The Malay archipelago* (New Clarendon: Tuttle publishing (Reprint)), 238.
10. Frederik Sigismund Alexander de Clercq (1842–1906), served as *Resident* of Ternate from 1885 to 1888. More about De Clercq and about the book: Paul

Michael Taylor, 'Introduction' in: F.S.A. de Clercq, *Ternate. The residency and its sultanate.* Translated from the Dutch by Paul Michael Taylor and Marie F. Richards. Smithsonian Institute 1999. pp. i–xviii. The translation and the Dutch original can be consulted at: http://www.sil.si.edu/DigitalCollections/Anthropology/Ternate/ (last visited 24 March 2014).

11. De Clercq, *Bijdragen tot de kennis der residentie Ternate*, 202–23.
12. This observation is made by De Clercq, *Bijdragen*, 207, Note 1.
13. De Clercq, *Bijdragen*, 338–46.
14. De Clercq, *Bijdragen* , 204.
15. De Clercq, *Bijdragen*, 209–10.
16. For Buniu's story see: De Clercq, *Bijdragen* , 204 and 213. The Moluccan islands formed the major destinations for political exiles in the nineteenth century.
17. De Clercq, *Bijdragen* , 205.
18. See, for example, Anna A. Akasoy, 'Interpreting earthquakes in medieval Islamic texts', 184–6, in: Christof Mauch and Christian Pfister (eds), *Natural disasters, cultural responses. Case studies towards a global environmental history* (Plymouth: Lexington).
19. Leonard Andaya (1993), *The world of Maluku. Eastern Indonesia in the early modern period* (Honolulu: University of Hawaii Press).
20. De Clercq, *Bijdragen*, 217.
21. De Clercq, *Bijdragen*, 207. See for the translation, De Clercq, *Bijdragen*,149. *Bobatos* are landlords.
22. De Clercq, *Bijdragen*, 153. For the original quote see De Clercq, *Bijdragen*.
23. De Clerq, *Bijdragen*, 203.
24. De Clerq, *Bijdragen*, 218–21.
25. Nationaal Archief Den Haag, Verenigde Oostindische Compagnie (VOC), entrance number 1.04.02, inv.nr. 3411, Ternate: 'Kort verhaal der Ternaatsche onheilen. 8 Juli 1774'.
26. Hoffman, 'Monster and the mother.'
27. Anthony Oliver Smith (2002), 'Introduction: why anthropologists should study disasters'in Smith and Hoffman (eds), *Catastrophe and culture. The anthropology of disaster* (Santa Fe, NM: School of American Research Press) 3–23.
28. Ulbe Bosma and Remco Raben, *De oude Indische Wereld*, 160.
29. Anna Akasoy, 'Interpreting earthquakes in medieval Islamic texts', 185 and Franz Maulshagen (2009), 'Disasters and political culture in Germany since 1500', 63. in: Christof Mauch and Christian Pfister (eds), *Natural disasters, cultural responses. Case studies towards a global environmental history* (Plymouth: Lexington).

Another example would be the role of religion in disaster understanding in modern Italy: D. K. Chester, A.M. Duncan and C.R.J. Dibben (2008). 'The importance of religion in shaping volcanic risk perceptions in Italy, with special reference to Vesuvius and Etna', *Journal of Volcanology and Geothermal Research* 172, 216–28.

30. De Clercq, *Bijdragen*, 177, Note 1.
31. Original text is found in ANRI, resentiearchief Menado 126: 'Stukken omtrent de uitbarsting van de vuurberg op de Sangie eilanden, 1856.'
32. *Tijdschrift voor Nederlandsch Indië*, 1856–II, 269.
33. David Henley (2005), *Fertility, food and fever: population, economy and environment in North and Central Sulawesi, 1600–1930* (Leiden: KITLV Press) 153. Henley gives a lowest estimate of 15,000 inhabitants, and a highest estimate of 30,000 inhabitants.
34. *Tijdschrift voor Nederlandsch Indië*, 1856–II, 268–72 and 403–5. Philodystenus (pseud.) (1856), *De uitbarsting van den vulkaan Awoe, op het eiland Groot-Sangir in Nederlandsch Oost-Indie, in Maart 1856: een gedicht, uitgegeven ten voordeele der overgeblevene bevolking van gezegd eiland* (The Hague).
35. *Tijdschrift voor Neerlandsch Indië*, p. 271. Exact term that is used in Dutch for temporary lease is 'tijdelijke verhuring.'
36. Ibidem, 271.
37. Gillen D'Arcey Wood (2008), 'The volcano lover: climate, colonialism, and the slave trade in Raffles's history of java (1817)', in: *Journal for Early Modern Cultural Studies*, Volume 8, Number 2, 33–55. Peter Boomgaard (2004), 'Human capital, Slavery and low rates of economic and population growth in Indonesia 1600–1910', in Gwyn Campbell (ed.), *The structures of slavery in Indian Africa and Asia* (London.: Cass) 83–95, 88–90. James Warren (2013) 'Ransom, escape and debt repayment in the Sulu Zone, 1750–1898' in: Gwyn Campbell and Alessandro Stanziani, *Bonded Labour and Debt in the Indian Ocean World* (London: Pickering & Chatto).
38. James Warren (1981), *The Sulu zone, 1768–1898: the dynamics of external trade, slavery, and ethnicity in the transformation of a Southeast Asian maritime state* (Singapore: NUS Press). In fact Menado's *resident* quotes a refugee who 'managed to stay out of the hands of the pirates' and had heard that Awu's ashes had come down as far as Mindanao.
39. ANRI, residentie archief Menado, 126.
40. These were most likely the black-and-white striped tricot pants popular at the time in the Netherlands and among newcomers in the Indies. Although they seem rather hot for the climate, under these circumstances they will surely have been useful one way or the other.

41. Perhaps the Book of Matthew was chosen on purpose as it could have helped people to give meaning to the disaster, see, for example, Matthew 24.7 on earthquakes. And the book could have served as a warning against competitive Islamic interpretations: in Matthew 24.5, there is a warning against a 'false messiah' making apocalyptic claims after disaster. It is more likely though that it was simply a coincidence, because there was a very recent translation of Matthew into Malay at hand in the region(Information via Th. Van den End). J.L. Swellengrebel (1974), *In Leijdeckers Voetspoor: Anderhalve eeuw bijbelvertaling en taalkunde in de Indonesische talen, deel I, 1820–1894*(Haarlem: Nederlands Bijbel genootschap), 175 and 177.
42. Franz Mauelshagen (2009), 'Disaster and political culture in Germany since 1500', in: Christof Mauch and Christian Pfister, *Natural disasters, cultural responses. Case studies toward a global environmental history* (Plymouth: Lexington books). 41–76. For present-day Italy as example of the persistence of religious interpretations of disasters, see: D.K. Chester, A.M. Duncan and C.R.J. Dibben, 2008. 'The importance of religion in shaping volcanic risk perceptions in Italy, with special reference to Vesuvius and Etna', *Journal of Volcanology and Geothermal Research* 172, 216–28.
43. 'Vulkanische uitbarsting op groot-Sangir', *Algemeen Handelsblad*, 20–8 1892.
44. When in 1829 the Tankuban Prahu in Krawang, North Java, started to smoke, farmers ran away from the volcano. The local official (*assistant-resident*) was annoyed by this, because he would be unable to collect taxes from the farmers. So he ordered the peasants to remain on their fields, despite the perceived threat of an eruption. See: Nationaal Archief Den Haag, Ministerie van Koloniën 20.10.01. Inv. Nr 2527 and 2824: 'Register der besluiten van den Gouverneur Generaal', date 29-4 1829, no.1.
45. This problematic relationship between people management and politics in the aftermath of disaster is not unique to colonial contexts: after the Mt Pinatubo eruption of 1991 local politicians did not cooperate in resettlement schemes, as they feared losing their constituencies. Greg Bankoff (1999), 'A history of poverty. The politics of natural disasters in the Philippines, 1985–1995', in: *The Pacific Review* 12:3, 381–420. 406.
46. Wim van den Doel, *Zo ver de wereld strekt* and Marieke Bloembergen and Remco Raben (2009), 'Introduction', in: Bloembergen and Raben (eds) *Het koloniale beschavingsoffensief: wegen naar het nieuwe Indië, 1890–1950* (Leiden: KITLV Press).
47. See, for example: Multatuli (1861), *Wijs mij de plaats waar ik gezaaid heb!* (Rotterdam). Johannes Quack (1862), *Gedenkboek van den watersnood in 1861, waardoor een groot gedeelte van Nederland en Nederlandsch-Indië geteisterd werd* (Dordrecht). Johannes Hageman and J. van Laar (1867), *Beknopte beschrijving, met 4 groote photographien, van de aardbeving, die*

het eiland Java in den ochtend van den 10den Junij 1867 heeft geteisterd: benevens eenige historische herinneringen aan den oorsprong van Passar-Gedée Semarang, 1867.

48. *Verslag van de verdeeling der gelden, door de particuliere liefdadigheid in Indië en Nederland bijeengebragt tot leniging der door den watersnood van 1861 in Midden-Java veroorzaakte rampen.* *Comité van gedelegeerden* (*Batavia*) (Batavia, 1862).
49. Alicia Schrikker, 'Natural disasters in the Dutch East Indies in the 19th century' paper presented at KNHG conference, 'A New Dutch imperial history' 1 October 2010. For an example of cooperation between Smeroe foundation and the Dutch Oranjekruis, see the report on the floods in the Netherlands and the earthquake in Padang: C.J. Mijnlieff (1926), *Verslag omtrent de bij de watersnoodramp* (*1925/1926*) *en bij de Padangramp verleende hulp.*
50. *Verslag van het centraal comité voor de noodlijdenden door de uitbarsting op Krakatau. 2 vols, Batavia 1884–85.* For the polemical discussion in the colonial newspapers on victims and receivers of funds see Gerard Termorshuizen and Anneke Scholten (2001), *Journalisten en heethoofden: een geschiedenis van de Indisch-Nederlandse dagbladpers, 1744–1905* (Leiden: KITLV Press); For an up-to-date analysis on the academic discussion on colonial racial categorization, see Susie Protschky (2011): 'Race, class, and gender: debates over the character of social hierarchies in the Netherlands Indies, circa 1600–1942', in: *KITLV bijdragen*, Vol.167 (4), 543–56.
51. This would relate to more general, enduring ambiguities regarding humanitarianism, see: Michael Barnett (2011), *Empire of humanity. A history of humanitarianism* (New York: Cornell University Press), especially. 1–18.
52. Andrew Goss (2011), *Floracrats:state-sponsored science and the failure of enlightenment in Indonesia* (Madison: University of Wisconsin Press). See also Peter Boomgaard, 'The making and unmaking of tropical science. Dutch research on Indonesia, 1600–2000',in: KITLV Bijdragen 162(2/3) (2006) 191–217.
53. Andreas Weber (2012), *Hybrid ambitions: science, governance, and empire in the career of Caspar C.G. Reinwardt* (*1773–1854*)(Amsterdam: Leiden University Pres).
54. Natuurkundige vereeniging voor Nederlandsch Indië. Journal title: *Natuurkundig Tijdschrift voor Nederlandsch Indië.*
55. Gregory Clancey (2006), *Earthquake nation, the cultural politics of Japanese seismicity* (Berkeley: University of California Press), 63–90.
56. In the 1875, the overflow of the crater lake was caused by a crack in the crater wall and not by an eruption.

57. N. Wing Easton (1923), *Vulcanology. The history and present state of scientific research in the Dutch East Indies* (Amsterdam). KNAW. 1–11.
58. N. Wing Easton, *Volcanology.* 1–11.
59. L.F. Dingemans (1923), *Nota in zake de maatregelen getroffen op Verkeers-Hygiënisch- en economisch gebied in de door de kloeteruptie van 19 en 20 mei 1919 in het gewest kediri geteisterde streken.* (Kediri) (transl: memorandum of measures taken in the field of infrastructure, hygiene and economics in the regions of Kediri that were hit by the eruption of the Kelud on 19 and 20 May 1919).
60. http://www.statengeneraaldigitaal.nl/ In December 1918 the Zuiderzeewet was discussed in parliament and approved. On 3 June we find the first mention of the Kelud project. Last visited 24 March 2014.
61. M.Neumann van Padang (1983), 'History of the volcanology in the former Netherlands East Indies', in: *Scripta Geol.*, 71: 1–76.
62. George Kemmerling (1923), *De vulkanen van den Sangi-archipel en van de Minahassa. Vulkanologische mededelingen 5* (Weltevreden).
63. Following the advice of volcanologist Kemmerling, when Mahangetang [island] (part of the Sangihe archipelago) was deemed dangerous for inhabitants, they were transferred to Manganitu and later to Bolaang Mongondow: Nationaal Archief The Hague, Ministerie van Koloniën 2.10.39, inventaris nr. 305: 'Memorie van Overgave van het Bestuur van den Res van M, J. Tideman (1926)'. In the 1930s inhabitants of the Merapi were forced to move to Lampung, which coincided with existing transmigration programmes.

Bibliography

Akasoy, Anna A. 2009. Interpreting earthquakes in medieval Islamic texts. In *Natural disasters, cultural responses. Case studies towards a global environmental history*, eds. Christof Mauch, and Christian Pfister. Plymouth: Lexington.

Andaya, Leonard. 1993. *The world of Maluku. Eastern Indonesia in the early modern period.* Honolulu: University of Hawaii Press.

ANRI, resentiearchief Menado 126: 'Stukken omtrent de uitbarsting van de vuurberg op de Sangie eilanden, 1856'.

Bankoff, Greg. 1999. A history of poverty. The politics of natural disasters in the Philippinnes, 1985–1995. *The Pacific Review* 12(3): 381–420.

Bankoff, Greg. 2004. In the eye of the storm: the social construction of the forces of nature and the climatic and seismic construction of god in the Philippines. *Journal of Southeast Asian Studies* 35(1): 91–111.

Barnett, Michael. 2011. *Empire of humanity. A history of humanitarianism.* New York: Cornell University Press.

Bloembergen, Marieke, and Remco Raben. 2009. Introduction. In *Het koloniale beschavingsoffensief: wegen naar het nieuwe Indië, 1890–1950*, eds. Marieke Bloembergen, and Remco Raben. Leiden: KITLV Press.

Boomgaard, Peter. 2004. Human capital, slavery and low rates of economic and population growth in Indonesia 1600–1910. In *The structures of slavery in Indian Africa and Asia*, ed. Gwyn Campbell, 83–95. London: Cass.

Boomgaard, Peter. 2006. The making and unmaking of tropical science. Dutch research on Indonesia, 1600–2000. *KITLV Bijdragen* 162(2/3): 191–217.

Boomgaard, Peter, D. Henley, and F. Colombijn. 1998. *Paper landscapes, explorations in the environmental history of Indonesia*. Leiden: KITLV Press.

Bosma, Ulbe, and Remco Raben. 2003. *De oude Indische Wereld, 1500–1920*. Amsterdam: Bakker(Translated into English in 2008 as *Being Dutch in the Indies*. Singapore: NUS Press).

Carey, Peter. 2007. *The power of prophecy: Prince Dipanagara and the end of an old order in Java, 1785–1855*. Leiden: KITLV.

Chester, D.K., A.M. Duncan, and C.R.J. Dibben. 2008. The importance of religion in shaping volcanic risk perceptions in Italy, with special reference to Vesuvius and Etna. *Journal of Volcanology and Geothermal Research* 172: 216–228.

Clancey, Gregory. 2006. *Earthquake nation, the cultural politics of Japanese seismicity*, 63–90. Berkeley: University of California Press.

de Clercq, F.S.A. 1890. *Bijdragen tot de kennis der residentie Ternate*. Brill: Leiden.

Dingemans, L.F., Nota in zake de maatregelen getroffen op Verkeers- Hygiënisch- en economisch gebied in de door de kloeteruptie van 19 en 20 mei 1919 in het gewest kediri geteisterde streken (Kediri, 1923). http://www.statengeneraal-digitaal.nl/. Last visited 24 Mar 2014.

Goss, Andrew. 2011. *Floracrats: State-sponsored science and the failure of enlightenment in Indonesia*. Madison: University of Wisconsin Press.

Hageman, Johannes, and J. van Laar. 1867. *Beknopte beschrijving, met 4 groote photographien, van de aardbeving, die het eiland Java in den ochtend van den 10den Junij 1867 heeft geteisterd: benevens eenige historische herinneringen aan den oorsprong van Passar-Gedée Semarang*.

Henley, David. 2005. *Fertility, food and fever:Population, economy and environment in North and Central Sulawesi, 1600–1930*. Leiden: KITLV Press.

Hoffman, Susannah. 'The monster and the mother', in Susannah Hoffman and Anthony Oliver-Smith, *Catastrophe and culture. The anthropology of disaster* (Santa Fe: School of American Research Press, 2002): 113–143.

Juneja, Monica, and Franz Mauelshagen. 2007. Disasters and pre-industrial societies: Historiographic trends and comparative perspectives. *The Medieval History Journal* 10: 1–31.

Kemmerling, George. 1923. *De vulkanen van den Sangi-archipel en van de Minahassa. Vulkanologische mededelingen 5.* Weltevreden: Landsdrukkerij.

Maulshagen, Franz. 2009. Disasters and political culture in Germany since 1500. In *Natural disasters, cultural responses. Case studies towards a global environmental history*, eds. Christof Mauch, and Christian Pfister. Plymouth: Lexington.

Meier, Astrirf. 2007. Natural disasters? Droughts and epidemics in pre-colonial Sudanic Africa. *The Medieval History Journal* 10: 209–236.

Mijnlieff, C.J. 1926. *Verslag omtrent de bij de watersnoodramp (1925/1926) en bij de Padangramp verleende hulp.*

Multatuli. 1861. *Wijs mij de plaats waar ik gezaaid heb!*Rotterdam, Nijgh.

Nationaal Archief Den Haag. Ministerie van Koloniën 20.10.01. Inv. Nr 2527 and 2824: 'Register der besluiten van den Gouverneur Generaal', date 29-4 1829, no.1; Ministerie van Koloniën 2.10.39, inventaris nr. 305: 'Memorie van Overgave van het Bestuur van den Res van M., J. Tideman (1926)'.

Nationaal Archief Den Haag, Verenigde Oostindische Compagnie (VOC), entrance number 1.04.02, inv. nr. 3411, Ternate: 'Kort verhaal der Ternaatsche onheilen. 8 Juli 1774'.

Natuurkundig Tijdschrift voor Nederlandsch Indië.

Philodystenus (pseud.). 1856. *De uitbarsting van den vulkaan Awoe, op het eiland Groot-Sangir in Nederlandsch Oost-Indie, in Maart 1856: een gedicht, uitgegeven ten voordeele der overgeblevene bevolking van gezegd eiland.* The Hague. 'S Gravenhage: C.W. Mieling

Protschky, Susie. 2011. Race, class, and gender: Debates over the character of social hierarchies in the Netherlands Indies, circa 1600–1942. *KITLV Bijdragen* 167(4): 543–556.

Quack, Johannes. 1862. *Gedenkboek van den watersnood in 1861, waardoor een groot gedeelte van Nederland en Nederlandsch-Indië geteisterd werd.* Dordrecht: J.P. Revers.

Schrikker, Alicia. 2010. Natural disasters in the Dutch East Indies in the 19th century. paper presented at KNHG conference *A New Dutch imperial history*, 1 October 2010.

Smith, Anthony Oliver. 2002. Introduction: Why anthropologists should study disasters. In *Catastrophe and culture. The anthropology of disaster*, eds. Susannah Hoffman, and Anthony Oliver-Smith, 3–23. Santa Fe: School of American Research Press.

Swellengrebel, J.L. 1974. *In Leijdeckers Voetspoor: Anderhalve eeuw bijbelvertaling en taalkunde in de Indonesische talen, deel I, 1820–1894.* Haarlem: Nederlands Bijbel genootschap.

Taylor, Paul Michael. 1999. Introduction. In *Ternate. The residency and its sultanate*, ed. F.S.A. de Clercq, i–xviii. Smithsonian Institute (translated from the Dutch by Paul Michael Taylor and Marie F. Richards, Washington).

Termorshuizen, Gerard, and Anneke Scholten. 2001. *Journalisten en heethoofden: een geschiedenis van de Indisch-Nederlandse dagbladpers, 1744–1905*. Leiden: KITLV Press.

Tijdschrift voor Nederlandsch Indië, 1856 – II.

van den Doel, Wim. 2011. *Zo ver de wereld strekt. De Geschiedenis van Nederland oversee vanaf 1800*. Amsterdam: Bert Bakker.

van Padang, M. Neumann. 1983. History of the volcanology in the former Netherlands East Indies. *Scripta Geol.* 71: 1–76.

Verslag van de verdeeling der gelden, door de particuliere liefdadigheid in Indië en Nederland bijeengebragt tot leniging der door den watersnood van 1861 in Midden-Java veroorzaakte rampen. Batavia: Lange & Co, 1862.

Verslag van het centraal comité voor de noodlijdenden door de uitbarsting op Krakatau. 2 vols, Batavia 1884-85.

'Vulkanische uitbarsting op groot-Sangir', *Algemeen Handelsblad*, 20-8, 1892.

Wallace, A.R. 2000. *The Malay archipelago*. New Clarendon: Tuttle Publishing (Reprint).

Warren, James F. 1981. *The Sulu zone, 1768–1898:The dynamics of external trade, slavery, and ethnicity in the transformation of a Southeast Asian maritime state*. Singapore: NUS Press.

Warren, James F. 2013. Ransom, escape and debt repayment in the Sulu Zone, 1750–1898. In *Bonded labour and debt in the Indian Ocean world*, eds. Gwyn Campbell, and Alessandro Stanziani. London: Pickering & Chatto.

Weber, Andreas. 2012. *Hybrid ambitions: Science, governance, and empire in the career of Caspar C.G. Reinwardt (1773–1854)*. Amsterdam: Leiden University Press.

Wing Easton, N. 1923. *Vulcanology. The history and present state of scientific research in the Dutch East Indies*. Amsterdam, KNAW.

Wood, Gillen D'Arcey. 2008. The volcano lover: Climate, colonialism, and the slave trade in Raffles's history of java (1817). *Journal for Early Modern Cultural Studies* 8(2): 33–55.

Yudhoyono 2012. Disasters are "greatest threats to our national security". 5th Asian Ministerial Conference on Disaster Risk Reduction 2012. http://www.unisdr.org/archive/29378. Last visited 15 Mar 2014).

CHAPTER 10

Cyclones, Drought, and Slavery: Environment and Enslavement in the Western Indian Ocean, 1870s to 1920s

Matthew S. Hopper

Introduction

A confluence of environmental, epizootic, economic, and political factors contributed to a surge in slave trading in the Western Indian Ocean in the late nineteenth century, increasing both the supply of enslaved Africans and demand for their labor. Historians of East Africa have long recognized the importance of environmental factors in the sporadic growth of the slave trade.[1] They have demonstrated how droughts, epidemics, and climatic disaster joined the disruptive forces of colonialism and international trade to create crises of resource scarcity that fueled the supply of vulnerable people who could be exploited by the slave trade. However, less well known are the environmental factors that influenced demand for slave labor in the Arabian Gulf. Climatic events affected Arabia as they affected East Africa. Severe cyclones hit Oman in 1885 and 1890 and Zanzibar in 1872. In addition, severe drought hit East Africa between

M.S. Hopper
History Department, California Polytechnic State University, San Luis Obispo, CA, USA

G. Bankoff, J. Christensen (eds.), *Natural Hazards and Peoples in the Indian Ocean World*, DOI 10.1057/978-1-349-94857-4_10

1888 and 1892. These climatic events were timed such that they played off one another and fueled a temporary increase in slave trading in the final decades of the nineteenth century and first decades of the twentieth century. This chapter suggests that the concurrence of environmental factors in East Africa and the Arabian Gulf acted synergistically to create an environment conducive to growth in the Indian Ocean slave trade despite Western political and diplomatic efforts to end it. Climate—specifically cyclones and drought—influenced both East Africa and the Arabian Gulf in the late nineteenth and early twentieth century, and consideration of these climatic influences can help us better understand how the confluence of supply in Africa and demand in Arabia contributed to growth of the slave trade in the Western Indian Ocean in this period.

Cyclones

On April 15, 1872, a devastating hurricane hit the island of Zanzibar and devastated the island's clove trees.[2] Robert Lynn's 1905 history of Zanzibar recounted that the hurricane 'began to blow at eleven o'clock on the night of the 14th from the south or southwest, and continued until 1.30 p.m. the next day, when there was a lull of half an hour. The storm then suddenly burst upon the island in greater fury from the north, and raged for about three hours.'[3]Descriptions of the storm are consistent with a 'direct hit' on the island, with the eye of the storm passing directly through Zanzibar harbor. The hurricane destroyed most of the island's productive clove trees as well as its shipping capacity. 'Every ship and dhow in the harbor was driven ashore save one, an English steamship, the *Abydos*, Captain Cumming, which by steaming at full speed was able to keep her moorings,' Lynn recalled. 'The town was wrecked; the clove and cocoanut plantations leveled, and many people were killed and drowned.... Natives, who remember the cyclone, declare that when the hurricane was over they could see for miles, as everything was blown flat.'[4]

Historians have long noted that the Zanzibar hurricane affected labor demands on the island. Abdul Sheriff argued that additional slave labor was required to replant the lost clove crop and that the hurricane drove up demand for slaves on the island in the years following the storm's destruction.[5] Frederick Cooper noted that the prices of slaves rose sharply after the hurricane, and slave owners moved their slaves to nearby Pemba, which was spared the brunt of the storm, to ensure a complete harvest on that island and to expand production there. The entire clove crop had to be replanted on Zanzibar, and since a clove tree takes six years to reach maturity and become productive, Cooper argued, 'the real burden fell on the slaves around 1880.'[6] Cooper noted that the loss of more than half of the productive capacity of Zanzibar's clove plantations drove prices up from $2 per frasila to about $9, leading to the rapid build-up of production on Pemba.[7] Even as late as the 1890s, two-thirds to three-fourths of the cloves shipped from Zanzibar were still imported from Pemba. Cooper also suggested that an increase in the slave trade in 1884 may have contributed to the larger harvests in the 1880s and 1890s. 'What is certain,' Cooper explains, 'is that the largest clove crops in the history of Zanzibar and Pemba were being gathered in the early 1890s.'[8] Sir John Kirk, the British Consul in Zanzibar at the time, noted the demand for slaves was particularly high in 1880, the year the first trees planted after the 1872 hurricane came into bearing.[9]

The expansion of clove production on Pemba and the re-creation of Zanzibar's clove crop essentially from scratch following the hurricane undoubtedly intensified demand for slave labor on both islands. Seth Pratt, the American Consul at Zanzibar, observed in 1890 that clove production was still dependent on slave labor and the island's economy was largely dependent on clove production (the sultan's annual revenue was nearly $400,000 from his 30 % levy on clove exports, which that year were estimated at 13 million pounds). 'Up to the present time the plantations have been worked with slave labor and at a comparatively small expense; but with stoppage of slave supplies from the mainland great difficulty will be experienced by the planters during harvest time,' Pratt wrote. 'One result will be an increase in expenses; but what the planters have most to fear is that the curtailment of the labor supply will entail a direct loss by rendering it impossible to harvest the crop until after it has blossomed, when it would be unfit for the uses of commerce.'[10]

The reason for the increased demand in slave labor following the hurricane was linked to the intensive labor required for both the replanting of clove trees and for harvesting them. Pratt described the process of planting in 1890 as follows:

> In the cultivation of the clove the first thing to be done is the starting of the shoot. The seeds are planted in long trenches and are kept well watered until and after sprouting. In the course of forty days the shoots appear above ground. They are carefully watered and looked after for the space of two years, when they should be about 3 feet in height. They are then

> transplanted, being set about 30 feet apart, and are kept watered till they become well rooted. From this time on the young trees require only ordinary care, though the best results are obtained when the ground about the trees is well worked over and kept free from weeds.... The growth of the tree is very slow, and five or six years are required for it to come into bearing, at which time it is about the size of an ordinary pear-tree and is usually very shapely. It is a pretty sight to see a young plantation just coming into bearing... For the past 15 years the cultivation of cloves has been the chief occupation of the Arab planters, and has always netted good returns...consumption of the article appears to keep pace with the inevitable increase of production.[11]

Replanting Zanzibar's clove trees and expanding Pemba's clove production required significant amounts of labor, and much of this came from the mainland of East Africa.

The 1872 hurricane was not the only severe storm to pound the Western Indian Ocean in the late nineteenth century. Cyclonic storms also hit Madagascar; a severe cyclone also hit Mauritius in 1892; and Aden and Somalia were pounded in 1885.[12] Although the Zanzibar hurricane is well known to historians of East Africa and the wider Indian Ocean world, considerably less attention has been paid to two devastating hurricanes that hit Oman in the late nineteenth century. These hurricanes made landfall in populated and productive agricultural regions of Oman roughly five years apart. The severity and timing of these storms made them particularly devastating. And, as the Zanzibar hurricane affected demand for labor for that island's lucrative clove industry, the same appears to have been true for Oman's profitable date industry. The Oman cyclones of 1885 and 1890, like the Zanzibar hurricane of 1872, reinforced demand for slave labor.

Cyclone is the meteorological name for a hurricane in the Indian Ocean, just as typhoon is the name for the same phenomenon in the Pacific.[13] The weather patterns in the Arabian Sea largely shield Arabia and the Gulf from tropical storms and cyclones, and only very rarely is Oman hit with a true cyclonic storm.[14] According to Fisher and Membery, the tropical storms and cyclones that hit Arabia most often develop over the southwestern quadrant of the Arabian Sea and tend to move northwesterly toward the Arabian Peninsula from there; but they often curve northeastward toward Gujarat or northwestward toward the Gulf of Aden. Only one in three tropical storms formed in the Arabian Sea actually approaches the Arabian Peninsula, and only once every three years on average does one of these storms actually cross the coast of Oman or Yemen. Even then, only half of these storms are likely to be true cyclones with winds of force 12 or higher, and most of these storms cross the coast in sparsely populated regions of Oman between Masirah Island and Salalah, where they dissolve over open desert. Only very rarely (perhaps once or twice a century) does an actual cyclone hit the more populated regions of northern coastal Oman where the region's date production is concentrated. The cyclone of June 4, 1890, brought nearly 300 mm of rain on Muscat and led to the deaths of at least 700 people.

David Membery explains that there are 'monsoon tropical cyclones' in the Arabian Sea—cyclones that 'form preferentially during the four-week period from mid-May to mid-June as a response to changes imposed on the region by the developing Asian monsoon.'[15] A devastating cyclone hit Masirah Island in June 1977 with winds of 90 knots and 431 mm of rainfall in 24 hours. Another hit Salalah in May 1963. But cyclones are almost unknown along the Arabian coast during the height of the monsoon during the summer months because the sea surface temperature (SST) in the Arabian Sea falls below 27 degrees Celsius. They tend to form instead during the transitional periods between monsoons in May/June and October/November (see Table 10.1).[16]

Like the 1890 and 1977 storms, the two cyclones to hit Oman in 2007 and 2010 also arrived in early June. Cyclone *Gonu* made landfall on June 6, 2007, and caused 50 deaths and $4.2 billion in damage in Oman. It also set records as the strongest tropical cyclone ever to hit Arabia and ever recorded in the Arabian Sea.[17] But *Gonu* was not a direct hit on Oman. It only made landfall on the easternmost tip of Oman before heading north into Iran and being downgraded to a tropical depression. Even so, it dumped 610 mm of rain on coastal Oman and thrust winds of up to 195 mph at its peak in the Arabian Sea.[18] Three years later, Cyclone *Phet* made landfall just north of Masirah Island on June 4, 2010, and passed over Oman directly through the port city of Sur before heading west and back out into the Arabian Sea. The most populous and productive regions of Oman were again spared, but Phet still killed 21 people and pelted Oman with 75 mph winds and 300+ mm of rain that caused more than $700 million in damage.[19] By comparison, Hurricane Katrina, which devastated New Orleans in 2009, dropped 380 mm of rain and sustained winds of 125 mph (Fig. 10.1).

Together, the two cyclones in 1885 and 1890, occurring so close in succession and with nothing nearly equivalent for a century before or since, caused significant damage to the region's agriculture, particularly its date plantations. The cyclone of 1885 hit the Batinah coast of Oman,

Table 10.1 Number of tropical storms > 34 knots and number of > 64 knots affecting the coastline of Oman and Yemen from 1890 to 1996

	J	*F*	*M*	*A*	*M*	*J*	*J*	*A*	*S*	*O*	*N*	*D*	*Total*
Storms + cyclones	0	0	0	0	9	8	1	1	1	8	7	1	36
Cyclones	0	0	0	0	6	4	0	0	0	3	3	1	17

Source: Martin Fisher, and David A. Membery, 'Climate,' in Shahina A. Ghazanfar and Martin Fisher (eds.), *Vegetation of the Arabian Peninsula*. Norwell, MA: Kluwer Academic Publishers, 1998. pp. 5–38.

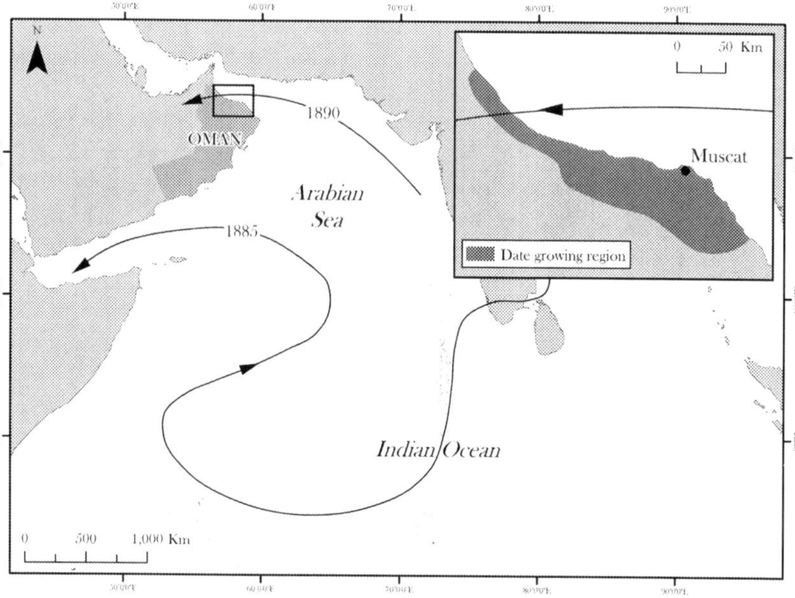

Fig. 10.1 Approximate paths of the cyclones of 1885 and 1890, with inset showing date growing region affected by 1890 cyclone

the country's most productive agricultural region, damaging much of the area's rich date crop and creating demand for labor to replant trees. The June 1890 cyclone hit the entire coast of Oman and destroyed much of the country's date-producing capacity and laid to waste much of the country's commercial capitals of Muscat and Mutrah. That storm killed at least 50 people in Muscat who were buried by debris. The storm destroyed more than 100,000 date trees valued at over $1.5 million at the time, and eliminated much of the region's productive capacity. Another cyclone came close to Muscat on June 3, 1898, causing further devastation in that city as well as in Jask in Baluchistan, but no storm affected Muscat and Batinah as severely again until *Gonu* more than a century later.[20]

The 1885 and 1890 cyclones were particularly devastating because they hit at a time in which much of Oman's economy was largely dependent on date production. Omani dates had grown in global demand throughout the nineteenth century, with their most lucrative markets in India and the USA. The eastern Arabian date industry, which fed local and regional markets for centuries, expanded to global proportions in the late nineteenth century, serving far-reaching ports in Singapore, Europe, Canada, and the USA. The American market in particular had become increasingly important in the second half of the nineteenth century, until the USA became Arabia's best foreign customer, importing nearly 80 million pounds of dates a year by 1925.[21] Following the seasonal monsoon winds of the Western Indian Ocean, American ships visited Arabia to exchange cotton cloth, piece goods, and specie for coffee, hides, and dates.[22] For most of the nineteenth century, Muscat was the center for Arabian date exports. Oman was home to particularly hardy varieties of dates, which could survive lengthy sea voyages, and which ripened earlier than most dates on account of Oman's southern latitude and intense summer heat. The *fardh* variety of dates would ripen in August, which would allow American ships enough time to load dates at Muscat, trade at Zanzibar, catch the monsoon winds, and make the 100-day journey home in time for the winter holidays. As American ships frequently returned from their voyages in the autumn, the arrival of Arabian dates in New York before Thanksgiving became an American tradition.[23] By 1871, the pressed *fardh* dates preferred by American merchants were selling at $30 a bag, and 15,000 bags were exported annually.[24]

Although *fardh* dates were not highly regarded in Arabia, Americans loved the sweet, sticky variety, and grocers stocked and sold millions of pounds of the fruit annually as a holiday confection. Grocers would use an ice pick to chisel off portions of blocks of sticky dates and sell them by the pound. The first American export company in Muscat, W.J. Towell Company, was established in 1866 specifically for shipping Omani *fardh* dates.[25] With the arrival of steam ships in the mid-nineteenth century, and the opening of the Suez Canal in 1869, the voyage from the Gulf to New York was cut to 60 days, and merchants began to add varieties of dates directly from Basra to their annual imports of *fardh* dates from Muscat. By 1885, Americans imported more than ten million pounds of dates annually, valued at $382,267.[26] A British observer in Oman noted that it was 'from the labors of the date cultivator that the country derives most of such wealth as it has.'[27] Although Basra would eventually surpass Muscat as the leading center for Gulf date exports, Muscat was still regularly visited by American barques well into the late nineteenth century and was still the major outlet for Omani dates, which were in high demand into the twentieth century (Fig. 10.2).

Not coincidentally, the primary area of nineteenth-century date expansion in Oman, Batinah—the 150-mile stretch of coast on the Gulf of Oman north of Muscat beginning around Seeb and continuing north of Sohar—became home to one of the region's largest populations of enslaved Africans. As the densest area of vegetation in eastern Arabia, Batinah had a bigger population and more agricultural production than any other part of the Arabian Gulf south of Iraq. But Batinah differed from the other date-producing areas in the region in that it required intense

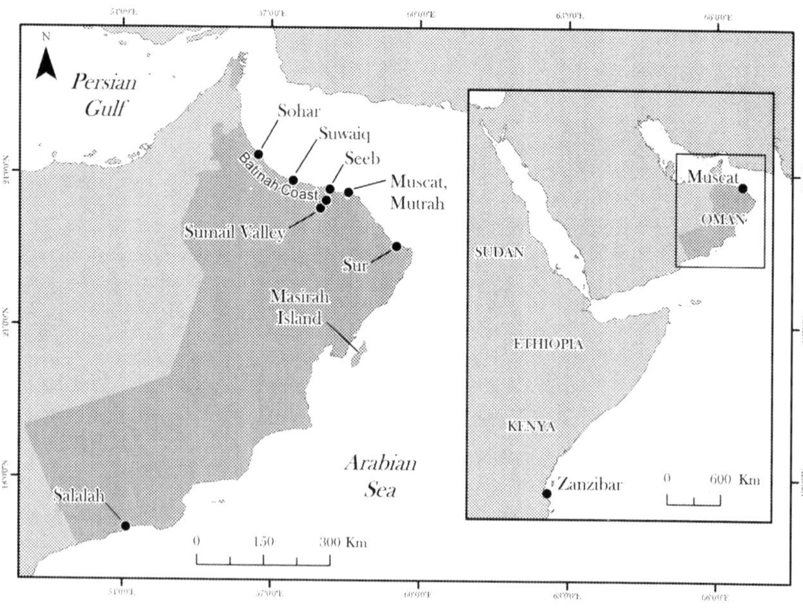

Fig. 10.2 Oman, showing places mentioned in chapter

human effort to irrigate its palms. Although it has some of the richest soil in Oman, Batinah receives no consistent flow of water from the inland mountains and relies entirely on ground water for irrigation through the use of wells.[28]Batinah farmers employed the *zijrah* (*zaygra*), or elevated well system used throughout the Arabian Peninsula and in Mesopotamia, consisting of a massive wooden framework, usually constructed of palm trunks, and a crossbar holding a rotating pulley wheel called a *manjur*, connected by rope to a bull (or in rare cases a mule) to lift water from 20 feet below the surface using leather bags which poured into cement (*saruj*)-coated holding tanks which further drained into irrigation channels, *aflaj*, to water several acres of date palms. The labor-intensive process used a male laborer, a *bidar*, to work in shifts around the clock in order to water approximately every 100 trees. The work of irrigation frequently fell on the shoulders of enslaved Africans. Bertram Thomas explained that runaway slaves were often punished with long hours in chains working in irrigation: 'The metallic chink of ankle-chains, heard, perhaps, from the bull-pit of a well within the date grove, is an indication of some such ill-fated escapade.'[29]

Like Zanzibar's cloves, Oman's dates required vast amounts of labor to produce the cash crop for global markets. In addition to irrigation, Batinah date plantations required constant maintenance. Date palms have to be pollinated by hand, offshoots (suckers) removed, dead branches cut off, extra date bunches removed, stocks kept clean, and—when the fruit is ripe—there is the enormous task of harvesting. In addition, in the shade of the date palms, farmers grew alfalfa and other animal feed in addition to vegetables for domestic consumption. Large parties of workers were required to boil the *maseybili* and *khameyzi* varieties, common in Batinah, in large copper cauldrons and dry them in the sun turning them frequently prior to shipping these popular varieties to India.[30] Dates also needed to be packed or pressed and conveyed overland or by sea to ports of export. Palm frond bags had to be woven to hold the dates, and once the dates were ready to export, they had to be loaded onto boats, conveyed to their destination and unloaded. Much of this work was performed by enslaved Africans.

Storms of the magnitude of the 1885 and 1890 hurricanes had not struck Oman in centuries and did not hit again until 2007 and 2010. When the first storm hit Batinah on March 30, 1885, it caused flooding that destroyed thousands of date palms in the region as far inland as Rostaq and significantly decreased the date crop exported to the USA.[31] A March cyclone is a rare event for Oman, but the mid-1880s were a tumultuous time for the world's climate. Just as Oman's date industry was beginning to recover, the second and more severe storm struck the entire coastline between Sur and Suwaiq on June 5, 1890. It began about midnight on the morning of the fifth and lasted for 24 hours, subjecting the coast to torrential rains and severe winds. Muscat received more than 279 mm of rain in 24 hours, and Mutrah, which was not protected from the northeast winds, suffered tremendous wreckage. Numerous large ships washed ashore, small local craft valuing at least $25,000 were destroyed, and houses collapsed and continued to collapse in the following weeks from weakened foundations. At least 50 people were drowned or buried by falling houses in Muscat and Mutrah alone. Within a week the sultan had received reports of at least 727 people killed across the country, and there were undoubtedly many more. The date crop was severely hit, with more than 100,000 date palms valued at over $1 million destroyed around the country. The majority of palms destroyed were in Batinah and several thousand were washed away due to flooding in the Sumail Valley, home to the largest production of the valuable *fardh* date variety.[32] Blount, Fritz, and Al-Harthy have argued convincingly that the unique bathymetry and the coastal track of the storm made it particularly severe. By employing a high resolution storm surge model and applying historical data from the 1890 storm with available data from *Gonu* through computer simulations, the climatologists derived a hypothetical track for the cyclone which followed the Batinah coastline much more closely than was previously assumed. The coastal track of the storm doubtless contributed to the severity of damage to the region's agriculture.[33] The damage to the Omani date industry as recorded by the British political agent in Muscat in 1890 was severe (Table 10.2).

The Oman cyclones of 1885 and 1890 destroyed significant portions of the region's date-producing capacity. Loss of this vital cash crop at a time of surging demand threatened to cripple the region's economy and created demand for more slave labor to replant destroyed date groves (Fig. 10.3).

Abdul Sheriff and Frederick Cooper note that the Zanzibar hurricane of April 1872 influenced demand for slaves in Zanzibar and Pemba. Likewise additional labor needs in Batinah and the Sumail Valley following the 1885 and 1890 hurricanes contributed to the increased slave traffic to Oman.[34] In fact, British officials noted an increase in the slave trade from East Africa beginning around 1885 but largely attributed the rise to problems of supply rather than demand. They correctly assumed that the discontinuation

Table 10.2 Statement showing the number of date trees with their prices, destroyed by the cyclonic storm of June 5, 1890

Name of place	*Number of date trees*	*Price*
Fanja	6,000	$120,000
'Amka'la	5,000	$100,000
Masra' bint Sa'ad	1,000	$20,000
Al Khowd	3,000	$60,000
Batineh Coast (From Seeb to Sohar)	63,000	$630,000
Wadi al Ma'awal	4,000	$100,000
Hajar and Khafeijee	3,000	$60,000
Kurayyat	2,000	$20,000
'Afa'	1,000	$20,000
Hail Ga'f	2,000	$40,000
Daqmar	1,000	$10,000
Miza'ra', Abayeh and Mahayeh	1,000	$20,000
Duba'b	500	$5000
Wadi al Arabiy'yain	500	$10,000
Suwakum	1,000	$20,000
Al Haboobiyyeh, A'rkee, Siya', Baldan Beni Alee, Al Hajar	2,000	$40,000
Fa'een [Fassn?]	1,500	$30,000
Wadi Dima', Haifad, Al Fiyyad, Al Haniineh, Hkabbeh, Simayysh {Buldan as Siabiyeen}	2,000	$40,000
Wadi al Miyeh	8,000	$160,000
Yiti	2,000	$20,000
Total	109,500	$1,525,000

Source: 'Statement Showing the Number of Date Trees with Their Prices, Destroyed by the Cyclonic Storm of 5th June 1890,' Surgeon Major A.S.G. Jayakar to Col. EC Ross (Sept. 21, 1890), IOR R/15/6/20

of the patrols of HMS *London*, the primary Royal Navy ship used for anti-slavery patrolling around Zanzibar, in 1884, and drought and growing famine on the coast of East Africa around the same time had created additional populations vulnerable to enslavement.[35] Six months after the storm, the British Political Resident in the Persian Gulf, declared that the slave trade was now being 'briskly carried on all over Oman and during the past year to such an extent as to constitute a grievous scandal requiring serious notice.'[36] The Royal Navy subsequently captured large shipments of slaves from East Africa en route to centers of Omani date production.

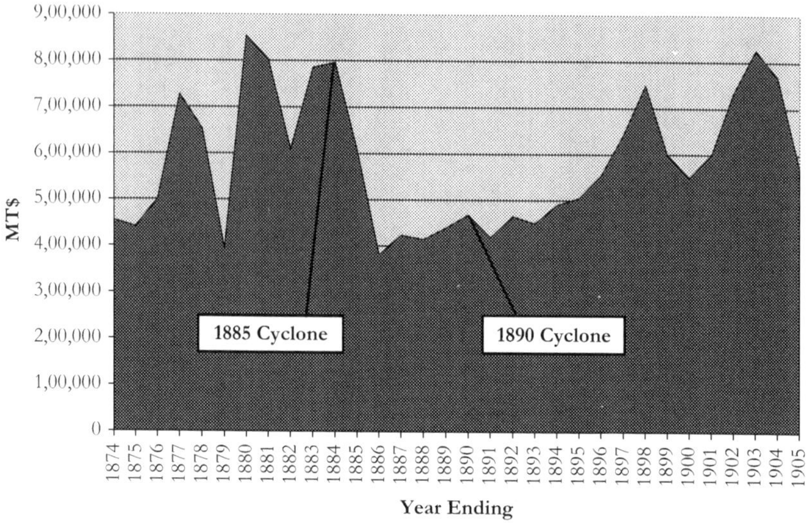

Fig. 10.3 Muscat date exports, 1874–1905 (Source: *Persian Gulf Administration Annual Reports, 1873–1905*) [MT$ = Maria Theresa dollar]

The first capture after the 1885 cyclone was made in September by the HMS *Osprey*.[37] Commander Herbert W. Dowding reported the incident as follows:

> On the 12th Sept a dhow was sighted off Ras Madraka coming from the Southward by boats detached from HMS *Osprey* to cruise off that point for the suppression of the slave trade. Lt. Valentine dispatched the gig to board her on which she wore and stood for the beach but a rifle being fired across her bow she lowered her sail and offered no resistance. Forty nine male and twenty four female slaves were found on board besides nine slave owners and nine crew. A prize crew of five men was put on board under the orders of Herbert Holyoake, Chief Quartermaster, who brought her to Ras al Hadd where she joined HMS *Osprey* on the morning of the 18th. The case was this day placed in the hands of Lt. Col Miles PA and Consul Muscat and condemned as a lawful prize, the slaves remaining in his charge for disposal.[38]

The same week, HMS *Ranger* seized a dhow that had just landed ten slaves at Sur.[39] Despite a drawdown in the British antislavery presence in the Western Indian Ocean after 1884, further captures continued to be made throughout the 1890s. The increased demand for labor happened to coincide with increasingly vulnerable populations in East Africa also due, in part, to climate.

Drought

Demand for labor to rebuild Oman's date plantations coincided with ecological disaster and an increase of a population susceptible to enslavement in Africa. The Oman cyclones corresponded with climatic events in East Africa that contributed to a rise in the supply of slaves on the coast. The many parts of East Africa that suffered from famine in the 1880s and 1890s have been well documented by historians. Scholars of Northeast Africa have demonstrated that environmental factors were not the only causes of the great famine of 1888–92. Richard Pankhurst, Douglas Johnson, Alex de Waal, and Steven Serels have identified multiple factors that contributed to separate famines in Ethiopia and Sudan.[40] Political factors such as the rise of the Mahadist state and British efforts to crush it contributed to famine, as did an outbreak of rinderpest virus in the region's cattle.[41] Likewise, scholars of East Africa have identified multiple factors that contributed to famine beginning in the 1880s. James Giblin's work on the Zigua ethnic group of northeastern Tanzania indicates that political and economic factors contributed alongside drought in both the 'lugala' famine of 1884–85 as well as the early colonial famine of 1894–96 (which was accompanied by a rinderpest outbreak and German military campaigns).[42] Juhani Koponen notes that these famines were preceded by earlier famines in the 1830s and 1860s.[43] Large parts of Kenya suffered from famine in the mid-1880s, in 1890–91, and again in 1898–99. Nile flood gauge records clearly indicate a significantly low rainfall in 1888 accompanying the East and Northeast African droughts. The same gauges support a preceding multi-year drought occurring in the 1830s.[44]

Drought was a driving factor, although certainly not the only factor, in several of East Africa's historical famines. East Africa's rainfall is highly variable with wide fluctuations year to year above and below annual averages. These conditions make the region particularly susceptible to drought. As Simon Nieuwalt explains, droughts occur in East Africa in periods in which rainfall is significantly below normal levels. Agricultural droughts are typically defined by rainfalls below one-half of annual averages. But the timing of rains is vital to East African agriculture, and examining annual totals alone can be deceiving. Droughts are often related to the late arrival or the failure of the rains. As Nieuwolt explains, 'Once a certain amount of rainfall has been received during a month, most crops are well supplied with moisture, and any further falls of rain are only useful as far as they can be stored in the soil for later use by the plants during dry months.' In East Africa the monthly amount is around 100 mm and 'rains above this limit are of little value to agriculture, as storage of moisture in the soil is only possible during the beginning of the rainy periods, the soils being saturated with water after one or two months of good rains.' A common problem in East Africa is soil erosion, flooding, and young crops being washed away in intense rain events. When the opposite problem occurs and rains fail to arrive or come too late, young plants whither and crop harvests fail. Maize in particular is more susceptible to drought than East African cereal grains such as millet and sorghum.[45] In other words, in addition to the volume of rain received in a given year, the timing of the rain is vitally important to African farmers as too little rain early in a growing season often cannot be made up with later heavy rains.

Drought conditions are well documented for East and Northeast Africa in the 1880s and 1890s. In fact, the mid-1880s were a tumultuous time climate globally and a time of severe drought and famine in numerous places, including Russia,[46] Bengal, Madras, and the Great Plains of North America.[47] Rainfall in Mauritius was significantly below annual averages between 1884 and 1887; on that island, 1886 marked the lowest recorded rainfall between 1862 and 1898 at 35.3 inches below average. India experienced famines in 1879–81, 1884–87, and 1890–92.[48] Likewise, Japan experienced widespread famine as a result of crop failures in 1884–85.[49] The great Ethiopian famine of 1888–92 coincided with drought as well as rinderpest and smallpox epidemics.[50] In East Africa, environmental disruptions and accompanying famines created vulnerable populations who could be exploited by the slave trade. Frederick Cooper explains that in East Africa a substantial revival of the slave trade took place in 1884 in part because of an increase in supply of slaves brought about by a devastating famine on the mainland of Kenya and Tanzania. Cooper notes that 'people were desperate enough to sell their neighbors, their children, and even themselves in order to survive.'[51] Consul Kirk and other observers on Zanzibar noted a marked increase in slave traffic from famine-stricken Tanganyika, particularly from Zaramo territory. Likewise, in Kenya, Cooper notes that in 1884 'large numbers of Mijikenda pawned themselves or their children for grain to the coastal plantation owners, who had been spared drought afflicting the immediate hinterland. Most likely the Mijikenda expected that such people could be redeemed, but this rarely happened.'[52] Timothy Fernyhough demonstrated that destitution in Ethiopia during the Great Famine of 1890–91 drove whole families into the hands of slavers.[53] Mike Davis has argued that the famines that devastated East Africa, Ethiopia, and Sudan in the 1880s were linked to back-to-back El Niños. The accompanying rinderpest, which 'killed off 90 percent of domestic and wild ruminants in the Horn of Africa before spreading south through the Rift Valley' left Ethiopia in turmoil. The resulting breakdown of Ethiopia's feudal system turned provincial governors and their soldiers into 'marauding bands.'[54] Davis further argues that the weakening of Ethiopia and Sudan by the droughts and famines of the 1880s made these regions more susceptible to European colonization. The Mahdist revolt was ultimately defeated by famine, Davis argues.[55] However, Steven Serels has recently shown that the famine in Sudan was in part created by British efforts to suppress the Mahdist revolt by deliberately shutting down grain shipments from the Red Sea on which parts of Sudan had become dependent.[56]

Coincidentally, famine in East Africa and destructive cyclones in Arabia were timed with a drawdown of the British antislavery efforts in the Indian Ocean. In 1884, the British Royal Navy's crusade against the slave trade in the Indian Ocean appeared to be at an end. A quartercentury of naval patrols, diplomacy, and political maneuvering seemed to have at last succeeded in ending the brutal East African slave trade. HMS *London*, flagship of British antislavery patrols in east Africa for the preceding decade, was deemed no longer necessary and was sold for scrap. For seven straight years, the Royal Navy witnessed declining shipments of enslaved Africans from the coast of East Africa, and admirals who led the antislavery crusade reversed the pessimism of their predecessors, declaring the slave trade to be 'paralyzed.' The celebrated treaty between the Sultan of Zanzibar and the British government signed under pressure from Bartle Frere and John Kirk in 1873 combined with the diligent work of the Royal Navy appeared to have finally succeeded in ending the traffic. In his classic history of the Royal Navy and the slave trade, Christopher Lloyd places the date for 'the end of the Arab slave trade' in 1883.[57] Other historians place the termination date in 1880 or even earlier.[58] In reality, the slave trade to Arabia continued well into the twentieth century, but the drawdown of the royal navy in 1884 and the British declaration of the protectorate over Zanzibar in 1890 thrust the problem of slavery out of the forefront of British priorities.[59]

The convergence of environmental, epidemic, and political factors in the 1880s created conditions particularly suited for exploitation of vulnerable populations. The slave trade increased in 1885 and continued into the 1890s. One of James Giblin's Zigua informants in a 1983 interview described the vulnerability of the 1884–85 drought period as follows: 'If you went travelling on the road with your wife and your possessions you could be robbed of everything.... If you were weak you would be robbed.... That's the way the *lugala* famine was.... The number of slaves increased [during 1884–1885].'[60] As late as 1901, British officials paid a visit to Indian subjects living in Sur—Oman's easternmost port and the former center of slave imports—and learned that at least a thousand slaves were still imported to that port annually.[61] Decades later, as the British administration under pressure from the League of Nations began to systematically grant manumission certificates to freed slaves in British-controlled ports in the Gulf, testimony from freed slaves attested to continued slave trading through the second decade of the twentieth century. More than half of the enslaved East Africans in the Gulf whose manumissions were recorded between 1921 and 1943 described being captured between 1895 and 1910, well after the supposed end of the slave trade.[62] Much of the demand for African labor in this period was connected to the Gulf's pearling industry. But another reason for the continued demand for African labor in the region was the need to replant Oman's destroyed date palms, for which there was still a lucrative and growing market in the closing decades of the nineteenth century.

A Perfect Storm

Recent climatological research suggests that the convergence of drought in East Africa with severe tropical cyclones in the Arabian Sea may not have been mere coincidence. In fact, all of these environmental events may have been related. Atmospheric scientists have linked some of the climatic events of the 1880s to disruptions in the transparency of the atmosphere resulting from dense veils of volcanic emissions following the Krakatoa eruption of 1883, which coincided with very strong El Niño/Southern Oscillation (ENSO) conditions in the Indo-Pacific basin. The pioneering work of meteorological historian H.H. Lamb in 1970 traced five centuries of volcanic eruptions to short-term climate change by assembling a 'Dust Veil Index' to estimate the amount of volcanic ash dispersed into the atmosphere and how long it stayed there. Lamb's work was followed shortly after by John D. Post's studies of Eurasia, which found similar links to disruptions in climate corresponding to years with high dust veil index. Post argued that high dust veil years lowered surface temperatures. La Marche and Hirschboeck confirmed severe frosts in North America in 1884 using tree-ring data to identify frost damage in bristlecone pines. More broadly, historians have linked the 'Krakatoa Effect' to climatic disruptions as varied as a rare summer snowfall in Korea, severe rainstorms and flooding in northern China, and poor harvests in Japan.[63] Handler and Andsager have shown that the volcanism associated with the eruption of Mt. Pinatubo in the Philippines coincided with cooler surface temperatures, an El Niño event, decreased Indian monsoon rainfall, fewer tropical storms in the North Atlantic, and higher winter rainfall in California in 1991–92, conditions parallel to previous El Niño events such as the 1982–83 event, which followed the El Chichon eruption of April 1982.[64]

More recently, a surge in interest in cyclones in the Arabian Sea following cyclones Gonu and Phet has produced much new and intriguing research on connections between climate in the Indian Ocean world and elsewhere.[65] Meteorological scientists now understand that 'warm' ENSO events (El Niño) and 'cold' ENSO (La Niña) events often produce consistently similar conditions in various geographies, although these events can be in interplay with more complex factors such as the Indian Ocean Dipole (IOD) and short-term anomalous variations in SST, which is now making possible predictions for severe storms, drought, and even epidemic diseases.[66] Emily Black has suggested that East African rainfall is associated with SSTs in the Pacific and Western Indian Oceans and as well as the Eastern Indian Ocean. Her suggestion that an 'anomalous zonal wind field that develops as a result of the perturbation in zonal SST gradient that characterizes the Indian Ocean Dipole causes excessive rainfall in coastal, equatorial East Africa,' has relevance to related work by climate scientists working to model cyclone formation in the Arabian Sea.[67] Scientists have had a keen interest in forecasting cyclones in the Northern Indian Ocean, as eight of the ten most severe cyclones (calculated in loss of life) in recorded history were in the region (Bay of Bengal and Arabian Sea).[68] Recent research has linked Indian Ocean atmospheric conditions to events in the Pacific. Most importantly, scientists now understand that anomalous warming events different from El Niño occur in the central equatorial Pacific in a horseshoe pattern around the equator flanked by colder SST anomalies on both sides. The resulting climatic event, known as El Niño Modoki (from a classical Japanese word meaning 'a similar but different thing') shares parallels with El Niño events but is a separate phenomenon and influences temperatures and precipitation around the world.[69] Atmospheric scientists have only very recently linked El Niño Modoki to the two severe storms to hit Oman in 2007 and 2010. It is clear that El Niño Modoki events that coincide with positive IOD events can generate more storms in the northwest Pacific.[70] Sumesh and Kumar suggest a similar confluence of factors may influence cyclogenesis in the northern Indian Ocean as well.[71] These studies suggest that the climatic disruptions of the 1880s and 1890s may well have been related to one another. The colder SSTs precipitated by the Krakatoa volcanic event of 1883, may have triggered an ENSO event or El Niño Modoki event which could have coincided with an IOD event or SST anomaly in the Indian Ocean which may have contributed to both the failure of rainfalls in East Africa and the cyclones in Oman and Aden in 1885. It is also possible that the ecological crises in East and Northeast Africa between 1888 and 1892 and the severe Oman cyclone of 1890 were precipitated by a similar climatic event not yet fully understood, but perhaps linked to another high volcanic dust index year, 1888. In that year eruptions of Bandai San (Japan) and Ritter Island (Bismarck Archipelago) combined may have rivaled Krakatoa according to Lamb's Dust Veil Index.[72] Preliminary corroboration of the African rainfall anomalies compiled in a remarkable historical dataset by Sharon Nicholson suggest that unusually low rainfall in East Africa in the 1880s during the drought and famines mentioned above corresponded with above-average rainfall in the Sahel. Further examination of this new and extensive source, which includes historical data from more than 400 rain gauges across Africa as well as lake levels and which is now archived at the Paleoclimate Data Center in Boulder, Colorado, may yield additional support for this possibility.[73]

Conclusion

Although this study is preliminary, evidence suggests a connection between climatic events that affected both East Africa and Arabia in the 1880s and 1890s. The growth of demand for labor in Arabia in the late nineteenth century may have been related, in part, to increased demand created by climactic catastrophes of the two cyclones of 1885 and 1890. Further research is needed to identify the extent to which enslaved Africans were put to use in the areas most affected by the storms of 1885 and 1890. By the turn of the twentieth century, the Batinah region of Oman was widely known as the center of the Gulf's largest African population, and Batinah was at the center of both of the late nineteenth-century cyclones. But additional investigation is necessary to connect the specific type of work imported African laborers performed in the years immediately following these storms. Preliminary evidence of the role of cyclones and drought suggests that environmental factors must also be considered alongside economic factors in the rise and demise of the slave trade to Arabia as well as in periods of fluctuation in its growth. To be sure, the environmental disasters in East Africa in the 1880s and 1890s were not entirely attributable to climate. Political and epidemiological factors also contributed. But drought appears to have precipitated famine in some places and contributed to upheaval in others. The increasing vulnerability of populations in East Africa in the midst of these crises doubtless contributed to the susceptibility of increasing numbers of people to enslavement. The convergence of these environmental events in these two regions already connected through the slave trade created an unfortunate perfect storm of supply and demand that warrants additional attention from historians.

Notes

1. See for example: Frederick Cooper, *Plantation Slavery on the East Coast of Africa* (Portsmouth: Heinemann, 1997), 126.
2. Mohammed RedaBhacker, *Trade and Empire in Muscat and Zanzibar: Roots of British Domination* (New York: Routledge, 1992), 128.
3. Robert Nunez Lynn, *Zanzibar in Contemporary Times* (London: Hurst and Blackett, 1905), 286.
4. Ibid.
5. Abdul Sheriff, *Slaves, Spices* & *Slavery in Zanzibar* (London: James Curry, 1987), 234–235.
6. Cooper, 132–137.
7. Ibid., 137.
8. Ibid., 132.
9. Ibid., 133.
10. Ibid.
11. Seth A. Pratt, 'Clove Culture in Zanzibar,' *U.S. Department of State Reports of the Consuls of the United States* 32, no. 112–115 (Jan.–Apr. 1890): 687–688.
12. David Membery, 'Monsoon Tropical Cyclones: Part 2,' *Weather* 57 (Jul. 2002): 246–255. For a list of cyclones to hit eastern Africa in the twentieth

century, see Colin Buckle, *Weather and Climate in Africa* (Essex: Longman, 1996), 164.

13. Robert Korty, 'Hurricane (Typhoon, Cyclone),' in *Encyclopedia of Natural Hazards*, P. Bobrowsky, ed. (Springer, 2013), 481–493.
14. Martin Fisher and David A. Membery, 'Climate,' in Shahina A. Ghazanfar and Martin Fisher (eds.), *Vegetation of the Arabian Peninsula* (Norwell: Kluwer Academic Publishers, 1998), 5–38.
15. David A. Membery, 'Monsoon Tropical Cyclones: Part 1,' *Weather* 56 (2001): 431–438.
16. Ibid., 17.
17. Mohammad Dibajnia, Mohsen Soltanpour, Rob Nairn, and Mohammadreza Allahyar, 'Cyclone Gonu: The Most Intense Tropical Cyclone on Record in the Arabian Sea,' in Y. Charabi (ed.), *Indian Ocean Tropical Cyclones and Climate Change* (New york: Springer, 2010), 147–157.
18. 'Oman pummeled as cyclone passes' *BBC News* (Jun. 6, 2007).
19. Saleh al-Shaibany, 'Cyclone Phet damages may cost $780 mln,' *Reuters* (Jun. 7, 2010).
20. Great Britain, Hydrographic Office, Admiralty, *The Red Sea and Gulf of Aden Pilot*, 5th ed. (London, 1900), 16.
21. Matthew S. Hopper, 'The Globalization of Dried Fruit: Transformation of the Eastern Arabian Economy, 1860s–1920s,' in James L. Gelvin and Nile Green (eds.), *Global Muslims in the Age of Steam and Print* (Berkeley: University of California Press, 2014), 158–182.
22. Bhacker, 109–113, 136–137.
23. Calvin Allen, 'Sayyids, Shets and Sultāns: Politics and Trade in Masqat Under the Al BūSa'īd, 1785–1914' (PhD dissertation, University of Washington, 1978), 140–156.
24. *Administration Report of the Political Agency Muscat for the year 1871–72*, India Office Records, British Library, London, UK (hereafter IOR) R/15/6/5.
25. Allen, 140–153.
26. Oscar Willoughby Riggs, 'The Fruit-Ships at New York,' *Frank Leslie's Popular Monthly* XXI, no. 5 (May 1886): 599.
27. Bertram Thomas, *Alarms and Excursions in Arabia* (Indianapolis: Bobbs-Merrill, 1931), 142.
28. J. C. Wilkinson, *Water and Tribal Settlement*, 47–51; Charles C. Graf, 'The Batinah Hydrologic Area,' in Sultanate of Oman, Public Authority for Water Resources, The Hydrology of the Sultanate of Oman: A Preliminary Assessment, PAWR 83-1 (Muscat, Sultanate of Oman: Public Authority for Water Resources, 1983), 28–43; Robert Dale, 'The Water Resources of Oman' (Report for TetraTech International, c. 1980), James H. Critchfield Papers, Georgetam University Special Collection, Box 77.

29. Bertram Thomas, *Alarms and Excursions* (Indianapolis: Bobbs-Merrill, 1931), 238.
30. *Administration Report of the Political Agency, Muscat, for the year 1876–77, 79–82*, Centre for Documentation and Research, Abu Dhabi, UAE (hereafter CDR) ND1/H.
31. *Administration Report of the Muscat Political Agency for 1884–85*. IOR R/15/6/17.
32. Surgeon Major A.S.G. Jayakar to Col. EC Ross, (Sept. 21,1890), IOR R/15/6/20.
33. Chris Blount, Hermann M. Fritz, and Ahmed Hamoud Mohammed Al-Harthy, 'Coastal Vulnerability Assessment Based on Historic Tropical Cyclones in the Arabian Sea,' in Y. Charabi (ed.), *Indian Ocean Tropical Cyclones and Climate Change* (New York: Springer Science, 2010), 207–214.
34. Sheriff, 234–235.
35. J.S. Saldanha, *Precis on Slave Trade in the Gulf of Oman and the Persian Gulf, 1873–1905 (With a Retrospect into previous history from 1852)* [Secret], 37–41, IOR L/PS/20/C246.
36. E. C. Ross, Political Resident in the Persian Gulf and Consul General for Fars to WJ Cuningham Sect to Govt of India Foreign Dept Calcutta (29 Dec. 1890), IOR R/15/6/20.
37. Herbert W. Dowding HMS Osprey to Rear Admiral Sir Fred'k W. Richards, Commander in Chief (19 Sept. 1885), National Archives, UK, Admiralty Records, (hereafter ADM) 1/6758.
38. Ibid.
39. Commander Ripon HMS Ranger to H.W. Dowding HMS Osprey, Senior Naval Officer, Persian Gulf Division (October 19, 1885), National Archives, UK, Admiralty Records, ADM 1/6758.
40. Richard Pankhurst and Douglas Johnson, 'The Great Drought and Famine of 1888–92 in Northeast Africa,' in *The Ecology of Survival; Case Studies from Northeast African History*, ed. Douglas Johnson and David Anderson (Colorado: Westview Press, 1988), 47–73; Alex de Waal, *Famine that Kills: Darfur Sudan* (Oxford: Oxford University Press, 2005); Steven Serels, 'Famines of War: The Red Sea Grain Market and Famine in Eastern Sudan, 1889–1891,' *Northeast African Studies* 12, no. 1 (2012): 73–94. See also, Steven Serels, *Starvation and the State: Famine, Slavery, and Power in Sudan, 1883–1956* (New York: Palgrave Macmillan, 2013).
41. Abdussamad H. Ahmad, 'Peasant Conditions in Gojjam During the Great Famine 1888–1892,' *Journal of Ethiopian Studies* 20 (Nov. 1987): 1–18; John A. Rowe and Kjell Hødnebø, 'Rinderpest in the Sudan, 1888–1890,' *Sudanic Africa* 5 (1994): 149–178.

42. James Giblin, 'Famine and Social Change During the Transition to Colonial Rule in Northeastern Tanzania, 1880–1896,' *African Economic History* 15 (1986): 85–105.
43. Juhani Koponen, 'War, Famine, and Pestilence in Late Precolonial Tanzania: A Case for a Heightened Mortality,' *International Journal of African Historical Studies* 21, no. 4 (1988): 637–676. For other related epidemics, see: Myron Echenberg, *Africa in the Time of Cholera: A History of Pandemics from 1817 to the Present* (New York: Cambridge University Press, 2011).
44. Gerald W. Hartwig, 'Demographic Considerations in East Africa During the Nineteenth Century,' *International Journal of African Historical Studies* 12, no. 4 (1979): 653–672.
45. S. Nieuwolt, 'The East African Monsoons and Their Effects on Agriculture,' *GeoJournal* 3, no. 2, (1979): 193–200.
46. Isaac Hourwich, 'The Crisis in Russian Agriculture,' *The Yale Review: A Quarterly Journal of History and Political Science* 1 (May 1892–Feb. 1893): 411–433.
47. Mike Davis, *Late Victorian Holocausts: El Niño Famines and the Making of the Third World* (Verso, 2001).
48. Norman and W.J.S. Lokyer, 'On Solar Changes and Variations of Rainfall in the Region Surrounding the Indian Ocean,' *Proceedings of the Royal Society of London* 67 (Jun.–Dec., 1900): 409–431.
49. Mikiso Hane, *Peasants, Rebels, Women, and Outcastes: The Underside of Modern Japan*, 2nd ed. (Lanham, MD: Rowman& Littlefield, 2003), 27.
50. Bahru Zewde, *A History of Modern Ethiopia, 1855–1974* (Athens: Ohio University Press, 1991), 71–72.
51. Cooper, 126.
52. Ibid., 128.
53. Timothy Derek Fernyhough, *Serfs, Slaves and Shifta: Modes of Production and Resistance in Pre-revolutionary Ethiopia* (Addis Ababa: Shama Books, 2010), 87.
54. Davis, 127–133.
55. Davis, 133–138.
56. Serels, 'Famines of War,' 73–92.
57. Christopher Lloyd, *The Navy and the Slave Trade* (London: Frank Cass & Co., 1968), 271.
58. Erik Gilbert, *Dhows and the Colonial Economy of Zanzibar, 1860–1970* (Oxford: James Currey, 2004), 63.
59. As was the case elsewhere in the British Empire (e.g. Northern Nigeria), once an area came under formal British control, the problem of slavery conveniently disappeared or became a minor problem that required slow, judicious handling. In Zanzibar, for instance, the horrors of slavery, which

ostensibly led to the declaration of the British protectorate over the island, were conveniently ignored once the region became part of the British Empire. When British officials eventually began to move toward the abolition of slavery on the island, the institution looked remarkably more benign than it had a few decades before. J.T. Last remarked, 'There are many reasons why the slaves of Zanzibar and Pemba should not be given their freedom in a mass, and even why the whole body of slaves should not be allowed to acquire their freedom in a short space of time. The slaves are generally quite satisfied with their position, the majority of them have no wish to be free, and some absolutely refuse the freedom offered to them.' See: J.T. Last, *Zanzibar Government commissioner, Mangapwani, Report by the Commissioner Last on the Working of the Decree for the Abolition of the Legal Status of Slavery* (Jan. 25, 1898) enclosure no. 2 in Sir A. Hardinge to the Marquess of Salisbury, Mombasa, Mar. 23, 1898, in *Correspondence Respecting the Abolition of the Legal Status of Slavery in Zanzibar and Pemba, Slave Trade*, vol. 95 (1898), 495.

60. Giblin, 97.
61. Marques of Lansdowne to Sir E. Monson (Apr. 16, 1902), IOR R/15/1/552.
62. Matthew S. Hopper, 'The African Presence in Arabia: Slavery, the World Economy, and the African Diaspora in Eastern Arabia, 1840–1940,' PhD dissertation (UCLA, 2006), 83.
63. William S. Atwell, 'Volcanism and Short-Term Climatic Change in East Asian and World History, c. 1200–1699,' *Journal of World History* 12, no. 1 (Spring, 2001): 29–98.
64. Paul Handler and Karen Andsager, 'El Niño, Volcanism, and Global Climate,' *Human Ecology* 22, no. 1 (Mar. 1994): 37–57.
65. 'Climate Change and Southern Hemisphere Tropical Cyclones: International Initiative – Progress since the First International Conference on Indian Ocean Tropical Cyclones and Climate Change,' in U.C. Mohanty et al. (eds.), *Monitoring and Prediction of Tropical Cyclones in the Indian Ocean and Climate Change* (New Delhi: Capital Publishing, 2014), 18–32.
66. See for example: Assaf Anyamba, Jean-Paul Chretien, Jennifer Small, et al., 'Prediction of a Rift Valley Fever Outbreak,' *Proceedings of the National Academy of Sciences of the United States of America* 106, no. 3 (Jan. 20, 2009): 955–959.
67. Emily Black, 'The Relationship Between Indian Ocean Sea-Surface Temperature and East African Rainfall,' *Philosophical Transactions: Mathematical, Physical and Engineering Sciences* 363, no. 1826, Atmosphere-Ocean-Ecology Dynamics in the Western Indian Ocean (Jan. 15, 2005): 43–47.

68. Secretariat of the World Meteorological Organization, *Tropical Cyclone Operational Plan for the Bay of Bengal and the Arabian Sea*, Report No. TCP-21 (2011): 5.
69. Karumuri Ashok et al., 'El Nino Modoki and its Possible Teleconnection,' *Journal of Geophysical Research* 112 (2007): 1–32.
70. P. K. Pradhan et al., 'Modoki, Indian Ocean Dipole, and western North Pacific typhoons: Possible implications for extreme events,' *Journal of Geophysical Research* 116 (2011): 1–12.
71. K. G. Sumesh and M. R. Ramesh Kumar, 'Tropical Cyclones over North Indian Ocean During El-Nino Modoki Years,' *Natural Hazards* 68 (2013): 1057–1074.
72. H. H. Lamb, 'Volcanic Dust in the Atmosphere; with a Chronology and Assessment of Its Meteorological Significance,' *Philosophical Transactions of the Royal Society of London* 266, no. 1178 (Jul. 2, 1970): 425–533.
73. Sharon E. Nicholson, Douglas Klotter and Amin K. Dezfuli, 'Spatial Reconstruction of Semi-quantitative Precipitation Fields over Africa During the Nineteenth Century from Documentary Evidence and Gauge Data,' *Quaternary Research* 78 (2012): 13–23. See also: Sharon E. Nicholson, Amin K. Dezfuli and Douglas Klotter, 'A Two-Century Precipitation Dataset for the Continent of Africa,' *Bulletin of the American Meteorological Society* 93, no. 8 (Aug. 2012): 1219–1231.

Bibliography

Administration report of the political agency Muscat for the year 1871–72, India Office Records, British Library, London, UK, R/15/6/5.

Administration report of the political agency, Muscat, for the year 1876–77, 79–82, Centre for Documentation and Research, Abu Dhabi, UAE, ND1/H.

Administration report of the Muscat political agency for 1884–85. IOR R/15/6/17.

Ahmad, Abdussamad H. 1987. Peasant conditions in Gojjam during the Great Famine 1888–1892. *Journal of Ethiopian Studies* 20: 1–18.

Allen, Calvin 1978. Sayyids, Shets and Sultāns: Politics and trade in Masqat under the Al Bū Sa'īd, 1785–1914. PhD dissertation, University of Washington.

al-Shaibany, Saleh. 2010. Cyclone Phet damages may cost $780 mln. *Reuters* Jun. 7.

Anyamba, Assaf, Jean-Paul Chretien, Jennifer Small, et al. 2009. Prediction of a rift valley fever outbreak. *Proceedings of the National Academy of Sciences of the United States of America* 106(3): 955–959 .Jan. 20, 2009

Ashok, Karumuri, et al. 2007. El Nino Modoki and its possible teleconnection. *Journal of Geophysical Research* 112: 1–32.

Atwell, William S. 2001. Volcanism and short-term climatic change in East Asian and world history, c. 1200–1699. *Journal of World History* 12(1): 29–98.

Bhacker, Mohammed Reda. 1992. *Trade and empire in Muscat and Zanzibar: Roots of British domination*. New York: Routledge.

Blount, Chris, Hermann M. Fritz, and Ahmed Hamoud Mohammed Al-Harthy. 2010. Coastal vulnerability assessment based on historic tropical cyclones in the Arabian Sea. In *Indian Ocean tropical cyclones and climate change*, ed. Y. Charabi, 207–214. New York: Springer Science.

Buckle, Colin. 1996. *Weather and climate in Africa*. Essex: Longman.

Cooper, Frederick. 1997. *Plantation slavery on the East Coast of Africa*. Portsmouth: Heinemann.

Dale, Robert. *The water resources of Oman* (Report for Tetra Tech International, c. 1980), James H. Critchfield Papers, Georgetum University Special collection.

Davis, Mike. 2001. *Late Victorian Holocausts: El Niño famines and the making of the third world*. New York: Verso.

de Waal, Alex. 2005. *Famine that kills: Darfur Sudan*. Oxford: Oxford University Press.

Dibajnia, Mohammad, Mohsen Soltanpour, Rob Nairn, and Mohammadreza Allahyar. 2010. Cyclone Gonu: The most intense tropical cyclone on record in the Arabian Sea. In *Indian Ocean tropical cyclones and climate change*, ed. Y. Charabi, 147–157. New York: Springer.

Dowding, Herbert W. 1885. HMS Osprey to Rear Admiral Sir Fred'k W. Richards, Commander in Chief (Sept. 19, 1885), National Archives, UK, Admiralty Records, 1/6758.

Echenberg, Myron. 2011. *Africa in the time of cholera: A history of pandemics from 1817 to the present*. New York: Cambridge University Press.

Fernyhough, Timothy Derek. 2010. *Serfs, slaves and shifta: Modes of production and resistance in pre-revolutionary Ethiopia*. Addis Ababa: Shama Books.

Fisher, Martin, and David A. Membery. 1998. Climate. In *Vegetation of the Arabian Peninsula*, eds. Shahina A. Ghazanfar, and Martin Fisher, 5–38. Norwell: Kluwer Academic Publishers.

Giblin, James. 1986. Famine and social change during the transition to colonial rule in northeastern Tanzania, 1880–1896. *African Economic History* 15: 85–105.

Gilbert, Erik. 2004. *Dhows and the colonial economy of Zanzibar, 1860–1970*. Oxford: James Currey.

Graf, Charles C. 1983. The Batinah hydrologic area. In *The hydrology of the Sultanate of Oman: A preliminary assessment*, PAWR 83-1, ed. Sultanate of Oman, Public Authority for Water Resources, 28–43. Muscat: Public Authority for Water Resources.

Great Britain, Hydrographic Office, Admiralty. 1900. *The Red Sea and Gulf of Aden pilot*, 5th edn. London.

Handler, Paul, and Karen Andsager. 1994. El Niño, volcanism, and global climate. *Human Ecology* 22(1): 37–57.

Hane, Mikiso. 2003. *Peasants, rebels, women, and outcastes: The underside of modern Japan*, 2nd edn. Lanham: Rowman & Littlefield.

Hartwig, Gerald W. 1979. Demographic considerations in East Africa during the nineteenth century. *International Journal of African Historical Studies* 12(4): 653–672.

Hopper, Matthew S. 2006. The African presence in Arabia: Slavery, the world economy, and the African diaspora in Eastern Arabia, 1840–1940. PhD dissertation, UCLA.

Hopper, Matthew S. 2014. The globalization of dried fruit: Transformation of the Eastern Arabian economy, 1860s–1920s. In *Global Muslims in the age of steam and print*, eds. James L. Gelvin, and Nile Green, 158–182. Berkeley: University of California Press.

Hourwich, Isaac. The crisis in Russian agriculture. *The Yale Review: A Quarterly Journal of History and Political Science* 1 (May 1892–Feb. 1893): 411–433.

Jayakar, Surgeon Major A.S.G. 1890. (in charge Pol. Agency Muscat) to Col. EC Ross (Pol Res in the Persian Gulf and Consul Gen for Fars, etc.) Muscat, (Sept. 21,1890), IOR R/15/6/20.

Koponen, Juhani. 1988. War, famine, and pestilence in late precolonial Tanzania: A case for a heightened mortality. *International Journal of African Historical Studies* 21(4): 637–676.

Korty, Robert. 2013. Hurricane (typhoon, cyclone). In *Encyclopedia of natural hazards*, ed. P. Bobrowsky, 481–493. New York: Springer.

Kuleshov, Yuriy. 2014. Climate change and southern Hemisphere tropical cyclones: International i Initiative – Progress since the first international conference on Indian Ocean tropical cyclones and climate change. In *Monitoring and prediction of tropical cyclones in the Indian Ocean and climate change*, eds. U.C. Mohanty et al., 18–32. New York: Springer Publishing.

Lamb, H.H. 1970. Volcanic dust in the atmosphere; with a chronology and assessment of its meteorological significance. *Philosophical Transactions of the Royal Society of London* 266(1178): 425–533. (Jul. 2, 1970).

Last, J.T. *Zanzibar Government commissioner, Mangapwani, Report by the commissioner last on the working of the decree for the abolition of the legal status of slavery* (Jan. 25, 1898) enclosure no. 2 in Sir A. Hardinge to the Marquess of Salisbury, Mombasa, Mar. 23, 1898, in *Correspondence respecting the abolition of the legal status of slavery in Zanzibar and Pemba, Slave Trade*, vol. 95 (1898), 495.

Lloyd, Christopher. 1968. *The navy and the slave trade*. London: Frank Cass & Co..

Lockyer, Norman, and W.J.S. Lockyer. 1900. On solar changes and variations of rainfall in the region surrounding the Indian Ocean. *Proceedings of the Royal Society of London* 67: 409–431.

Lynn, Robert Nunez. 1905. *Zanzibar in contemporary times*. London: Hurst and Blackett.

Marques of Lansdowne to Sir E. Monson (Apr. 16, 1902), IOR R/15/1/552.

Membery, David A. 2001. Monsoon tropical cyclones: Part 1. *Weather* 56: 431–438.

Membery, David A. 2002. Monsoon tropical cyclones: Part 2. *Weather* 57: 246–255.

'Oman pummeled as cyclone passes' *BBC News* (Jun. 6, 2007).

Nicholson, Sharon E., Amin K. Dezfuli, and Douglas Klotter. 2012a. A two-century precipitation dataset for the continent of Africa. *Bulletin of the American Meteorological Society* 93(8): 1219–1231.

Nicholson, Sharon E., Douglas Klotter, and Amin K. Dezfuli. 2012b. Spatial reconstruction of semi-quantitative precipitation fields over Africa during the nineteenth century from documentary evidence and gauge data. *Quaternary Research* 78: 13–23.

Nieuwolt, Simon. 1979. The East African monsoons and their effects on agriculture. *Geo Journal* 3(2): 193–200.

Pankhurst, Richard, and Douglas Johnson. 1988. The great drought and famine of 1888–92 in northeast Africa. In *The ecology of survival;case studies from northeast African history*, eds. Douglas Johnson, and David Anderson, 47–73. Colorado: Westview Press.

Pradhan, P.K., et al. 2011. Modoki, Indian Ocean Dipole, and western North Pacific typhoons: Possible implications for extreme events. *Journal of Geophysical Research* 116: 1–12.

Pratt, Seth A. 1890. Clove Culture in Zanzibar. *U.S. Department of State Reports of the Consuls of the United States* 32(112–115): 687–688.

Riggs, Oscar Willoughby. 1886. The Fruit-Ships at New York. *Frank Leslie's Popular Monthly* XXI(5): 599.

Ripon, Commander. 1885. HMS Ranger to H.W. Dowding HMS Osprey, Senior Naval Officer, Persian Gulf Division (Oct. 19, 1885), National Archives, UK, Admiralty Records, ADM 1/6758.

Ross, EC Political Resident in the Persian Gulf and Consul General for Fars to WJ Cuningham Sect to Govt of India Foreign Dept Calcutta (Dec. 29, 1890), IOR R/15/6/20.

Rowe, John A., and Kjell Hødnebø. 1994. Rinderpest in the Sudan, 1888–1890. *Sudanic Africa* 5: 149–178.

Saldanha, J. S. *Precis on slave trade in the Gulf of Oman and the Persian Gulf, 1873–1905* (*with a retrospect into previous history from 1852*) [Secret], 37–41, IOR L/PS/20/C246.

Secretariat of the World Meteorological Organization. 2011. *Tropical cyclone operational plan for the Bay of Bengal and the Arabian Sea*, Report No. TCP-21, 5.

Serels, Steven. 2012. Famines of war: The Red Sea grain market and famine in eastern Sudan, 1889–1891. *Northeast African Studies* 12(1): 73–94.

Serels, Steven. 2013. *Starvation and the state:famine, slavery, and power in Sudan, 1883–1956*. New York: Palgrave Macmillan.

Sheriff, Abdul. 1987. *Slaves, spices &slavery in Zanzibar*. London: James Curry.

Sumesh, K.G., and M.R. Ramesh Kumar. 2013. Tropical cyclones over north Indian Ocean during El-Nino Modoki years. *Natural Hazards* 68: 1057–1074.
Thomas, Bertram. 1931. *Alarms and excursions in Arabia*. Indianapolis: Bobbs-Merrill.
Wilkinson, J.C. 1977. *Water and tribal settlement*. Oxford: Oxford University Press.
Zewde, Bahru. 1991. *A history of modern Ethiopia, 1855–1974*. Athens: Ohio University Press.

CHAPTER 11

Their Inescapable Portion? Cyclones, Disaster Relief, and the Political Economy of Pearlshelling in Northwest Australia, 1865–1935

Joseph Christensen

The start of the 1935 season could not come soon enough for the pearlers of Broome, an isolated port on Australia's remote Northwest coast. Markets for mother-of-pearl had crashed during the Great Depression, and increasing competition from Japanese sampans operating in the nearby Timor and Arafura Seas was now threatening the viability of what had once been the world's largest pearlshelling fleet. The West Australian Pearlers' Association had spent the past summer's 'lay-up' season lobbying for assistance to avert the industry's collapse, and a one-off grant of £2500 announced by Australia's Commonwealth Government in early March was the inducement the pearlers needed to put to sea for another year.[1] The bulk of the fleet had been out for barely a week when the morning of 26 March broke with a strong south-easterly and a haze across the horizon, weather that Broome's residents feared signalled the approach of a late-season tropical storm. The coastal trader SS *Koolinda*, north of Broome on the coast of the Dampier Peninsula, was soon transmitting

J. Christensen
Asia Research Centre, Murdoch University, Perth, Australia

G. Bankoff, J. Christensen (eds.), *Natural Hazards and Peoples in the Indian Ocean World*, DOI 10.1057/978-1-349-94857-4_11

reports of decreasing air pressure and worsening weather. Around noon the town's postmaster received a telegram from the Australian Bureau of Meteorology advising that an atmospheric disturbance was imminent. There could be little doubt that a cyclone was now bearing down from the north. Storm-warning flags were posted at the harbour, and Broome's residents began preparing for the storm.[2]

Little, however, could be done for the pearling luggers already at sea. Few vessels in the fleet were equipped with engines, and none carried radios. At least 36 luggers lay in the storm's path around the Lacapede Islands and Beagle Bay (see Fig. 11.1).[3] Past experience had taught the pearlers that no anchor could hold against a Northwest cyclone, so as the barometer began falling and dark clouds closed in over the horizon, each vessel's master faced a simple yet fraught decision: to make for the mainland and seek sanctuary amongst the inlets and mangrove creeks lining the shoreline; or put out to open waters for sea room to ride out the storm. The fleet scattered, and by mid-afternoon 'wind of hurricane force, almost continuous wind and rain squalls, high mountainous seas [and] heavy confused swell' had, as the *Koolinda*'s logbook recorded, set in along the coast.[4] *Koolinda* succeeded in manoeuvring around the cyclone and was immediately diverted to search for the missing pearling fleet. Some luggers had found havens along the mainland, and some others had ridden out the storm at sea, but in the confusion that followed the cyclone's passage at least 20 vessels were reported missing. During a three-day cruise through wreckage-strewn waters *Koolinda* rescued 18 men. An aircraft sent out from Broome on 30 March counted 15 luggers afloat, although most were dismasted and making a slow passage home under makeshift rigs. The first rescue boats to reach the Lacapedes observed wreckage and bodies up to seven miles out from the Islands, but only a single survivor was found ashore.[5]

It was more than a week before the full scale of losses could be tallied. The barometer had fallen as low as 29.30 inches and winds of 65 miles (105 km) per hour had been recorded in Broome, making it a Category 2 cyclone by the time it reached the town. To the north, the Catholic Mission at Beagle Bay had borne its full fury and lay in ruins. At sea, 21 luggers had sunk and a further 15 were damaged, with most remaining afloat only after having their masts cut away. The number of dead was eventually fixed at 142. All but one were Asian divers and crewman employed under regulations that exempted the pearling industry from the Commonwealth's *Immigration Restriction Act*, or the 'White Australia Policy': 60 Japanese,

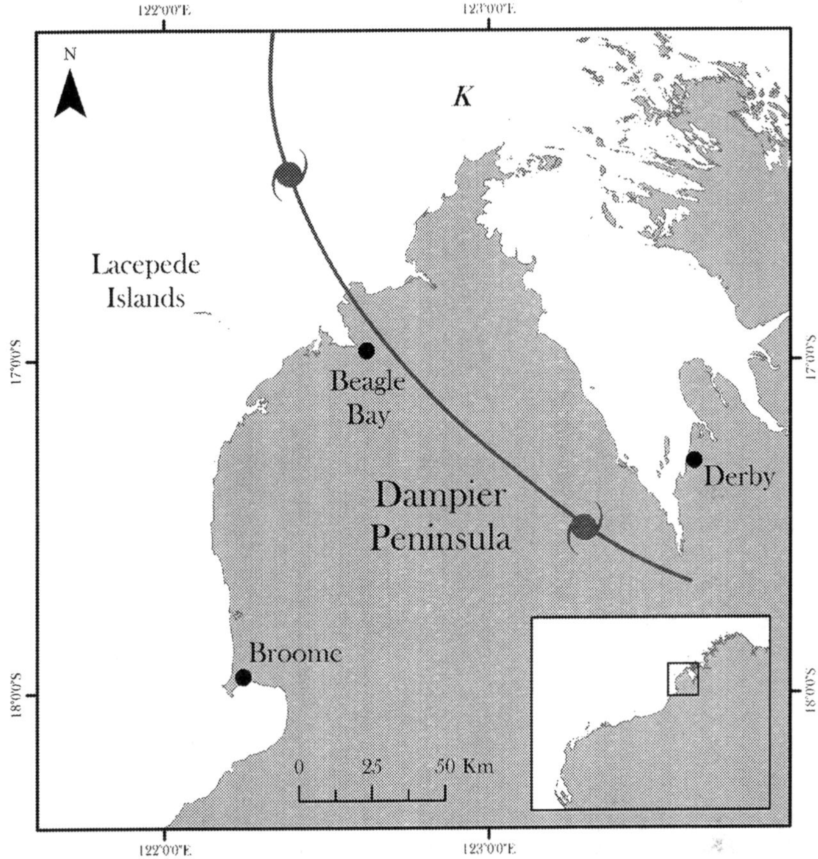

Fig. 11.1 Broome and the Dampier Peninsula, showing the track of the March 1935 cyclone and the approximate position of the *Koolinda* (marked a 'K') when the cyclone was first reported

34 Malays (Indonesians), 30 Koepangers (Timorese), and 17 Chinese died in the storm. A sole Aboriginal crewman was lost; two European 'shell-openers' at sea with the fleet survived. The Pearlers' Association estimated that losses totalled £37,000.[6] It was the single worst disaster to befall an industry that had weathered more than 60 tropical storms since its inception in 1865, and in terms of lives lost, the second most costly cyclone on the Northwest Australian coast after the death of 150 passengers and crew

in sinking of the SS *Koombana* on 20 March 1912. It remains today as one of Australia's worst weather-related disasters.[7]

The cyclone of 27 March 1935 is also significant because it marks an important milestone in the development of Australian disaster relief policy. In the wake of the cyclone the Commonwealth Government built on its £2500 payment with a £5000 grant to assist the industry's recovery, supplementing a Western Australian Government contribution of £7000 for disaster relief.[8] This milestone has been overlooked in the past. In *Bushfire Disaster: An Australian Community in Crisis* (1975), Roger Wettenhall suggests that a £1000 grant to relieve Tasmanian communities destroyed by bushfire in February 1939 inaugurated Commonwealth provision of financial aid for disaster relief, an area of public policy not specified in Australia's Constitution and thus remaining a residual function of State Governments.[9] Historians of other natural hazards, particularly drought, have followed Wettenhall in highlighting the significance of the 1939 Tasmanian bushfires.[10] Yet four years earlier, Broome's pearlshellers had received five times this amount from the Commonwealth, at a time when their industry was only marginally economical, and where the main victims of the cyclone had been 'alien' or non-European workers whom were denied, as a matter of Government policy, from ever settling permanently in Australia.

This chapter examines the March 1935 disaster and its contribution to the development of Australian disaster relief policy. It is divided into three parts. The first section reviews the pearling industry's historical experience with extreme tropical weather on the Northwest Australian coast. The second section considers the early meteorological study of Australian cyclones and the problems with forecasting and storm warning that contributed to the March 1935 disaster. The third section examines the factors that influenced the decision to assist the industry to rebuild after the storm. In the early twenty-first century, the Commonwealth's Natural Disaster Relief and Recovery Arrangements (NDRRA) provide hundreds of millions of dollars annually in assistance to State Governments for relief and recovery from storms, droughts, floods, fires, and other natural hazards, reflecting the increasing co-operation and coordination between governments that developed in the second half of the twentieth century.[11] This chapter suggests that a central root of such co-operation lies in the unique political economy of Australian pearlshelling, which mandated Government support for the industry because of its crucial strategic role in maintaining European settlements along the continent's isolated northern coasts against a backdrop of growing concern over Japanese designs in the region. The cruel irony of the March 1935 disaster is that its Japanese, Chinese, and Indonesian victims died sustaining a vision for a 'White Australia' from which they were themselves excluded.

Cyclones and the Development of Pearlshelling in Northwest Australia

Tropical cyclones (also known as hurricanes or typhoons) are low-pressure systems that form over warm waters and produce gale-force winds of at least 63 km/h. A cyclone is classified as 'severe' when it produces hurricane-force winds of 118 km/h, although some cyclones are known to produce winds exceeding 200 km/h and be accompanied by torrential rain and storm surges that swamp low-lying coastlines. Australia's Northwest coast, lying south of the equator in the eastern Indian Ocean, receives around 9 % of the world's tropical cyclones and three-quarters of Australia's severe tropical cyclones, making it the most cyclone-prone region in the Southern hemisphere.[12] The cyclone season coincides with the summer monsoon or 'wet season', running from November to late April, although most cyclones occur between late December and the end of March. An average of five cyclone form off the coast each year, and an average of two pass across it (one of which will be classified as 'severe'), mostly in the area between Broome and Onslow (see Fig. 11.2). However, there is considerable annual variation in the frequency of storm systems, and the paths taken by cyclones can be erratic, with several since the mid-nineteenth century passing south to Shark Bay and beyond.[13] Northwest cyclones are also affected by variations in the El Niño Southern Oscillation (ENSO), with a tendency to be less common in El Niño years and more common in La Nina years, although the frequency of severe cyclones is similar in either climatic phase.[14]

European settlement in the 'Northwest' (the area north of the 26th parallel, comprising the Gascoyne, Pilbara, and Kimberley regions) began at Cossack and nearby Roebourne in the early 1860s, before extending westward to the Ashburton and Gascoyne Rivers and northeast into the West Kimberley during the 1870s and 1880s.[15] The *Australian Encyclopaedia* (1958) records that over 300 ships and more than 1000 people were lost as a result of tropical storms in this region during the century ending in 1954.[16] Most of these losses were sustained by the pearling industry, the

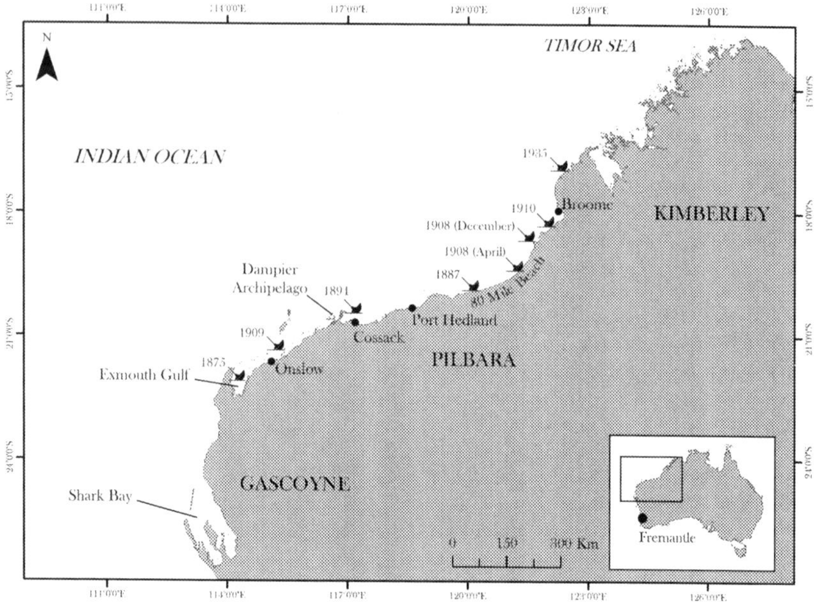

Fig. 11.2 Northwest Australia, showing places mentioned in the chapter and the locations of the severe cyclones of 1875, 1887, 1908 (April and November), 1910, and 1935

most extensive maritime activity carried out along the coast in this period. Pearlshelling began around 1865 when the first settlers in the district around Cossack began to collect the large mother-of-peal shell *Pinctada maxima* as a sideline to their pastoral activities. As the profitability of the new industry was demonstrated, the Northwest coast began to attract men and boats from the colony's south, other Australian colonies, or via Southeast Asian ports such as Koepang or Singapore, and the first dedicated pearlshelling operations began. By 1870, there was a fleet of around 30 vessels and more than 300 men working on the coast.[17]

When pearlshelling began, operations were based around the collection of shell by hand, a practice aided by the ample tides that daily exposed large stretches of shoreline. As the coast around Cossack became denuded the industry transitioned to free or naked diving, employing Aboriginal divers working from small dinghies attached to larger vessels under the command of a European master. Divers from the South Pacific and later the Malay Archipelago were brought to the coast from about 1869 to replace Aboriginal labour that was becoming scarce due to the ravages of smallpox and other exotic diseases and the high attrition of divers that resulted from the shift to deeper waters. Malays, in particular, were favoured because of their ability to reach greater depths, and their employment grew rapidly; in 1876, they accounted for 800 of the 1200 divers working in the Northwest. A second and more profound transformation took place in the mid-1880s through the introduction of diving suits (or the 'diving dress'), comprising of a copper or brass helmet, rubber air hose, and canvas suit. The new technology opened up vast new beds of shell in deeper waters north of Cossack, off the Eighty Mile Beach and West Kimberley coast. A new class of skilled labour, originating mainly from Japan or the Philippines archipelago, entered the industry in place of free divers, and the lugger, a two masted ketch suitable for drifting slowly over the beds with a diver underneath, became standard in the fleet. The industry gravitated north towards Broome, founded in 1883, and underwent a rapid expansion. By the early 1900s, the industry employed more than 2000 men, over 900 of whom were Japanese divers and tenders, aboard more than 350 European-owned boats, and produced more than 1000 tons annually, approximately half the global supply of high-quality *maxima* shell.[18]

Figure 11.3 lists the tropical storms that affected the industry between its establishment and the 1935 cyclone. Several points should be made about this table. It lists only events that resulted in the loss of five or more lives and/or three or more vessels; a number of lesser events are not included. The table also records all 'tropical storms' irrespective of whether or not it would meet current criteria for cyclone classification. This is partly because accurate data only dates from the late nineteenth century; although a recording station was established at Cossack in 1881, more systematic records were only kept following the opening of the telegraph line between Perth and Broome in 1889, and even then, the number of recording stations was few.[20] Another problem surrounds local nomenclature. By the 1880s, pearlshellers typically distinguished two types of tropical storm: a 'cock-eyed Bob', referring to a sudden squall at sea or a tornado that originated on land before crossing the coast, with localised effects only; and a 'willy-willy', or true cyclone, which was marked by sustained strong winds and which, with a life span of two to three days, could cause considerable damage over a wide area.[21] Figure 11.2 covers both cock-eyed bobs and willy-willies. By focussing on damage to the pearling industry,

Date	Location/s	Losses
4 Jan 1868	Cossack	3 vessels
25 Dec 1870	Cossack, Port Walcott	2 lives; 8 vessels
20 Mar 1872	Cossack, Port Headland	2 lives; 8 vessels
24 Dec 1875	Exmouth Gulf, Cossack	56 lives; 3 vessels
4 Mar 1878	Cossack	8 lives; 6 vessels
24 Jan 1879	Lacapede Is, Cossack	19 lives; 3 vessels
7 Jan 1881	Cossack	6 lives; 12 vessels
6 Mar 1882	Cossack, Pt Walcott	6 lives; 4 vessels
8 Feb 1883	Cossack, Condon	10 lives; 2 vessels
16 Jan 1884	Broome	5 lives; 5 vessels
11 Jan 1887	Cossack	6 lives; 5 vessels
22 Apr 1887	80 Mile Beach (Yammadery Ck)	140 lives; 20 vessels lost, 34 damaged
6 Feb 1889	King Sound	4 vessels
28 Feb 1889	Cossack	6 lives; 10 vessels
27 Jan 1890	Camden Sound	3 vessels
25 Feb 1893	Shark Bay, Exmouth Gulf	12 lives; 49 vessels
3 Jan 1894	Cossack	4 vessels
9 Jan 1894	Cossack	20 lives; 14 vessels
24 Jan 1898	Shark Bay	11 vessels
2 Apr 1898	Cossack	3 vessels
12 Jan 1899	Cossack and Broome	10 vessels
17 Apr 1904	Broome	8 lives; 4 vessels
13 Mar 1907	Broome to Shark Bay	3 vessels
27 Apr 1908	80 Mile Beach (La Grange)	117 lives; 44 vessels
8 Dec 1908	80 Mile Beach (Wallal)	50 lives; 39 vessels
5 Apr 1909	Onslow	24 lives; 4 vessels
19 Nov 1910	Broome	40 lives; 34 vessels lost, 67 damaged
6 Mar 1912	Broome	5 vessels
27 Mar 1935	Lacapede Is, Beagle Bay	142 lives; 21 vessels lost, 15 damaged

Fig. 11.3 Major losses through cyclones in the pearling industry, Northwest Australia, 1868–1935 (only losses of five or more lives and/or three or more vessels have been included; losses sustained by other industries, such as passenger or merchant shipping, are not counted) (Sources: Hunt H.A., S.S. Visher and D. Hodge [eds.] *Australian Hurricanes and Related Storms: With an Appendix on Hurricanes in the South Pacific* [Melbourne: Commonwealth Bureau of Meteorology, Bulletin No.16, 1925], and others)[19]

the table glosses over storms that caused extensive destruction on land, including cyclones that devastated Cossack in February 1893, Onslow in December 1897, and Broome in January 1926. Storms that affected the Shark Bay pearl fishery, a separate industry based on the exploitation of the pearl oyster *Pinctada albina*, have not been included, although severe cyclones were recorded here in February 1893 and February 1921. Finally, all figures should be treated as approximations. Due to its isolation (and the tendency for cyclones to disrupt coastal shipping and bring down telegraph lines), there was often a considerable lag in the transmission of information from the Northwest to the colonial capital of Perth, creating confusion and misinformation in the wake of a storm. Vessels reported as lost might later show up, vessels reported as beached or sunk could later be refloated, and it is sometimes unclear if a particular vessel was engaged in pearlshelling, in coastal trade, or another pursuit such as turtle fishing. This contributes to a tendency for losses to be exaggerated, sometimes wildly, in historical accounts.[22]

Despite these caveats, Fig. 11.3 captures the most costly disasters in the industry's history: those of December 1875; April 1887; January 1894; April and December 1908; April 1909; November 1910; and March 1935. The significance of these facts can be interpreted in light of the monthly frequency of Northwest cyclones. Figure 11.4 gives these figures for a 50-year period between 1872 and 1923. It clearly delineates the peak cyclone season of December to March, which accounts for 85% of storms during this period. The table also records only two cyclones in November and six in April, barely 10 % of the total. Yet these same cyclones were responsible for nearly 50 % of all deaths in pearlshelling recorded in Fig. 11.3.

These insights help to place the March 1935 disaster in context. During the industry's early years, losses were relatively slight. The free-diving phase was largely confined to the summer months because pastoral labour was idle at this time, and because the sea temperature was warmer; regulations under the *Pearl Shell Fishery Regulation Act 1875* restricted diving to the period between November and April in order to protect Aboriginal divers from the harmful effects of prolonged immersion in cold water. This meant that pearlshelling took place during the peak cyclone season, but as the industry was effectively limited by the depth that free divers could ascend to (at best, around six fathoms), the fleet was confined to the shallower waters around Cossack and the Dampier Archipelago, so that shelter could be readily sought if a cyclone approached.[23] The first

Month	No. of Tropical Storms
January	23
February	14
March	16
April	6
May	0
June	0
July	1
August	0
September	1
October	1
November	2
December	10

Fig. 11.4 Monthly frequency of tropical storms, Northwest Australia, 1872–1923 (Source: Adapted from Hunt H.A., S.S. Visher and D. Hodge (eds.) *Australian Hurricanes and Related Storms: With an Appendix on Hurricanes in the South Pacific* (Melbourne: Commonwealth Bureau of Meteorology, Bulletin No.16, 1925))

major losses took place only after the introduction of Malay divers and the denudation of the early pearlshelling grounds pushed the industry into the deeper and more exposed waters of Exmouth Gulf, where the cyclone of 24 December 1875 resulted in 56 deaths, of whom at least 46 were Malays, seven Europeans, and the remainder (probably including some uncounted) Aboriginals.[24] The introduction of diving suits, which offered protection against cold water, initiated a transition to winter diving and the implementation of a summer lay-up coinciding with the peak cyclone season between December and March. Diving suits allowed divers to reach 20 fathoms and beyond, so that the fleet could now venture out as far as 50 km from the coast.[25] Henceforth, the industry was considerably less exposed to the dangers of tropical storms at large, but in moving into offshore waters, considerably more exposed to the vagaries of the early- or late-season cyclones that occasionally bore down on the coast.

The dangers of late-season cyclones was revealed in 1887, only the second season after the widespread adoption of diving suits. The cyclone that crossed 80 Mile Beach on 22 April was the latest encountered since the industry's inception. Reports of its passage through the pearling fleet working along this exposed stretch of coast were brought into Cossack by the SS *Australind*, which had rescued a handful of survivors at sea, shocking the colonists with the news that as many as 500 men were missing; as the *Western Mail* reported, 'people were so unprepared for it, the season for such visitations having passed over without serious damage. No more dreadful disaster has, probably, occurred in the history of these colonies... there is every reason to fear that the disaster has been attended by a loss of life unparalleled in the history of Australian pearl fishing'.[26] It took almost three months before the final tally of 140 deaths was announced, making it Australia's worst weather-related disaster at the time (although this was later surpassed by the 400 pearlers lost in the 1899 Bathurst Bay cyclone and storm surge in Queensland).[27] Significantly, the cyclone may also have affected the industry's racial composition. When diving suits were introduced to the coast as many as 30 European or 'white' divers worked in the fleet, but a number of the divers were lost in the April 1887 cyclone, and their numbers never recovered as the industry came to rely increasingly on imported Japanese divers from the late 1880s on.[28]

The industry's vulnerability was demonstrated again by a cluster of early- and late-season storms during 1908–10. The cyclone of 27 April 1908 was a repeat of the 1887 disaster, catching pearlers by surprise on the exposed waters off Eighty Mile Beach. Losses might even have reached a similar level had not the SS *Paroo* been diverted from Broome to render assistance to the shattered fleet; the final tally was 117 deaths, 44 vessels lost, and damages estimated at £25,000.[29] A second cyclone on 8 December, reckoned at the time to be even stronger than April's cyclone, caught a group of luggers in waters near Lagrange Bay. As survivors' accounts accumulated and bodies washed ashore with the tide, it became clear that another tragedy was unfolding; losses were eventually calculated at 50 deaths, 39 vessels, and £20,000 in damage.[30] On 5 April 1909, a small fleet working off Onslow was hit by another late-season cyclone that resulted in 24 deaths and four vessels sinking.[31] A fourth cyclone in November 1910 approached Broome from the northwest before curving to follow the coast in a south-westerly direction, causing extensive damage in the town and catching the fleet unexpectedly not far from Roebuck Bay. One pearler with more than 40 years of experience on the coast reckoned it to be the strongest cyclone he had ever witnessed.[32] At least 67 vessels were blown ashore after having their masts cut away, helping their crews to survive, but another 34 luggers went down in the storm, at the cost of another 40 lives.[33]

Meteorology and Storm Warning on the Northwest Australian Coast

The sequence of disasters in 1908–10 forced pearlers to confront the dangers of cyclonic weather. Following the April 1908 cyclone the *Western Mail* published one industry veteran's suggestion that a closed season running from mid-December to mid-May should be fixed by law, and that storm anchors and barometers be made compulsory aboard every lugger.[34] After the December 1908 cyclone the newspaper ran an editorial calling for a more effective distribution of local weather forecasts to the fleet, and the *West Australian* proposed that rockets be supplied to coastal telegraph stations to warn shipping in the event of approaching storms.[35] The Pearlers' Association heeded these calls by including a notice on 'Hurricanes' in its annual reports from 1908 onwards. It cautioned against pearlshelling 'between 1st December and 30th April, except in those waters from which shelter can be easily reached'; advised masters to carefully monitor atmospheric pressure at sea, 'as it is an unfailing guide to weather in the tropics'; and reminded pearlers that 'one great lesson of the last two "blows"... has been not to work away from the place which shelter can be easily obtained, and secondly, not to seek the shelter of an open beach because the wind is in the east', as this would place vessels on a lee shore once the eye of the cyclone had passed by.[36] The cyclone of November 1910 then came as a particular shock. 'There seems to have been no indication that anything resembling a storm was approaching', exclaimed the *West Australian*, its sense of shock echoed by the *Western Mail*, which reported that 'not for many a year, perhaps never before, has there been a blow if such violence in November'.[37]

Such reports expose the lack of an effective storm-warning system on the Northwest coast. Tropical cyclones had been studied carefully during the second half of the nineteenth century in the Bay of Bengal, South China Sea, and the Gulf of Mexico, and during the 1870s and 1880s networks of local observers had been established around each of these seas to relay weather reports along the known paths of cyclonic storm systems via telegraph.[38] In the early 1900s the invention of wireless telegraphy promised to improve the dissemination of weather reports by allowing transmissions to vessels at sea.[39] Australia, and the continent's western side in particular, lagged a long way behind these developments. For example, Findlay's *Sailing Directions for the Indian Ocean* (1870), despite reporting at length on cyclone systems in the Bay of Bengal and the seas around Mauritius, stated only that '[o]ff the West coast of Australia storms of great violence, and these often cyclonic, occur to the North of Cape Leeuwin', thus describing more than 2000 miles of coast, where British settlement dated back to 1829.[40] There was some improvement by 1905, when the *Australia Directory* was able to delineate a cyclone season on the Northwest coast running from December to April, and offer practical tips for navigators caught in a storm.[41] However, although the Northwest coastal settlements had been connected via telegraph to Perth and Java in 1889, the region was too thinly settled, and lay too far away from major shipping routes, for a storm-warning system to be feasible.

Local meteorological studies remained neglected until the late 1890s. The first figure to develop a sustained interest in Australian tropical storms, the Government Meteorologist in Queensland, Clement Wragge, had by the mid-1890s come to believe that the Coral Sea was the origin of all Australian cyclones—that is, of both the northeast, northern, and northwest coasts. These ideas were challenged following the appointment of Western Australia's first Government Astronomer (with responsibility for meteorology), William Ernest Cooke, in 1896. Like many of his contemporaries, Cooke was no admirer of the colourful and ambitious Wragge, and after corresponding with meteorologists in Mauritius he established that cyclones in the Northwest formed in similar latitudes to cyclones elsewhere in the southern Indian Ocean, before following similar courses from the Timor and Arafura Seas to the Australian coast.[42] He carefully studied one cyclone that passed over Cossack and Roebourne on 2 April 1898, and, assisted by recordings made by the SS *Albany* as it was steaming south from the Dampier Peninsula, produced the first synoptic chart of a Northwest cyclone.[43] It was probably Cooke who supplied the information on cyclonic storms that was distributed by the Pearlers' Association from 1908 onwards.

In this context it is not surprising to find that pearlshellers developed their own strategies to deal with tropical storms during the industry's early decades. Aneroid barometers might provide as much as 12 hours' notice of an approaching cyclone, but this was scarcely any longer than visible changes in local weather conditions which experienced masters could recognise as foreshadowing extreme weather. During the free-diving phase, when pearling took place during the summer, some had come to recognise that the first signs of a cyclone could often be detected underwater. Severe cyclones are generally accompanied by storm surges created by the sea literally rising under an envelope of low atmospheric pressure, potentially affecting currents across a wide area of ocean.[44] There is some evidence to suggest that this effect could be felt by naked divers. Pearling master John Brockman recorded one instance in January 1881 where his Aboriginal divers refused to take to the water due to sudden temperature changes, prompting Brockman to seek shelter from what soon developed into a furious cyclone. His schooner *Sarah* sustained damage after being beached in a mangrove creek, but the crew survived the storm that cost six lives and damaged another 11 vessels in the vicinity of Cossack.[45] Arthur Bligh, in his memoir of pearling in the 1890s, also records that 'many hours before you could see a storm coming up, the water near the bottom was very rough, sometimes so rough that the divers could not stay below. Divers…could often tell 12 hours ahead'.[46] Yet the shift from free-diving to diving suits meant that divers were less exposed to changing water temperatures, and the growing use of Japanese divers increased the proportion of those who were unfamiliar with local conditions, probably worked against this method of forecasting. During 1912, the Western Australian Fisheries Department supported an investigation of sea temperatures involving measurements taken by passenger steamers traversing the Western Australian coast. The data yielded on temperatures and currents, emerging as it did in the aftermath of the *Koombana* tragedy, led to a resurgence of interest in the old tales about changing sea temperatures presaging the approach of cyclones.[47] When contacted on the matter, however, the Pearlers' Association had replied that 'no change in sea temperature at the bottom occurs prior to a 'cock-eye bob' or 'willy-willy'', based on the experience of divers, and the Commonwealth Meteorologist later dismissed the theory as misguided.[48]

The loss of the *Koombana* did however lead to some improvements in storm warning. The Court of Inquiry into the disaster recommended that vessels employed in coastal trade carry wireless radio sets to receive and relay weather conditions, that daily weather reports should be posted for public viewing in Northwest ports, and that in the event of poor forecasts, storm-warning signals should be displayed to warn mariners.[49] These recommendations provided the basis of the system that operated at Broome as the cyclone approached the town on 26 March 1935, and which proved unable to warn the luggers already at sea. Radios might have been used to warn the fleet, but the cost of carrying the cumbersome and expensive receivers was too high for an industry caught in straitened times, and the coast too sparsely settled to support a network of local observers to relay weather reports in good time. These limitations had been recognised by the American climatologist Stephen Sargent Visher in 1925 in the Commonwealth Bureau of Meteorology's first systematic appraisal of Australian tropical cyclones, *Australian Hurricanes and Related Storms.* On the 'Prediction of Hurricanes', Visher's report suggested that 'to the present, however, a completely satisfactory result has not been attained. It will, perhaps, be impossible at any time to do more than state the probabilities of a hurricane occurring somewhere in a rather large region.'[50]

Such constraints left the pearlshellers in much the same position they had been in before the recommendations of the *Koombana* inquiry. Equipped only with the means to make short-term forecasts with the aid of a barometer, each master needed to balance the prospects of an early- or late-season cyclone against the potential financial benefits of a reduced lay-up season, the prospect of being caught in an unfavourable position should a cyclone approach, and a captain's skill at handling his vessel in a storm. It was perhaps inevitable that the dangers of putting to sea in late March would eventually be overlooked. As one of the European shell-openers who survived the cyclone recalled, 'the fleet was caught unawares' on 27 March; although his vessel had cut away its mast and put out two sea anchors to avoid turning broadside to the storm, before drifting some 16 miles to the safety of Beagle Bay,[51] other luggers had simply been swamped, and the barometer had only begun falling in the hours before the cyclone began.[52]

Disaster Relief and the Political Economy of Pearlshelling

Historians of Australian pearling have not been blind to the impact of cyclones on the Northwest coast. In *Full Fathom Five* (1982), Mary Albertus Bain suggests that the 'slow destruction the pearling industry on the west coast had many and varied causes but the repeated effects of cyclones was an important factor'.[53] In his history of Broome, *Port of Pearls* (1983), Hugh Edwards observes that 'cyclones have been part of the way of life' in the town and that 'the pearling fleets have suffered severely'.[54] A degree of fatalism pervades this literature. Bain writes that cyclones 'made pearling the most dangerous occupation in Australia', and were 'the most dreaded fate' in the industry.[55] The question of risk was taken up by the *West Australian* in the aftermath of the March 1935 disaster:

> Of the thousands who, year in and year out, are exposed to perils on the sea, it is probably safe to say that none run greater risks than the pearling crews in our northern waters. It is all very well to say, as some have been saying in the past week, that the perilous period is so clearly demarked that prudence ought to dictate the luggers be laid up over this peculiarly dangerous interval in the year. But the truth is that the willy-willy season is not so well defined as many suppose. The idea has become prevalent that by suspension of operations from January 1 to March 31 the pearlers might prosecute their industry and enjoy practical immunity from hurricane risk. But this is a popular error. At any rate the records show that some of the worst of the disasters which have befallen the industry have occurred as early as mid-December and as late as the end of April. Partly for purposes of overhaul, but mainly, of course, to minimise the risks of attending the raising of shell in the hurricane season, the lugger fleets are laid up for eight or ten weeks and sometimes for longer periods every year. But for reasons, the more obvious in view of the present depressed condition of the industry, the lying-up period cannot be so greatly extended as to eliminate all chances of the fleets escaping an unusually early or late-coming willy-willy. Risk, and more or less grave risk, is their inescapable portion.[56]

Was the disaster the 'inescapable portion' of the pearlshellers? Greg Bankoff argues that there is no such thing as a 'natural disaster'; that 'what makes a hazard into a disaster depends primarily on the way a society is ordered', or in other words, that the origins of disasters lay 'in the political structures, economic systems and social orders of the societies in which they take place'. The *West Australian* may well have been correct to observe that risk was a fundamental element of the pearlshelling industry, but dismissing cyclones as fate (in a fashion echoed by historians) ignores the social factors that contributed to the March 1935 disaster. These factors are laid bare in the circumstances that led to this event becoming a landmark in the development of Australian disaster relief policy.

Recovery had been rapid after 1908–10 due to strong demand for mother-of-pearl. Prices had risen from £110 per ton in 1906 to over £190 in 1911, and a year later the Northwest pearlshellers had their best year on record, producing 1596 tons worth £421,609; the fleet of 401 vessels was comfortably the largest in the world.[57] But three successive crises were to reverse their fortunes. The first, the global restriction on trade during the Great War 1914–18, brought an end to the long period of growth that had sustained the industry through its heyday in the 1890s and early 1900s. The second was the Great Depression of the early 1930s, when the global market for mother-of-pearl collapsed. Pearlshell was a luxury product used mainly in the manufacture of consumer items such as buttons and cutlery and, as such, was highly susceptible to changes in discretionary spending or the use of substitute products such as glass and bone. Pearls could provide a valuable bonus, but were not a reliable source of revenue, and by the 1920s were no longer as plentiful as they had been during the industry's early years. Prices for shell fell to around £107 in 1933, less than the estimated costs of production.[58] The pearl-shellers themselves had contributed to their precarious financial position by meeting falling prices with increased production, creating over-supplies that kept prices low.[59] Before the Great War, the industry was famed for the riches it yielded, and many master pearlers did indeed amass considerable fortunes. In the 1920s and early 1930s, however, a constant struggle to remain solvent was a more frequent experience. The fleet fell below 300 vessels in 1921 and below 200 in 1927. The 87 vessels licensed in 1935, of which only half were operational, was the lowest number on the Northwest coast since the advent on the diving suit era in the mid-1880s.

It was against this backdrop that a third challenge to the industry's viability emerged in the form of rising competition from Japanese fleets operating beyond Australian territorial waters in the Timor and Arafura Seas.[60] The first Japanese pearlshelling operations had appeared off Australia's northern coast in 1931. By 1934, at least ten vessels were operating in the Arafura Sea, with additional boats based at Timor and the Aru Islands in the Dutch East Indies also venturing into waters between the Malay Archipelago and Australia. These vessels harvested both *maxima* and trochus shell, a cheaper substitute that placed further pressure on prices. Their presence had immediate ramifications. The Northwest pearlshellers abandoned plans that had been developed to limit their production in order to improve market conditions, even though their costs were now outstripping profit, and turned instead to government for assistance.[61] At a meeting held in Broome on 23 January 1935 to discuss the crisis, the leading pearling master Arthur Male claimed that the Japanese were not only exempt from the tariffs on equipment and supplies paid by local producers, but that they were also being subsidised in their operations by their own government. The meeting concluded with a resolution to 'implore the respective [i.e. state and federal] Governments to take swift and decisive action to save this outpost of the north from reverting to desolation, and the old-established industry of pearling from extinction. This meeting unanimously agrees that if action is not speedily taken the end of Broome and the industry is certain'.[62]

Behind these urgent concerns, moreover, lay more than two decades of anxiety over the role of Japanese in Australian pearlshelling. Although Japanese divers had become the mainstay of the industry in the 1890s, their presence had been jeopardised by the Commonwealth's *Immigration Restriction Act* of 1901, and only a series of exemptions during the early 1900s had allowed the Northwest pearlshellers to retain their imported Asian labour.[63] A scheme that trialled British divers at Broome failed in 1912 through illness, accidents, the harsh climate, and the rigours of sustained diving.[64] A Royal Commission into Australian pearlshelling eventually decided, in 1916, that the industry would be allowed to keep its imported divers in the interests of its European owners and the coastal communities they lived in. The Commission 'decided that diving for shell is not an occupation which our [white] workers should be encouraged to take', as a diver's arduous and dangerous occupation was 'incompatible with that a European worker is entitled to live'.[65] However, although Europeans retained exclusive rights to own vessels and be issued licences to collect shell, the practice of illegally leasing luggers to Japanese, known locally as 'dummying', was commonplace during the 1920s. Riots in Broome between Japanese and Timorese in 1920 exposed wider tensions amongst the town's racially diverse population.[66] Japan's government, dissatisfied with the handling of its territorial claims at the 1919 Paris Peace Conference, had also been increasingly critical of the White Australia Policy in the years following the Great War.[67] By the early 1930s, then, the expanding distant water fleet appeared to be part of Japan's broader imperial agenda in Southeast Asia and the Southwest Pacific, in waters off Australia's northern coast.

Male addressed such concerns when speaking in Broome on 23 January 1935. He argued that 'the pearling industry has made possible the establishment of many of the large stations of the North, established fairly large towns and supported them in a sparsely populated part of our continent, and effectively patrolled thousands of miles of our northern waters, which would otherwise be left open to foreigners'.[68] There was much truth to this statement—several pastoral stations had been established by pearlshellers, the industry supported the major coastal settlements between

Geraldton and Wyndham in the far northeast of the Kimberley region, helped to keep the State Shipping service viable, and was the main maritime presence along a stretch of the Australian coastline that neither the Australian customs agency nor the Royal Australian Navy were capable of patrolling.[69] This reality had been recognised by the Royal Commission in 1916, which 'admitted that the pearlshelling industry has an importance of its own, in that it maintains a European population in centres where no other industry would take place. Broome, for instance, is the only town of any importance between Geraldton and Darwin, a steaming distance of 1985 miles, and is entirely supported by this industry.'[70] In threatening the viability of pearlshelling, the Japanese fleet was therefore imperilling the principal coastal settlements on the stretch of the Australian coastline that lay closest to Southeast Asia.[71] No government would have been so careless as to openly telegraph such concerns, but it does help to explain why the Commonwealth acted so quickly to extend a grant of £2500 to Broome's pearlshellers in February 1935.[72]

These same factors explain why the Commonwealth joined the Western Australian Government in granting relief in the aftermath of the March 1935 disaster. Calls went immediately to the Prime Minister and Treasurer to extend assistance to the industry, and by mid-April an agreement had been reached; the Commonwealth would make £5000 available to Broome's pearlshellers,[73] and the Western Australian Government would contribute an additional £7000 and supervise the distribution of relief funds.[74] A committee was duly formed, tasked with allocating the money in such a way as to place each pearling master 'into the same position as he was in prior to the cyclone, enabling him to again put a boat to sea, thus preventing the collapse of the industry.'[75] No other rationale for the relief funds was ever publicly stated, although Western Australia's Minister for Agriculture defended the expenditure of state funds on the pearling industry at the same time as refusing requests for assistance from storm-affected orchardists in the state's agricultural districts by stating simply that 'Broome must be preserved'.[76] The industry had itself lobbied for the speedy distribution of funds by pointing to the impending loss of Broome's surviving experienced divers; 'Only the other day one pearler missed a very fine diver for his boat, because he could not give a definite order, and that skilled man went to a Japanese sampan where a job was offering. These and other instances would convince anyone, that if the full benefits are to be derived from the Grants they must be distributed quickly'.[77] By mid-June the relief funds had been expended, allowing a fleet of 90 vessels to be made ready for the start of the 1936 season.[78]

Conclusion: Pearlshelling, Disaster Relief, and European Settlement in Northwest Australia

The 1936 season marked a turnaround in the fortunes for the Northwest pearlshellers. The weather was benign, and the fleet collected 825 tons worth £102,817. The following season was even better, yielding shell worth £125,447, the best return since the beginning of the Great Depression. But it proved to be the industry's Indian summer. Japan's fleet had also drastically escalated its harvest, with 74 vessels collecting 750 tons outside Australian territorial waters and a further 1100 tons in the Arafura Sea in 1936, and 145 vessels taking an unknown but substantial harvest in 1937.[79] The combined Australian and Japanese production caused a glut on the market that would take years to clear, driving prices to record lows. The industry limped through to 1941 with the aid of additional State and Federal grants, loans and subsidies. Following the outbreak of war in the Pacific, the industry was effectively shut down as its Japanese divers were interred and the most of the fleet's luggers either pressed into military service or destroyed lest they be used to assist an invasion of Australia's north. The Japanese bombings of Broome in early 1942, following on attacks on Darwin and coming after Japan's rapid advance across Southeast Asia and the Pacific, dramatically exposed the vulnerability of the Northwest coast. Pearlshelling revived after the war, without ever reaching the same scale as the early twentieth century, before collapsing due to the rise of plastics in the 1960s. Broome eventually found a new future in the cultured pearl industry, and from the 1980s onwards, in the international tourist trade.

There has not been another cyclone in the Northwest comparable to the March 1935 disaster. The decline of pearlshelling removed the vast fleet of sail-powered vessels that once worked the coast, and satellite-based forecasting and modern radio communications now give advance warning of any approaching storm. Commonwealth disaster relief payments proceeded according to ad hoc arrangements until a policy of matching State spending dollar-for-dollar emerged in the early 1960s.[80] Cyclone Tracy, which devastated Darwin on 24 December 1974, prompted the Commonwealth Government to expend more than $150 million through the Darwin Reconstruction Commission, and although this took place in the Northern Territory (an Australian federal territory) it does reflect the increasing Commonwealth commitment to disaster relief that developed in the second half of the twentieth century. In the early 2000s, a more sophisticated policy framework was established through the NDRRA, a form of federal–state co-operation in which the Commonwealth funds three-quarters of all disaster relief payments in Australia.

The historian David Walker has examined the development of Australian anxieties over the continents 'empty north' and relative proximity to Asia, tracing their emergence in nineteenth-century ideas about climate and race which mandated tropical areas as unfit for European settlement and destined instead to support non-white populations, to the rise of strategic dilemmas in the twentieth century linked to population deficiencies, economic under-development in northern regions, and the rise of new economic and military powers in Asia.[81] Within this context, Japanese expansion in the 1920s and 1930s was often seen as a direct threat to Australian interests, and these concerns explain why pearlshelling received extensive government support during these decades, including the grants made to the industry in February and April 1935. And yet the March 1935 disaster has been overlooked by historians of Australian natural hazards. Disasters that affect the continent's more heavily settled southern and eastern regions, such as Victoria's Black Friday (1939), Ash Wednesday (1983), and Black Saturday (2009) bushfires, and the droughts and floods that regularly affect agricultural districts, are familiar stories in Australia, and influence expressions of national identity and understandings of the nation's historical development.[82] The Northwest coast has no place in narratives of a national character forged by resilience to climatic variability and extreme weather. Simple monuments in Broome's cemetery commemorate the Japanese and Chinese divers and crewmen who drowned in the waters off Beagle Bay and at the Lacapede Islands in March 1935—martyrs to an industry dependent on their labour, victims of a society that would expose them to risk in order to preserve its vision of a White Australia on the continent's remote Northwest coast.

Notes

1. *Northern Times*, 20 March 1935: 2; *Western Mail*, 31 January 1935: 18. Also: Bach, John P.S., *The Pearling Industry in Australia: An Account of Its Social and Economic Development* (Canberra: Department of Commerce and Agriculture, Commonwealth of Australia, 1956): 186–188.

2. Norman, John E. deB, G.V. Norman, *A Pearling Master's Journey*. (Strathfield: John E de B Norman, 2007): 204–205.
3. Edwards, Hugh, *Port of Pearls* (Adelaide: Rigby, 1983): 101.
4. *West Australian*, 3 April 1935: 16.
5. *Daily News*, 28 March 1935: 1; 1 April 1935: 2; 6 April 1935: 1; *West Australian*, 3 April 1935: 16.
6. See the list compiled by the Pearling Inspector, Broome, for the Chief Pearling Inspector, 18 April 1935. In: State Records Office of Western Australia. Fisheries Department file 193/48, 'Storm Damage to Pearling Fleet. Broome District. General File', Cons. 477.
7. Writer, Larry, *The Australian Book of Disasters* (Sydney: Murdoch Books Australia, 2011): 70–76; for the *Koombana*, see Boyd, Annie, *Koombana Days* (Fremantle: Fremantle Press, 2013).
8. See correspondence in: State Records Office of Western Australia. Premier's Department file 218/35, 'Relief to Pearling Industry for losses caused by Cyclone 1935', Cons. 1496; and *West Australian*, 5 July 1935: 23.
9. Wettenhall, Roger L., *Bushfire Disaster: An Australian Community in Crisis* (Sydney: Angus and Robertson, 1975): 46–47.
10. Heathcote, R.L., B.G. Thom (eds.), *Natural Hazards in Australia* (Canberra: Australia Academy of Science, 1979); Botterill, L.C., 'Uncertain Climate: The Recent History of Drought Policy in Australia', *Australian Journal of Politics and History*, 49:1 (2003): 61–73.
11. Middleman, Miriam H. (ed.), *Natural Hazards in Australia: Identifying Risk Analysis Requirements* (Canberra: Geoscience Australia, 2007): 10–21.
12. Wright, Bill (ed.) *Drought, Dust and Deluge: A Century of Climate Extremes in Australia* (Canberra: Australian Bureau of Meteorology, 2004): 41–48.
13. Hanstrum, Barry, 'A History of Tropical Cyclones in the Southwest of Western Australia 1830–1992', *Early Days: Journal of the Royal Western Australian Historical Society*, 10:4 (1992): 396–407.
14. *Drought, Dust and Deluge*: 41–48.
15. See: De La Rue, Kathy, *Pearl Shell and Pastures: The Story of Cossack and Roebourne and Their Place in the History of the North West, from the Earliest Explorations to 1910* (Roebourne: the Cossack Project Committee, 1978); Forrest, Kay, *The Challenge and the Chance: The Colonisation and Settlement of North-West Australia 1861–1918* (Perth: Hesperian Press, 1996); Powell, Alan, *Northern Voyagers: Australian Monsoon Coast in Maritime History* (Melbourne: Australian Scholarly Publishing, 2010); Withnell Taylor, Nancy E., *Yeera-muk-a-doo:A Social History of the Settlement of North-West Australia Told Through the Withnell and Hancock Families, 1861 to 1890* (Perth: Hesperian Press, 2002).

16. 'Storms', in Chisolm, A.H. (ed.), *The Australian Encyclopaedia* (Sydney: Angus and Robertson, 1958).
17. *The Pearling Industry in Australia*: 3–9; also Streeter, Edwin W. 1886. *Pearls and Pearling Life* (London: George Bell and Sons).
18. *The Pearling Industry in Australia*: *An Account of Its Social and Economic Development*: 85, 275–282.
19. This table was compiled from a large collection of documentary sources, including many of those listed in the bibliography.
20. Gentilli, Joseph, 'A History of Meteorological and Climatological Studies in Australia', *University Studies in History*, 5:1 (1967): 54–79.
21. For a contemporary descriptions, see: *Daily News*, 29 July 1913:3; 6 May 1937: 4.
22. A case in point is: Longshore, David, *Encyclopaedia of Hurricanes, Typhoons and Cyclones* (New York: Facts on File 1998), section on 'Australia' pp.22–25.
23. See: McCarthy, Mike, 'Before Broome', *The Great Circle*: *Journal of the Australian Association for Maritime History*, 16:2 (1995), 76–89; McCarthy, Mike, 'Naked Diving for Mother-of-Pearl'. *Early Days*: *Journal of the Royal Western Australian Historical Society*, 13:2 (2006): 243–259.
24. *The Inquirer and Commercial News* 12 January 1876, p.3; *The Western Australian Times* 14 January 1876, p.2; *The Herald* 29 January 1876, p.3.
25. *The Pearling Industry in Australia*: 28–33, 85.
26. *Western Mail*, 7 May 1887: 28. For more on the cyclone, see *West Australian*, 29 April 1887: 3; *Inquirer and Commercial News*, 18 May 1887: 2; *Western Mail*, 30 April 1887: 15; *Western Mail*,7 May 1887: 28.
27. *Northern Voyagers*: 167' *Inquirer and Commercial News*, 3 August 1887: 2.
28. *The Pearling Industry in Australia*: 28–33.
29. *Sunday Times* 3 May 1908, p.5; 14 June 1908, p.3; *Western Mail* 16 May 1908, p.51; *West Australian* 12 June 1908, p.5.
30. *West Australian*, 12 December 1908: 12; 14 December 1908: 7; 15 December 1908: 5; 16 December 1908: 7.
31. *West Australian*, 14 April 1909: 7.
32. *West Australian*, 24 November 1910: 7.
33. *West Australian*, 24 November 1910: 7; 29 November 1910: 5.
34. *Western Mail*, 16 May 1908, p.51.
35. *West Australian* 14 December 1908: 7; 19 December 1908: 12.
36. See the *Annual Reports* for 1908 and subsequent years in: West Australian Pearlers Association, 'Collection of papers on Broome and the pearling industry'. Battye Library of Western Australian History, MN 377, ACC 1935A, 5006A.
37. *West Australian*, 24 November 1910: 7; also *Western Mail*, 3 December 1910: 13.

38. Cushman, George T, 'The Imperial Politics of Hurricane Prediction: From Calcutta and Havana to Manilla and Galveston, 1839–1900', in: Lawrence, M., E. Bsumek, D. Kinkela (eds), *Nation-States and the Global Environment* (New York: Oxford University Press, 2013), 137–162.
39. P. Kevin Mackeown, *Early China Coast Meteorology: The Role of Hong Kong*. Royal Asiatic Society Hong Kong Study Series(Hong University Press,2010): 133–172.
40. Findlay, Alexander George, *Sailing Direction for the Indian Ocean* (London: Richard Holmes Laurie, 1870), 76.
41. *Australia Directory. Volume III* (London: Hydrographic Office, 1905), 76.
42. W.E Cooke to Sir John Forrest, 13 July 1898,in State Records Office of Western Australia. Premier's Department file 1898/0643, 'W.E. Cooke Esq. Government Astronomer, Perth. Worthlessness of Mr C. Wragge's warnings to shipping', Cons. 1496; and Hutchison, David, 'William Ernest Cooke, First Government Astronomer', *Early Days: Journal of the Royal Western Australian Historical Society*, 8:4 (1980): 93–113.
43. Cooke, William Ernest, *The Climate of Western Australia: From Meteorological Observations Made During the Years 1876–1899* (Perth: Government Printer, 1901): 11–14.
44. *Encyclopaedia of Hurricanes, Typhoons and Cyclones*: 291–295.
45. Brockman, John, *Pearling Days* (Perth: Hesperian Press, 2010), 57–62.
46. Bligh, Arthur C.V., *The Golden Quest* (Perth: Hesperian Press, 1984), 186. Similar accounts are also detailed by Kay Forrest in *The Challenge and the Chance*:267.
47. See the report by J.J. East to the Chief Inspector of Fisheries, 4 October 1912, in: State Records Office of Western Australia. Aborigines and Fisheries Department file 1917/2297, 'Ocean Temperatures: Information supplied by J.J. East re:', Cons. 477. Also: *Western Mail*, 27 December 1912: 54.
48. Stanley Piggot, Secretary West Australian Pearlers' Association, to Chief Inspector of Fisheries, 14 December 1912; and Hunt, H.A., Commonwealth Metereologist, 'Notes on Mr East's Ocean Temperature Investigations', and the accompanying letter, Prime Minister of Australia to the Premier of Western Australia, in: State Records Office of Western Australia. Aborigines and Fisheries Department file 1917/2297, 'Ocean Temperatures: Information supplied by J.J. East re:', Cons. 477.
49. State Records Office of Western Australia. Colonial Secretary's Department, Harbour and Light file 438/12, '"Koombana" S.S. Overdue at Broome – Total Loss of, Inquiry re:', Cons.
50. Hunt, H.A., S.S. Visher, D. Hodge (eds.) *Australian Hurricanes and Related Storms: With an Appendix on Hurricanes in the South Pacific*

(Melbourne: Commonwealth Bureau of Meteorology, Bulletin No.16, 1925): 9.

51. *Daily News*, 1 April 1935: 2.
52. *Northern Times*, 3 April 1935: 3.
53. Bain, Mary Albertus, *Full Fathom Five* (Perth: Artlook Books, 1982), 161.
54. *Port of Pearls*: 101.
55. *Full Fathom Five*: 155.
56. 'The Northern Cyclone', *West Australian* 1 April 1935, p.14.
57. *The Pearling Industry in Australia*: 274–283.
58. *The Pearling Industry in Australia*: 274–283.
59. Bach, John P.S., 'The Political Economy of Australian Pearlshelling', *The Economic History Review*, 14:1 (1961): 105–114.
60. *The Pearling Industry in Australia*: 274–283.
61. *The Pearling Industry in Australia*: 219–226.
62. *Western Mail*, 31 January 1935: 18.
63. Bach, John P.S., 'The Pearlshelling Industry and the "White Australia" Policy', *Historical Studies*, 10:38 (1962): 203–213.
64. *The Pearling Industry in Australia*: 219–226.
65. *Report and Recommendations of the Royal Commission into the Pearl-shelling Industry* (1916): 6.
66. *Port of Pearls*: 68–71.
67. 'The Pearlshelling Industry and the "White Australia" Policy': 201–213.
68. *Western Mail*, 31 January 1935: 18.
69. For a general context, see *Northern Voyagers*; and *The Pearling Industry in Australia*: 219–226.
70. *Report and Recommendations of the Royal Commission into the Pearl-Shelling Industry* (1916): 7.
71. Oliver, Pam, 'Interpreting "Japanese Activities" in Australia 1888–1945', *Journal of the Australian War Memorial*.
72. State Records Office of Western Australia. Premier's Department file 218/35, 'Relief to Pearling Industry for Losses Caused by Cyclone 1935', Cons. 1496.
73. See Assistant Secretary, Department of the Treasury, to The Secretary, Prime Minister's Department, 15 April 1935, and other correspondence in National Archives of Australia. Prime Minister's Department F393/1/1, 'Relief of Distress—Pearling Disaster—Broome', Series A461.
74. Prime Minister of Australia to Premier of Westrn Australia, 15 April 1935, in State Records Office of Western Australia. Premier's Department file 218/35, 'Relief to Pearling Industry for losses caused by Cyclone 1935', Cons. 1496.

75. State Records Office of Western Australia. Premier's Department file 218/35, 'Relief to Pearling Industry for losses caused by Cyclone 1935', Cons. 1496.
76. *Daily News*, 15 April 1935: 1, 4.
77. E.S. Reynolds, Secretary of Pearlers' Committee, to F.J. Wise, Minister for the North, 16 May 1935. In: State Records Office of Western Australia. Premier's Department file 218/35, 'Relief to Pearling Industry for losses caused by Cyclone 1935', Cons. 1496.
78. *West Australian*, 5 July 1935: 23.
79. *The Pearling Industry in Australia*: 283, 299.
80. Uncertain Climate: The Recent History of Drought Policy in Australia': 63.
81. Walker, D., *Anxious Nation: Australia and the Rise of Asia 1850–1939* (St Lucia: University of Queensland Press, 1999), 2-4-17; also Walker, D., 'The Curse of the Tropics'. In: Sherratt, T., T. Griffiths, L. Robin (eds) *A Change in the Weather: Climate and Culture in History* (Canberra: National Museum of Australia, 2005): 92–101.
82. For example, see Garden, Donald, *Droughts, Floods and Cyclones* (North Melbourne: Australian Scholarly Publishing, 2009).

Bibliography

Australia Directory. 1905. *Volume III, North, north-west, and west coasts, from Torres Strait to Cape Leeuwin:Being chiefly the result of various surveys made by order of the Lords Commissioners of the Admiralty.* London: Hydrographic Office.

Bach, John P.S. 1956. *The pearling industry in Australia: An account of its social and economic development.* Canberra: Department of Commerce and Agriculture, Commonwealth of Australia.

——— 1961. The political economy of pearlshelling. *The Economic History Review* 14(1): 105–114.New Series

——— 1962. The pearlshelling industry and the "White Australia" policy. *Historical Studies* 10(38): 203–213.

Bain, Mary Albertus. 1982. *Full fathom five.* Perth: Artlook Books.

Bankoff, Greg. 2003. *Cultures of natural disaster: Society and Natural Hazard in the Philippines.* London: Routledge Curzon Press.

———. 2004. The historical geography of natural disaster. In *Mapping vulnerability: Disasters,development and people*, eds. Greg Bankoff, G. Frerks, and D. Hilhorst. London: Earthscan.

———. 2010. No such things as "natural disasters: Why we had to invent them. *Harvard International Review*,24 August 2010. http://hir.harvard.edu/no-such-thing-as-natural-disasters.

Bligh, Arthur C.V. 1984[1958]. *The golden quest: Roaring days of the West Australian gold rushes, and life in the pearling industry.* Perth: Hesperian Press.

Botterill, L.C. 2003. Uncertain climate: The recent history of drought policy in Australia. *Australian Journal of Politics and History* 49(1): 61–73.

Boyd, Annie. 2013. *Koombana days.* Fremantle: Fremantle Press.

Brockman, John. 2010[1918]. *Pearling days: The pearling voyage of the Sarah to the North West and Kimberley in 1880 and 1881, with an appendix on the death of W.H. Lowe during an expedition N.E. of the Gascoyne in late 1881.* Perth: Hesperian Press.

Lynne, Cairns, and Graeme Henderson. 1995. *Unfinished voyages: Western Australian shipwrecks,1881–1900.* Crawley: The University of Western Australia Press.

Commonwealth of Australia, *Report and recommendations of the Royal Commission into the pearl-shelling industry.* Melbourne: Government Printer, 1916.

Cooke, William Ernest. 1901. *The climate of Western Australia: From meteorological observations made during the years 1876–1899.* Perth: Government Printer.

Cushman, George T. 2013. The imperial politics of hurricane prediction: From Calcutta and Havana to Manilla and Galveston, 1839–1900. In *Nation-states and the global environment*, eds. M. Lawrence, E. Bsumek, and D. Kinkela. New York: Oxford University Press.

Daily News (Perth).

De La Rue, Kathy. 1978. *Pearl shell and pastures:The story of Cossack and Roebourne and their place in the history of the north west, from the earliest explorations to 1910.* Roebourne: The Cossack Project Committee.

Edwards, Hugh. 1983. *Port of pearls.* Adelaide: Rigby.

Findlay,Alexander George. *Sailing direction for the Indian Ocean[A directory for the navigation of the Indian Ocean, with descriptions of its coasts, islands, etc. from the Cape of Good Hope to the Strait of Sunda and Western Australia, including also the Red Sea and the Persian Gulf:The winds, monsoons, and currents, and the passages from Europe to its various ports*] (London: Richard Holmes Laurie, 1870)

Forrest, Kay. 1996. *The challenge and the chance:The Colonisation and settlement of North-West Australia 1861–1918.* Perth: Hesperian Press.

Garden, Donald. 2009. *Droughts,floods and cyclones: El Niños that shaped our colonial past.* North Melbourne: Australian Scholarly Publishing.

Gentilli, Joseph. 1967. A history of meteorological and climatological studies in Australia. *University Studies in History* 5(1): 54–79.

Hanstrum, Barry. 1992. A history of tropical cyclones in the southwest of Western Australia 1830–1992. *Early Days: Journal of the Royal Western Australian Historical Society* 10(4): 396–407.

Heathcote, R.L., and B.G. Thom, eds. 1979. *Natural hazards in Australia.* Canberra: Australia Academy of Science.

Henderson, Graeme, and Kandy-Jane Henderson. 1988. *Unfinished voyages: Western Australian shipwrecks 1851–1880*. Nedlands: University of Western Australia Press.

Herald (Fremantle).

Hunt, H.A., S.S. Visher, and D. Hodge, eds. 1925. *Australian hurricanes and related storms: With an appendix on hurricanes in the South Pacific*. Melbourne: Commonwealth Bureau of Meteorology Bulletin No. 16.

Hutchison, David. 1980. William Ernest Cooke, first Government Astronomer. *Early Days: Journal of the Royal Western Australian Historical Society* 8(4): 93–113.

Inquirer and Commercial News (Perth).

Longshore, David. 1998. *Encyclopaedia of hurricanes,typhoons and cyclones*. New York: Facts on File.

Mackeown, P. Kevin. 2010. *Early China coast meteorology: The role of Hong Kong*, Royal Asiatic Society Hong Kong Study Series. Hong Kong University Press, 2010.

McCarthy, Mike. 1995. Before broome. *The Great Circle: Journal of the Australian Association for Maritime History* 16(2): 76–89.

———. 2006. Naked diving for mother-of-pearl. *Early Days: Journal of the Royal Western Australian Historical Society* 13(2): 243–259.

National Archives of Australia. Prime Minister's Department F393/1/1, 'Relief of Distress – Pearling Disaster – Broome', Series A461.

Norman, John E. deB, and G.V. Norman. 2007. *A pearling master's journey*. Strathfield: John E de B Norman.

Northern Times (Carnarvon).

Oliver, Pam. 2012. Interpreting 'Japanese activities' in Australia, 1888–1945. *Journal of the Australian War Memorial* 36, https://www.awm.gov.au/journal/j36/oliver.asp (accessed 1 June 2016).

Powell, Alan. 2010. *Northern voyagers: Australian monsoon coast in maritime history*. Melbourne: Australian Scholarly Publishing.

State Records Office of Western Australia. Aborigines and Fisheries Departmentfile 1917/2297, 'Ocean Temperatures: Information supplied by J.J. East re:', Cons. 477.

State Records Office of Western Australia. Colonial Secretary's Department, Harbour and Light file 438/12, '"Koombana" S.S. Overdue at Broome – Total Loss of, Inquiry re:', Cons. 1066.

State Records Office of Western Australia. Fisheries Department file 193/48, 'Storm Damage to Pearling Fleet. Broome District. General File', Cons. 477

State Records Office of Western Australia. Premier's Department file 1898/0643, 'W.E. Cooke Esq. Government Astronomer, Perth. Worthlessness of Mr C. Wragge's warnings to shipping', Cons. 1496.

State Records Office of Western Australia. Premier's Department file 218/35, 'Relief to Pearling Industry for losses caused by Cyclone 1935', Cons. 1496.

Storms. In *The Australian Encyclopaedia*, ed. AHChisolm. Sydney: Angus and Robertson, 1958.

Streeter, Edwin W. 1886. *Pearls and pearling life*. London: George Bell and Sons.

Sunday Times (Perth).

Walker, D. 1999. *Anxious nation: Australia and the rise of Asia 1850–1939*. St Lucia: University of Queensland Press.

——— 2005. The curse of the tropics. In *A change in the weather: Climate and culture in history*, eds. T. Sherratt, T. Griffiths, and L. Robin. Canberra: National Museum of Australia.

West Australian (Perth).

West Australian Pearlers Association, Collection of papers on Broome and the pearling industry. Battye Library of Western Australian History, MN 377, ACC 1935A, 5006A.

Western Australian Times (Perth).

Western Mail (Perth).

Wettenhall, Roger L. 1975. *Bushfire disaster: An Australian community in crisis*. Sydney: Angus and Robertson.

Withnell Taylor, Nancy E. 2002. *Yeera-muk-a-doo:Asocial history of the settlement of north-west Australia told through the Withnell and Hancock families, 1861 to 1890*. Perth: Hesperian Press.

Wright, Bill, ed. 2004. *Drought,dust and deluge: A century of climate extremes in Australia*. Canberra: Australian Bureau of Meteorology.

Writer, Larry. 2011. *The Australian book of disasters*. Sydney: Murdoch Books Australia.

Index

A

Abu-Lughod, Janet, L., 4–5
accretion, 63–7, *63*, *66*
Adamson, George, 17, 18, **79–104**
adaptive measures, 18, 92–4
agriculture, 20, 39–41
  India, 79–104, 106
  Oman and Zanzibar, 256–68, *265*, *266*, 270
  Philippines, 18, 127–30
  Vietnam, 60–1, 69–70
Agung (Sultan), 44–5
Algué, José (Fr), 108
Al-Harthy, Ahmed Hamoud Mohammed, 264
Alpers, Edward, 3
Anderson, William, 201
Andsadler, Karen, 271
Annales of the Propagination of the Faith (letters of French missionaries), 170, 172, 181, 186
Arabian Gulf, 7, 10–12, 20, 255–82, *262*
archives, 8, 20, 80–1, *82*, 91
Armitage, David, 3
Aubréville, André, 153
Australia, 10, 20–1, 43–4, 283–311, *288*
Australian Directory, 295
Australian Hurricanes and Related Storms (Visher), 297

B

Babad Dipanegara (Diponegoro), 226
Bain, Mary Albertus, 297
Bali, 39
Bangladesh, 7–9, 14
Bankoff, Greg, **1–30**, 19, 132, **199–224**, 226, 298
Bay of Bengal, 7–9, 11–13
Berg, C.C., 45
Black, Emily, 271
Bligh, Arthur, 296
Blount, Chris, 264

Page references in *italics* refer to tables.
Page references in **bold** refer to the main contribution from each author.

G. Bankoff, J. Christensen (eds.), *Natural Hazards and Peoples in the Indian Ocean World*, DOI 10.1057/978-1-349-94857-4

Bombay Presidency, 18, 79–104
Boomgaard, Peter, 37
Bose, Sugata, 1–2
Bowling Alone (Putnam), 206
Bowman, David, 147
Braudel, Fernand, 2, 3
Brockman, John, 296
Broome, 283–311, *285*
Bui Quang Tung, 177
Bushfire Disaster (Wettenhall), 286
bushfires, 18–19, 143–68, 286

C
Carey, Peter, 226
Carnac, J.R., 86
Cham rice, 60–1
charity, 180, 210–11
  India, 84, 85, 87
  Indonesia, 19, 229, 233, 236–8, 240–1, 245
Chaudhuri, Kirti N., 2–3
Christensen, Joseph, **1–30**, 20, **283–311**
Christie, Jan, 41
civic community culture, 19, 199–224
climate change, 18, 121, 126, 131, 132, 213, 270–2
climatic hazards, 79–104. *See also* hazards
colonialism, 18, 19, 128–9, 128–30, 149–54, 225–54, 255
conflict, 35, 38, 82–3
Confucian doctrine, 169, 176, 184–7, 191
Cong V. Mai, 67
conservation, 18, 151–3, 155–9
Cooke, Nola, 181
Cooke, William Ernest, 295
Cooper, Frederick, 256–7, 264, 268–9
cross-cultural cooperation, 21, 234–41, 244
cultural adaptation, 200–1, 212–13
cyclones and cyclonic storms, 7–10, 21
  Australia, 283–311, *285*, *288*, *290*, *292*
  Oman and Zanzibar, 20, 255–82, *260*, *265*
  Philippines, 18, 105–42, *122*, *123*, 202

D
Dao The Tuan, 60
Davis, Mike, 125
de Clercq, Frederik S.A., 227–8
Delgado, Juan J. (Fr), 117
Democracy in America (Tocqueville), 206
Deppermann, Charles (Fr), 111–12
de Waal, Alex, 267
dikes, 17, 55–78, *57*, *59*
Dinh Nhi dike, 64–5
Diponegoro, 226
disaster response and mitigation, 225–54, 283–311
  Philippines, 18–19, 130–3, 199–224
  Vietnam, 20–1, 169–98
disasters, 1–30, 199–224
  Australia, 21, 285–6, 293, 298, 303
  Indonesia, 17, 31–4, 225–6, 232–4
  Vietnam, 20, 169–98, *173*
'disaster subculture', 200–2
disease, 20, 39, 86, 174, 188, 267–9, 289
Doan Ngoc Han, 70
Dong, Keqin, 123–4
Dowding, Herbert W., 266
droughts, 7, 18, 143
  India, 17, 79–104
  Oman and Zanzibar, 20, 255–82
  Philippines, 124, 125
  Vietnam, 174, 179
Du dia Chi (Nguyen Trai), 70
Dyt, Kathryn, 19–20, **169–98**

E
earthquakes, 10–15, 19, 33–5, 42–3, 45, 225–54
East Africa, 20, 255–82
East India Company (EIC), 80, 82–5, 89, 94
Edwards, Hugh, 297–8
Eastern delta, 55–78
Einrem, Johannes, 144
El Niño phenomena, 6–7, 18, 106, 121–7, *124*, *125*, 130, 270–2, 287
Elvin, Mark, 189
environmental factors, 20, 255–82
epidemics, 143, 174, 176, 180–4, 188
erosion, 65–8, *66*, 71

F
Fallow, James, 206
famines, 20, 265, 267–70
  India, 17–18, 79–104
  Vietnam, 62, 70, 175–6, 180, 181
Farrar, Caroline, 88
Findlay, Alexander George, 295
fire, 18–19, 143–68, 286
firestorms, 226
Fisher, Martin, 258
floods, 7–10, 143, 263
  Philippines, 19, 105, 110, *114*, 124, 125, 128, 130, 199–203, 209–12
  Vietnam, 62, 67, 172–7, 179
'Floods, Famine and Poverty among Northern Peasants in, 1857' (Nguyen Van Kiem), 177
Flores, 40
food shortages, 80, 84, 176
Frere, Bartle, 269
Fritz, Hermann M., 264
Froc S. J., Louis, 106
Full Fathom Five (Bain), 297

G
geology, 31–53
Giblin, James, 267, 270
global warming, 121, 126, 131, 132
Goss, Andrew, 241
Gourou, Pierre, 57, 60, 62, 68

H
Hageman, J., 242
Hai Hau, 66–8
Handler, Paul, 271
hazards, 5–6, 18, 20–1, 79–104, 143–68
  Australia, 286, 298
  Indonesia, 16, *36*
  Philippines, 105, 200–2
Helback, D.F.H., 227–34
Hirschboeck, 271
Hoa Binh hydropower dam, 67, 71
Hoàng Thu, 189–90
Hoffman, Susannah, 226, 233
Hollensteiner, Mary, 210
Homo floresiensis, 40
Hong Duc dike, 66, 71
Hopper, Matthew S., 20, **255–82**
Hostetter, Harvey Edgar, 210
households (Vietnam), *61*
'humane hospitals', 85 87
humanitarian aid, 19, 130–2, 239–41, 245
Humbert, Henri, 152
hurricanes. *See* cyclones and cyclonic storms

I
India, 7–9, 13–14, 17–18, 79–104, *81*, 106
Indian Ocean Dipole (IOD), 6–7
The Indian Ocean (Pearson), 3
The Indian Ocean: A History of People and the Sea (McPherson), 3

The Indian Ocean in World History (Alpers), 3
Indian Ocean World (IOW), 1–30, *4*, 16, 79
Indonesia, 11, 14–17, 19, 31–53, 225–54, *228*
Iran, 12–13
irrigation, 92–3, *93*, 154, 263

J

Java, 19, 32, 36–8, 41–6, 226, 240, *242*
Johnson, Douglas, 267

K

Kham dinh Dai Nam hoi dien su le (Imperial Vietnam compendium), 56
Kirk, John (Sir), 257, 268, 269
Koponen, Juhani, 267
Kull, Christian A., 18–19, **143–68**
Kumar, M.R. Ramesh, 272

L

laissez-faire policies, 18, 86, 87, 89, 94
La Marche, 271
Lamb, H.H., 270–2
land reclamation, 17, 55–78
La Niña phenomena, 6–7, 106, 121–7, *124*, *125*, 130, 271, 287
Le Ba Thao, 68
Lloyd, Christopher, 269
locust swarms, 86, 92, 143
Luzon, 36–7, 202, 203, 204, *205*
Lynn, Robert, 256

M

Madagascar, 10, 18–19, 143–68, *146*
Male, Arthur, 299, 300
Marsden, William, 42
Mascarene Islands, 7–8, 15
Maskrey, Andrew, 200, 212
Mauelshagen, Franz, 238
McPherson, Kenneth, 3
Membery, David, 258–9
migration, 70, 85–7, 91–2
missionaries, 169–98, 238
monsoons, 2–5, 7, 17, 79–80, 84, 113, 259
Moore, Harry, 200–1
Myanmar, 7, 10

N

Nash, D.J., 80
natural hazards. *See* hazards
Nepal, 14
Ngo Ngo Cat, 65
Nguyen Dong Chi, 177–8
Nguyen Trai, 70
Nguyen Van Kiem, 177
Niazi, Tarique, 132–3
Nicholls, Jasper, 84–5
Nicholson, Sharon, 272
Nieuwalt, S., 267–8
Ninh Viêt Giao, 177–8

O

Observatory Atlas of the Tracks of 620 Typhoons 1893–1918 (Froc), 106
Oman, 10, 20, 255–82, *262*

P

Pakistan, 9, 14
Pankhurst, Richard, 267
Pararaton (Javanese chronicle), 41
pearlshelling industry, 283–311
Pearson, Michael, 3, 5

Pellerin, François (Mgr), 174, 176, 181, 182
Pelley, Patricia, 177
Perrier de la Bâthie, Henri, 152
Philippines
  tsunamis, 12, 14–16, 32, 34
  typhoons, 7–8, 18, 19, 40, 105–42, *112*, *116*, *126*, 199–224, *205*
Pinker, Steven, 35
political conflict, 35, 38, 82–3
politics, 230–4, 233–6, 239–46
Ports of Pearls (Edwards), 297–8
Post, 271
Pratt, Seth, 257–8
Puginier, Paul-François (Fr), 173–4
Putnam, Robert, 206–7, 210, 212

R

Raffles, T.S., 37 38, 39
Ratu Kidul, 38, 44–5
Red River delta, 17, 55–78, *57*, *58*, *59*, *63*
refugees, 86, 89, 231, 236, 239, 245
Reid, Anthony, 16–17, **31–53**
religious traditions, 93–4
  Indonesia, 226, 229–30, 233–4, 237–41, 244–5
  Vietnam, 180–3, 186, 190
Retord, Pierre (Mgr), 170, 172–3, 175–7, 182, 183
Reyes, Joaquin, 204, 207–11
Ricklefs, Merle, 44
riots, 88–9
risks, 5–16, 21, 45, 200–2, 212, 227–34, 243–5, 298
rural economies, 83, 90, 92, 128–30

S

Sailing Directions for the Indian Ocean (Findlay), 295
saltwater intrusion, 56, *58*, 61–2, 65
San Isidro, 19, 199–224, *205*
Schrikker, Alicia, 19, **225–54**
science, 241–6, 245
sea dikes, 55–78, *57*, *59*
sea intrusion. *See* saltwater intrusion
Selga, Miguel (Fr), 204
Senopati, 44–5
Serels, Steven, 267
Sheriff, Abdul, 256, 264
slavery, 20, 255–82
social responses, 79–104
Southeast Asia, 31–53, *36*
Sri Lanka, 7, 11
storms, 199–224. *See also* cyclones and cyclonic storms; typhoons
storm surges, 7, 9, 10, 108, 110, 126, *127*, 130, 296
'The Succession of Thieu Tri' (Bui Quang Tung), 177
Sumatra, 2, 11, 32, 35, 36, 41, 42–5
Sumesh, K.G., 272
Super Typhoon Haiyan, 106, *107*, 108, 110, 120–1, 132–3, 201–4

T

Tagliacozzo, Eric, 5
Tana, Li, 17, **55–78**
taxpayers in Hai Duong and Nam Dinh (Vietnam), *69*
Thomas, Bertram, 263
Tocqueville, Alexis De, 206
trade, 2–3, 20, 80, 255
Trade and Civilization in the Indian Ocean (Chaudhuri), 2–3
transnational history, 1, 3–4
The Treasure of Vietnamese Folktales (ed. Nguyên Dŏng Chi), 177–8
tropical storms. *See* cyclones and cyclonic storms
Tsuboi Yoshiharu, 176

tsunamis, 2, 7, 10–12, 14, 16–17, 31–5, 42–5, 225–54
Tu Duc (Emperor), 20, 169–98
typhoons, 10
Philippines. *See under* Philippines
Vietnam, 62, 67–8, 71, 173–4
Tyrell, Ian, 4

V
Veritable Records (dynastic chronicles), 172, 174, 179, 183–4, 187, 189
Viet Chronicle, 62
Vietnam, 17, 19–20, 55–78, *57*, 169–98, *171*, *173*
'Vietnamese Politics and Confucianism in the Nineteenth-Century' (Tsuboi), 176
Visher, Stephen Sargent, 297
volcanic eruptions, 11, 14–17, 19, 31–42, 45–6, 225–54, 270–2

W
Walker, David, 303
Wallace, Alfred Russel, 227, 234
Warren, James Francis, 17–18, **105–42**
Weller, Jack, 201
Wenger, Dennis, 201
Western delta, 65–8, 70–1
Western Indian Ocean, 255–82
Wettenhall, Roger, 286
winds, 106–8, 116–21, *119–20*, 202
Woodside, A., 184
Wragge, Clement, 295

Z
Zain, Muhammad (Sultan), 227–34, 239, 245
Zanzibar, 255–82

Printed in the United States
By Bookmasters